Advanced Polymer Nanocomposites

Advanced Polymer Nanocomposites

Editors

Ting-Yu Liu
Yu-Wei Cheng

MDPI • Basel • Beijing • Wuhan • Barcelona • Belgrade • Manchester • Tokyo • Cluj • Tianjin

Editors
Ting-Yu Liu
Department of Materials
Engineering
Ming Chi University
of Technology
New Taipei City
Taiwan

Yu-Wei Cheng
Department of Chemical
Engineering
Ming Chi University
of Technology
New Taipei City
Taiwan

Editorial Office
MDPI
St. Alban-Anlage 66
4052 Basel, Switzerland

This is a reprint of articles from the Special Issue published online in the open access journal *Polymers* (ISSN 2073-4360) (available at: www.mdpi.com/journal/polymers/special_issues/Advanced_Polymer_Nanocomposites).

For citation purposes, cite each article independently as indicated on the article page online and as indicated below:

LastName, A.A.; LastName, B.B.; LastName, C.C. Article Title. *Journal Name* **Year**, *Volume Number*, Page Range.

ISBN 978-3-0365-5958-2 (Hbk)
ISBN 978-3-0365-5957-5 (PDF)

© 2022 by the authors. Articles in this book are Open Access and distributed under the Creative Commons Attribution (CC BY) license, which allows users to download, copy and build upon published articles, as long as the author and publisher are properly credited, which ensures maximum dissemination and a wider impact of our publications.

The book as a whole is distributed by MDPI under the terms and conditions of the Creative Commons license CC BY-NC-ND.

Contents

About the Editors vii

Preface to "Advanced Polymer Nanocomposites" ix

Chia-Hsin Zhang, Chia-Hung Huang and Wei-Ren Liu
Structural Design of Three-Dimensional Graphene/Nano Filler (Al_2O_3, BN, or TiO_2) Resins and Their Application to Electrically Conductive Adhesives
Reprinted from: *Polymers* **2019**, *11*, 1713, doi:10.3390/polym11101713 1

Weibing Zhong, Haiqing Jiang, Liyan Yang, Ashish Yadav, Xincheng Ding and Yuanli Chen et al.
Ultra-Sensitive Piezo-Resistive Sensors Constructed with Reduced Graphene Oxide/Polyolefin Elastomer (RGO/POE) Nanofiber Aerogels
Reprinted from: *Polymers* **2019**, *11*, 1883, doi:10.3390/polym11111883 13

Yanni Guo, Deliang He, Aomei Xie, Wei Qu, Yining Tang and Lei Zhou et al.
The Electrochemical Oxidation of Hydroquinone and Catechol through a Novel Poly-geminal Dicationic Ionic Liquid (PGDIL)–TiO_2 Composite Film Electrode
Reprinted from: *Polymers* **2019**, *11*, 1907, doi:10.3390/polym11111907 25

Yang-Yen Yu and Cheng-Huai Yang
Preparation and Application of Organic-Inorganic Nanocomposite Materials in Stretched Organic Thin Film Transistors
Reprinted from: *Polymers* **2020**, *12*, 1058, doi:10.3390/polym12051058 43

Eun Bin Ko, Dong-Eun Lee and Keun-Byoung Yoon
Electrically Conductive Nanocomposites Composed of Styrene–Acrylonitrile Copolymer and rGO via Free-Radical Polymerization
Reprinted from: *Polymers* **2020**, *12*, 1221, doi:10.3390/polym12061221 59

Chien-Hong Lin, Yu-De Zhuang, Ding-Guey Tsai, Hwa-Jou Wei and Ting-Yu Liu
Performance Enhancement of Vanadium Redox Flow Battery by Treated Carbon Felt Electrodes of Polyacrylonitrile Using Atmospheric Pressure Plasma
Reprinted from: *Polymers* **2020**, *12*, 1372, doi:10.3390/polym12061372 73

Hao-Hueng Chang, Yi-Ting Tseng, Sheng-Wun Huang, Yi-Fang Kuo, Chun-Liang Yeh and Chien-Hsin Wu et al.
Evaluation of Carbon Dioxide-Based Urethane Acrylate Composites for Sealers of Root Canal Obturation
Reprinted from: *Polymers* **2020**, *12*, 482, doi:10.3390/polym12020482 89

J. J. Encalada-Alayola, Y. Veranes-Pantoja, J. A. Uribe-Calderón, J. V. Cauich-Rodríguez and J. M. Cervantes-Uc
Effect of Type and Concentration of Nanoclay on the Mechanical and Physicochemical Properties of Bis-GMA/TTEGDMA Dental Resins
Reprinted from: *Polymers* **2020**, *12*, 601, doi:10.3390/polym12030601 109

Kai-Ting Hou, Ting-Yu Liu, Min-Yu Chiang, Chun-Yu Chen, Shwu-Jen Chang and San-Yuan Chen
Cartilage Tissue-Mimetic Pellets with Multifunctional Magnetic Hyaluronic Acid-Graft-Amphiphilic Gelatin Microcapsules for Chondrogenic Stimulation
Reprinted from: *Polymers* **2020**, *12*, 785, doi:10.3390/polym12040785 123

Nguyen-Phuong-Dung Tran and Ming-Chien Yang
The Ophthalmic Performance of Hydrogel Contact Lenses Loaded with Silicone Nanoparticles
Reprinted from: *Polymers* **2020**, *12*, 1128, doi:10.3390/polym12051128 . . . . **139**

Jen-Hung Fang, Che-Hau Liu, Ru-Siou Hsu, Yin-Yu Chen, Wen-Hsuan Chiang and Hui-Min David Wang et al.
Transdermal Composite Microneedle Composed of Mesoporous Iron Oxide Nanoraspberry and PVA for Androgenetic Alopecia Treatment
Reprinted from: *Polymers* **2020**, *12*, 1392, doi:10.3390/polym12061392 . . . . **153**

Chuan-Chih Hsu, Yu-Wei Cheng, Che-Chun Liu, Xin-Yao Peng, Ming-Chi Yung and Ting-Yu Liu
Anti-Bacterial and Anti-Fouling Capabilities of Poly(3,4-Ethylenedioxythiophene) Derivative Nanohybrid Coatings on SUS316L Stainless Steel by Electrochemical Polymerization
Reprinted from: *Polymers* **2020**, *12*, 1467, doi:10.3390/polym12071467 . . . . **167**

Yu-Kai Wang, Fang-Chang Tsai, Chao-Chen Ma, Min-Ling Wang and Shiao-Wei Kuo
Using Methacryl-Polyhedral Oligomeric Silsesquioxane as the Thermal Stabilizer and Plasticizer in Poly(vinyl chloride) Nanocomposites
Reprinted from: *Polymers* **2019**, *11*, 1711, doi:10.3390/polym11101711 . . . . **179**

Hyeon Il Shin and Jin-Hae Chang
Transparent Polyimide/Organoclay Nanocomposite Films Containing Different Diamine Monomers
Reprinted from: *Polymers* **2020**, *12*, 135, doi:10.3390/polym12010135 . . . . **193**

Manel Chihi, Mostapha Tarfaoui, Chokri Bouraoui and Ahmed El Moumen
Effect of CNTs Additives on the Energy Balance of Carbon/Epoxy Nanocomposites during Dynamic Compression Test
Reprinted from: *Polymers* **2020**, *12*, 194, doi:10.3390/polym12010194 . . . . **211**

Fei Kung and Ming-Chien Yang
Improvement of the Heat-Dissipating Performance of Powder Coating with Graphene
Reprinted from: *Polymers* **2020**, *12*, 1321, doi:10.3390/polym12061321 . . . . **227**

Sihao Yin, Xinlin Ren, Peichao Lian, Yuanzhi Zhu and Yi Mei
Synergistic Effects of Black Phosphorus/Boron Nitride Nanosheets on Enhancing the Flame-Retardant Properties of Waterborne Polyurethane and Its Flame-Retardant Mechanism
Reprinted from: *Polymers* **2020**, *12*, 1487, doi:10.3390/polym12071487 . . . . **243**

Wei Liu, Lutao Lv, Zonglin Yang, Yuqing Zheng and Hui Wang
The Effect of OMMT on the Properties of Vehicle Damping Carbon Black-Natural Rubber Composites
Reprinted from: *Polymers* **2020**, *12*, 1983, doi:10.3390/polym12091983 . . . . **257**

About the Editors

Ting-Yu Liu

Ting-Yu Liu achieved his PhD degree at the Department of Materials Science and Engineering, National Chiao Tung University, Taiwan, in 2008. He was a Visiting Researcher in the Department of Materials Science and Engineering, University of Pennsylvania, USA, for 1 year. Subsequently, he became a Post-Doc Fellow at the Institute of Atomic and Molecular Sciences, Academia Sinica, Taiwan, from 2009 to 2011, and then a Project Assistant Professor in the Institute of Polymer Science and Engineering, National Taiwan University, from 2011 to 2013. He is currently a Distinguished Professor at the Department of Materials Engineering, Ming Chi University of Technology, Taiwan. His research covers nanomaterials, biomaterials, polymer composites, and opto-electric (surface-enhanced Raman spectroscopy detection) and electrochemical sensing. He has published more than 100 SCI-indexed journal articles and has gained an h-index of 31, according to the citation report from Google Scholar.

Yu-Wei Cheng

Yu-Wei Cheng achieved his PhD degree at the Institute of Polymer Science and Engineering, National Taiwan University. He is currently an Assistant Professor at the Department of Chemical Engineering, Ming Chi University of Technology, Taiwan. His research interests include the synthesis of functionalized polymers and the fabrication of optimized Raman-enhanced nanoparticle arrays for small molecule detection. His recent research has focused on manipulated metal nanoparticle arrays grown on 2D and 3D substrates for multifunctional smart sensors and surface-enhanced Raman scattering detection, published in highly ranking journals of chemistry and physical fields. He has published 26 SCI-indexed journal articles and has gained an h-index of 12, according to the citation report from Google Scholar.

Preface to "Advanced Polymer Nanocomposites"

Polymer nanocomposites currently attract high levels of industrial interest in various scientific areas in the field of nanomaterials, because they can improve the performance of polymeric matrices and inorganic nanomaterials, such as enhancing light/magnetic behaviors, electrical/thermal conductivity, toughness, stiffness, and mechanical strength. Inorganic quantum dots/nanoparticles, nanorods/nanotubes, and 2D materials (such as graphene-based nanosheets) can be used to decorate the polymer matrix through chemical synthesis or physical blending to improve its performance. Thus, how to fabricate a homogeneous dispersion of fillers in the polymer matrix is a crucial technique in the nanomaterials field. This Special Issue, entitled "Advanced Polymer Nanocomposites", collects high-level original and review papers, focused on scientific discussions and practical applications in the field of functional polymer nanocomposites, including (a) optoelectronic materials (papers 1–6), (b) biomedical materials (papers 7–12), and (c) other functional polymer nanocomposites (papers 13–18). We hope that this Special Issue will promote academic exchanges, and identify and respond to tremendous challenges to disseminate research in this developing field.

Ting-Yu Liu and Yu-Wei Cheng
Editors

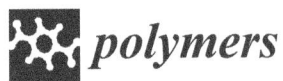

Article

Structural Design of Three-Dimensional Graphene/Nano Filler (Al₂O₃, BN, or TiO₂) Resins and Their Application to Electrically Conductive Adhesives

Chia-Hsin Zhang [1], Chia-Hung Huang [2] and Wei-Ren Liu [1,*]

[1] Department of Chemical Engineering, R&D Center for Membrane Technology, Center for Circular Economy, Chung-Yuan Christian University, Chungli 32023, Taiwan; cindy560559@gmail.com
[2] Metal Industries Research and Development Centre, Kaohsiung 81160, Taiwan; chiahung@mail.mirdc.org.tw
* Correspondence: WRLiu1203@gmail.com; Tel.: +886-3-2653315; Fax: 886-3-2653399

Received: 7 October 2019; Accepted: 17 October 2019; Published: 18 October 2019

Abstract: In this study, we designed a three-dimensional structure of electrically conductive adhesives (ECAs) by adding three different kinds of nano filler, including BN, TiO$_2$, and Al$_2$O$_3$ particles, into a few-layered graphene (FLG)/polymer composite to avoid FLG aggregation. Three different lateral sizes of FLG (FLG3, FLG8, and FLG20) were obtained from graphite (G3, G8, and G20) by a green, facile, low-cost, and scalable jet cavitation process. The corresponding characterizations, such as Raman spectroscopy, scanning electron microscopy (SEM), atomic force microscopy (AFM), and transmission electron microscopy (TEM), verified the successful preparation of graphene flakes. Based on the results of four-point probe measurements, FLG20 demonstrated the lowest sheet resistance value of ~0.021 Ω/■. The optimized ECAs' composition was a 60% solid content of FLG20 with the addition 2 wt.% of Al$_2$O$_3$. The sheet resistance value was as low as 51.8 Ω/■, which was a reduction of 73% compared to that of pristine FLG/polymer. These results indicate that this method not only paves the way for the cheaper and safer production of graphene, but also holds great potential for applications in energy-related technologies.

Keywords: graphene; electrically conductive adhesive; fillers; TiO$_2$; Al$_2$O$_3$; BN; resins

---

1. Introduction

The electronic industry is currently one of the most diversified industries. Electronic products can be seen everywhere in daily life, since the new generation of electronic products emphasizes the needs of personalization and portability [1,2]. Among the recent advances in electronic packaging technologies, electrically conductive adhesives (ECAs) have attracted most researchers' attention. The characteristics of ECAs are that they are environmentally friendly, bendable, have a high workability, and are simple to apply [3–10]. In ECAs, the electrically conducive fillers play a significant role in improving conductivity and strength. Different kinds of electrically conductive fillers, such as silver [11–13], copper [14,15], carbon black, carbon nanotubes, graphite, and graphene [16–21], have been widely reported. In recent years, carbon-based electrically conductive filler applications in ECAs have been universal because the cost of these materials is lower than that of metal fillers and they demonstrate a much better stability. Among these carbon-based materials, graphene has drawn much attention due to its exceptionally high crystallinity and electronic quality, and these features mean that graphene has high mechanical properties (>1060 GPa), high electrical conductivity (10^4 S/m), high thermal conductivity (~3000 W/m K), and a light weight [22–24]. These unique thermal, mechanical, optical, and electrical properties are better than those of other carbon materials, so this material has been widely used in energy storage materials, lithium ion battery materials, solar cells, super capacitors, and other applications [25–28].

There are a lot of approaches for preparing graphene-based materials, such as chemical reduction [29], pyrolytic graphite [30], Hummers method [31], and jet cavitation [32,33]. Nevertheless, strong acids, organic solvents, and oxidants are always used in an environmentally unfriendly way during the production process. Therefore, in our study, we propose a jet cavitation-assisted green process to synthesize few-layered graphene (FLG). This method is facile, low cost, green, and scalable to the production of few-layered graphene. Therefore, in the first part of our study, we used FLGs to construct an electrically conductive network in a polymer matrix to increase the electrical conductivity properties of ECAs.

The dispersion of FLGs in a polymer matrix is another important issue. In order to avoid the aggregation of FLGs in a polymer, nano-sized fillers are required to fill spaces between FLGs and the polymer matrix. He et al. [6] investigated graphene/MnO_2 composite networks as flexible supercapacitor electrodes, lowering the electrical conductivity of the graphene/MnO_2 composite due to the increase of MnO_2 with its low electronic conductivity of $10^{-5} \sim 10^{-6}$ S/cm [34]. Pu et al. investigated the application of N-GNSs (N-doped graphene nanosheets) in Ag-filled ECAs to reduce the resistivity with lower Ag loading ratios, and showed that adding merely 1 wt.% of N-GNSs can convert a non-conducting 30 wt.% Ag-filled polymer resin into an ECA with a resistivity of 4.4×10^{-2} Ω-cm [7]. Peng et al. advanced the weight ratio of SGNs to silver flakes to 20:80 (%), and the resistivity reached the lowest value of 2.37×10^{-4} Ω cm [35]. Ghaleb et al. demonstrated the effect of GNP (Graphene nanoplates) loading (0.05–1 vol%) on the tensile and electrical properties of GNP/epoxy thin-film composites, and the electrical conductivity of the 0.1 vol% GNP thin film increased from 4.32×10^{-7} to 1.02×10^{-3} S/m [36]. The above's cost is higher and CNTs also agglomerate easily. In order to overcome the possible drawback of graphene aggregation, we used alumina (Al_2O_3), titanium dioxide (TiO_2), and hexagonal boron nitride (BN) in a graphene/polymer composite. Alumina is abundant in the world and it is low cost, exhibiting characteristics of anti-oxidation, corrosion resistance, and chemical and thermal stability. It not only prevents carbon from being oxidized in the air, but also has a high stability when combined with materials and polymers [37]. Titanium dioxide seemed to be desirable when investigating the electrical properties of semiconducting crystals [38]. In terms of boron nitride, during the conduction process, the efficiency of the electronic product is lowered due to the heat release, so our study investigated the thermal conductivity of boron nitride. These fillers would effectively prevent graphene agglomeration and significantly reduce the cost [39,40].

In our study, we firstly synthesized three different sizes of FLGs by a low-temperature ultra-high pressure continuous homemade flow cell disrupter. Secondly, we incorporated Al_2O_3, TiO_2, and BN particles into the matrix resin to prepare ECAs in order to prevent FLG stacking. The corresponding characterizations, such as Raman spectroscopy, scanning electron microscopy (SEM), atomic force microscopy (AFM), and transmission electron microscopy (TEM), as four-point probe measurements of FLG and FLG/polymer and FLG/fillers/polymer composites, were carried out in this research.

2. Experimental Section

2.1. Preparation of Few-Layered Graphene (FLG)

First, three different lateral sizes of graphite, including KS-6 (Timcal®, d_{50} = 3 μm, namely G3), 8 μm graphite (KNANO®, d_{50} = 8 μm, namely G8)), and KS-44 (Timcal®, d_{50} = 20 μm, namely G20)), were dispersed in 500 mL of deionized (DI) water by sonication for 15 min (about solid content 5 wt.%). After dispersing them sufficiently, the solution was transferred into the tank of the low-temperature ultra-high pressure continuous homemade flow cell disrupter (LTHPD). As part of the work in designing the LTHPD, the suspension was poured into the device and the process was operated three times at each pressure (800, 1200, and 1500 bar). Therefore, the process was conducted nine times in total. Then, the graphite was separated by cavitation effects with different pressures.

The course operated in a circulation cooling water bath, which kept the temperature at 14–16 °C The whole exfoliated experiment was carried out in room conditions. The suspension of graphene was vacuum-filtered to obtain graphene cake and was transferred to an oven at a temperature of 40 °C for 24 h. Finally, the cake of FLG was milled into powder with a grinder.

2.2. Preparation of Electrically Conductive Adhesives (ECAs)

The ECAs were mainly composed of a resin matrix, FLG, and nano fillers (Al_2O_3, TiO_2, and BN). First, A-polymer and B-polymer were mixed at the mass ratio of 1:1. Then, ethyl acetate was added to the resin drop by drop over 30 min, with stirring. The graphene and filler samples were added to the resin with various mass ratios, such as 95:5, 90:10, and 85:15, and the total mass fraction was 50%. After stirring for 24 h, the graphene/filler composite was dispersed into the resin. The slurry was coated on the PET (Polyethylene terephthalate) and cured at 150 °C for 2 h in the oven. Finally, the FLG/nano filler content was increased to 55% and 60% in order to improve the ECAs' efficacy.

2.3. Characterizations

The morphologies of the sample were analyzed using scanning electron microscopy (SEM) by Hitachi S-4100 (Tokyo, Japan) and high-resolution transmission electron microscopy (HRTEM) JEOL-JEM2000FXII. The height profile of the as-synthesized few-layer graphene was measured by using atomic force microscopy (AFM, Bruker Dimension Icon, Berlin, Germany). Raman spectra were measured by a micro Raman spectroscopy system (Hsinchu, Taiwan), with a laser frequency of 532 nm as the excitation source. The electrical conductivities of the graphene adhesives were measured using a resistivity meter (KeithLink TG2, Shenzhen, China) with a four-point probe.

3. Results and Discussion

The morphologies of graphite and FLG were observed by SEM (Figure S1a–d in Supplementary Materials). Figure 1a,c,e present the different size graphite images of G3, G8, and G20, which were arranged two-dimensional material and comprised of micron-sized stackable sheet structures, and the lateral sizes of G3, G8, and G20 were in the range of 3–5, 8–10, and 20–22 µm, respectively. These were thus larger and had a greater thickness than FLG. Both FLGs were efficiently exfoliated to form separated thin sheets, as shown in Figure 1b,d,f, demonstrating that FLG3, FLG8, and FLG20 were obtained by LTHPD, respectively, and the lateral sizes were in the range of 2–3, 5–8, and 17–20 µm. In comparison, FLG was composed of thinner sheets and smaller sizes than graphite.

We used atomic force microscopy (AFM) to characterize the thickness and surface morphology of as-synthesized FLGs. Figure 2a,c,e depict the AFM images of FLG3, FLG8, and FLG20. Figure 2b,d,f show the thickness which was measured from the height profile of the AFM image, and the average thickness was about 15, 3, and 2 nm, respectively. Since the lateral size of FLG3 was small, the effect of being stripped was poor, and the inverted thickness did not decrease significantly. However, this is consistent with the data reported in the literature, indicating that the thickness of graphene sheets was about 2–4 nm, so FLG8 and FLG20 were high-quality few-layered graphene.

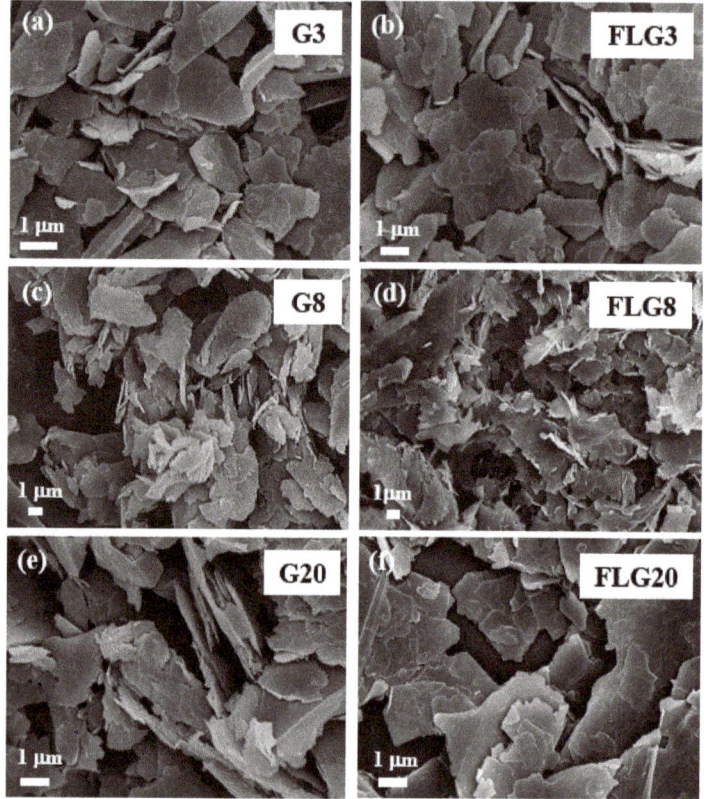

Figure 1. Scanning electron microscopy (SEM) images of (**a**) graphite (G)3, (**b**) few-layered graphene (FLG)3, (**c**) G8, (**d**) FLG8, (**e**) G20, and (**f**) FLG20.

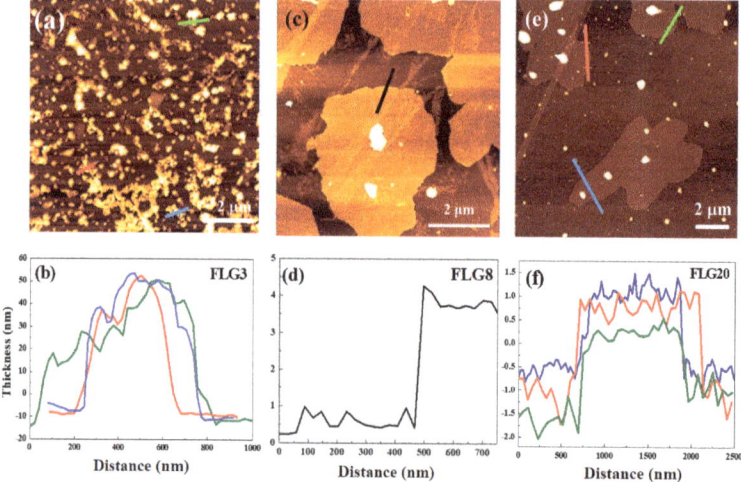

Figure 2. Atomic force microscopy (AFM) images of (**a**) FLG3, (**c**) FLG8, and (**e**) FLG20. Distribution of the thickness of (**b**) FLG3, (**d**) FLG8, and (**f**) FLG20 calculated from the obtained AFM analysis.

Figure 3a–c show the XRD patterns of graphite and FLG. The XRD patterns of graphite indicated the presence of two peaks at 25.0° and 43.5°, which corresponded to the inter-layer spacing of graphite d_{002} and the d_{101} reflection of the carbon atoms, respectively. Additionally, after exfoliating, the intensities of the diffraction peaks present a slight decrease in the few-layered graphene. The average crystallite size of G3, G8, and G20 was about 200,168, and 240 Å, respectively. However, the few-layered graphene obtained by LTHPD exhibited a slight decrease in crystallite sizes because the process wrecked the crystallinity. In Figure 3d, the picture shows a comparison of powder before and after manufacturing by LTHPD. The FLG produced by LTHPD was fluffier than that of graphite, so using this feature to prepare ECAs can reduce the amount of graphite and increase the conductive path to improve the conductivity.

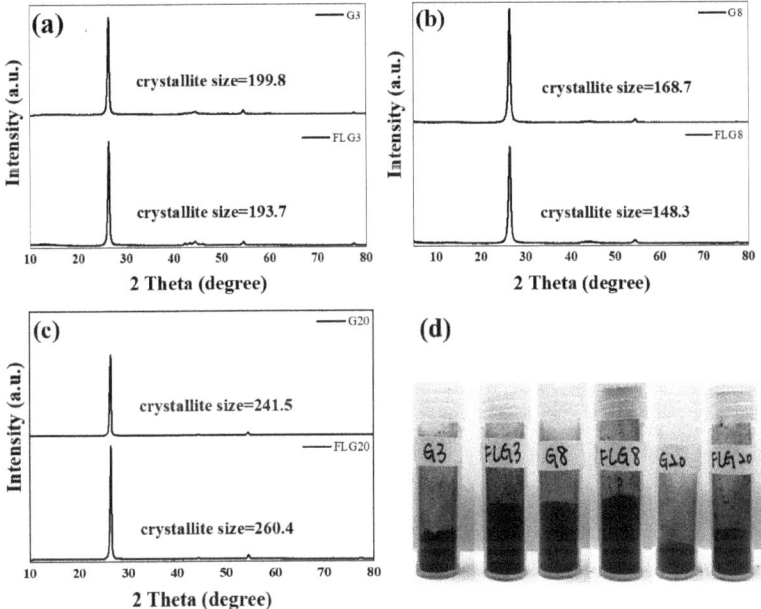

Figure 3. XRD patterns of (a) G3 and FLG3, (b) G8 and FLG8, and (c) G20 and FLG20, and a (d) comparison of those of powder before and after manufacturing by a low-temperature ultra-high pressure continuous homemade flow cell disrupter (LTHPD).

Figure 4a–c show the Raman spectra of graphite and FLG. The main features in the Raman spectra of carbons are the so-called G and D peaks, which lie at around 1560 and 1360 cm^{-1}, respectively, for visible excitation. The G peak is due to the doubly degenerate zone center E_{2g} mode, while the D peak is a breathing mode of κ-point phonons of A_{1g} symmetry [41]. The intensity ratio (I_D/I_G) of the D peak to G peak of the G3, G8, and G20 was about 0.226, 0.198, and 0.11, respectively. However, for FLG3, FLG8, and FLG20 which were obtained by LTHPD, the I_D/I_G was 0.231, 0.206, and 0.14, separately. The I_D/I_G ratio increased because the stripping process created defects. The results of graphite and FLG for testing the four-point probe are shown in Figure 4d. The sheet resistance of FLG was higher than that of graphite because the defects of FLG were increased; however, FLG was bulkier for graphite, which was beneficial for manufacturing the ECAs. The sheet resistance of FLG20 was lower than that of FLG. Therefore, we used FLG20 as the foremost material to export and apply to ECAs.

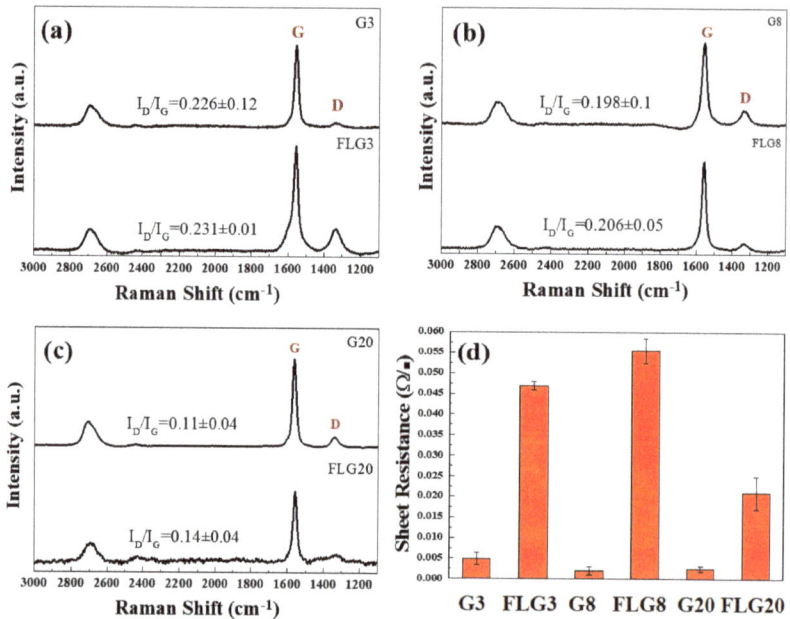

Figure 4. Raman spectra of (**a**) G3 and FLG3, (**b**) G8 and FLG8, and (**c**) G20 and FLG20, and (**d**) the sheet resistance of graphite and FLG.

The morphology of FLG20 were observed by TEM and HRTEM. Figure 5a shows that FLG20 appeared as a micro-size transparent sheet structure with a smooth surface and wrinkled pattern on the edge, which was typical of graphene. An HRTEM analysis of folding at the edges of sheets gave the number of layers by direct visualization as Figure 5b, and it was clear that the number of layers of FLG20 was about 7~10 layers.

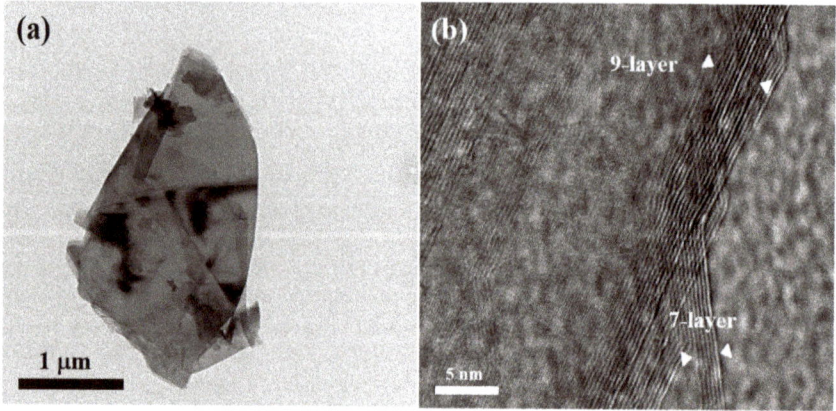

Figure 5. (**a**) Transmission electron microscopy (TEM) image of FLG20 and (**b**) high-resolution transmission electron microscopy (HRTEM) image of FLG20.

Figure 6a–c show SEM images of BN, TiO_2, and Al_2O_3, respectively. The morphology of BN was two-dimensional and its lateral size was in the range of 30–50 nm, as shown in Figure 6a. TiO_2 and

Al$_2$O$_3$ are granular materials, and had particle sizes of 20–30 and 10–15 nm, respectively (Figure 6b,c). In the study, we used these nanoparticles as nano fillers in ECAs, with the purpose of preventing the agglomeration of graphene lamellae and propping up the graphene to form a three-dimensional structure to increase conductive channels. The XRD patterns of BN, TiO$_2$, and Al$_2$O$_3$ are shown in Figure 6d–f. All these diffraction peaks match well with the standard values and are in agreement with the hexagonal structure of the Bragg positions in ICSD-241875, ICSD-9852, and ICSD-66559, respectively. In Figure 6d, the XRD patterns had a peak at 23.5°, showing that the type of TiO$_2$ was retile, and the Al$_2$O$_3$ was γ-phase Al$_2$O$_3$, as shown in Figure 6f.

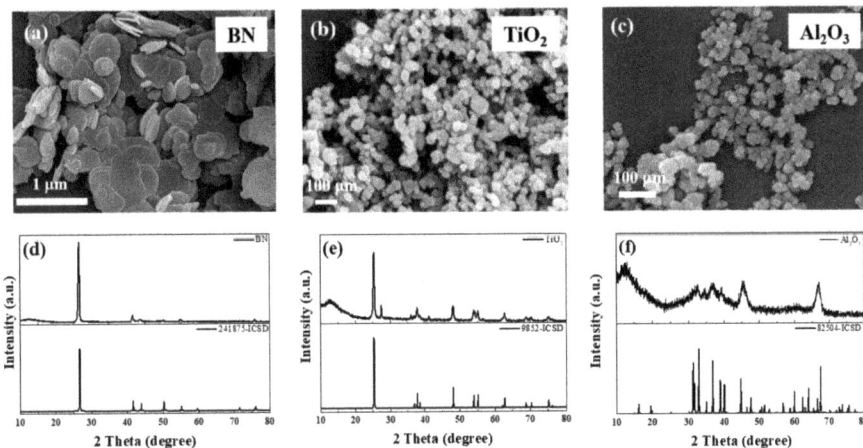

Figure 6. SEM images of (**a**) boron nitride (BN), (**b**) titanium dioxide (TiO$_2$), and (**c**) alumina (Al$_2$O$_3$); XRD patterns of (**d**) BN, (**e**) TiO$_2$, and (**f**) Al$_2$O$_3$.

Figure 7a–c and Table S1 compare the sheet resistance of ECAs with G20 and FLG20 for different nano filler mass ratios. The blank EACs manufactured by G20 and FLG20 had a sheet resistance of 192 Ω/■ and 190 Ω/■, respectively. When the mass ratio of nano fillers to G20 increased, the sheet resistance of ECAs increased; however, the mass ratio of nano fillers to FLG20 increased, and the sheet resistance of ECAs decreased first and then increased. This result shows that FLG20 had better consequences because FLG20 was fluffier, so it generated a continuous conductive pathway easily. When the mass ratio was 95:5 (FLG20/nano fillers), all the sheet resistances reached a minimum value of 135 Ω/■, 156 Ω/■, and 85.7 Ω/■, respectively, which were 20%~56% lower than the blank. This result can be explained by the fact that the content of nano fillers was too high, so the nano fillers could not disperse raggedly, resulting in an increase in the sheet resistance. The effect of FLG20/Al$_2$O$_3$ was the best, so we continued to explore the ratios of 99:1, 98:2, 97:3, and 96:4, as shown in Figure 7c. When 2 wt.% Al$_2$O$_3$ was added, the sheet resistance reached a minimum value of 79.3 Ω/■, which was 60% lower than the blank. With the FLG20 to Al$_2$O$_3$ mass ratio of 98:2, Al$_2$O$_3$ prevented the agglomeration of FLG20 and propped it up to form more electrically conductive networks. As shown in Figure 7d, we increased the solid content to 55% and 60%, and the sheet resistance reduced from 79.3 Ω/■ to 51.8 Ω/■, which was a reduction of 73% compared to without adding any nano fillers. This result shows that increasing the solid content obviously decreased the sheet resistance.

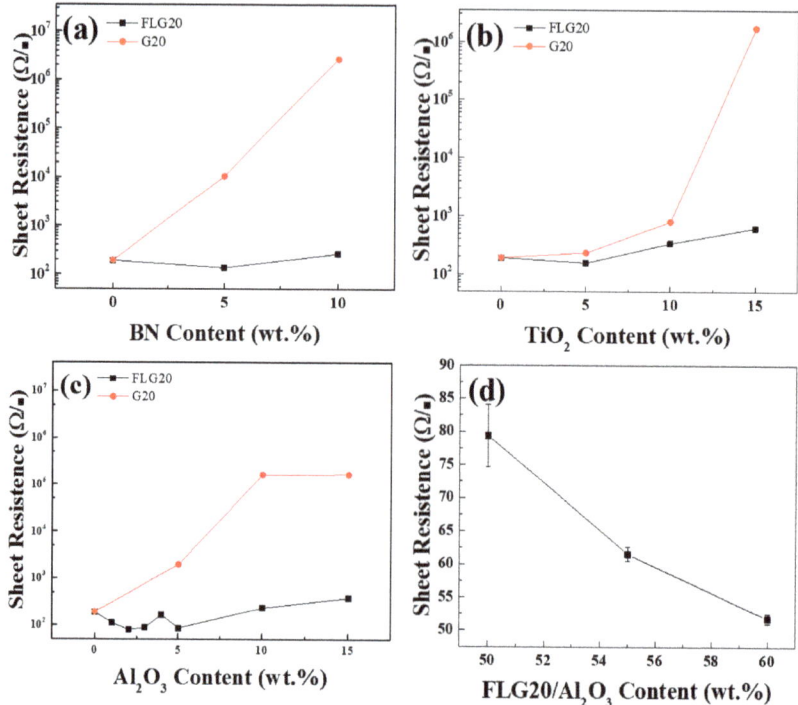

Figure 7. Sheet resistance of electrically conductive adhesives (ECAs) filled with different proportions of (**a**) BN, (**b**) TiO$_2$, and (**c**) Al$_2$O$_3$. (**d**) Different FLG20/Al$_2$O$_3$ solid contents.

The FLG20/Al$_2$O$_3$ composite also exhibits a great flexibility and mechanical strength after being coated on flexible plastic sheets. The electrical resistance of the FLG20/Al$_2$O$_3$ thin film on the PET film was revealed by the bending test. As shown in Figure 8a, there was a fairly small amount of variation after thousands of bending cycles, with an R/R$_0$ value retention value of 98%. This great mechanical strength makes the FLG20/Al$_2$O$_3$ a great conductive adhesive for flexible electronic applications. For demonstration, two pieces of printed FLG20/Al$_2$O$_3$ thin films on the PET film were connected to an LED. As shown in Figure 8c, the PET film with slight curvature could remain bright, showing that the EACs had a flexible property.

Figure 9 is a schematic diagram of the situation before and after adding nano fillers. In Figure 9a, since graphene is a two-dimensional material, a large number of graphene may generate much restacking without adding nano fillers, causing the electrical conductivity to decrease. Therefore, adding different kinds of nano fillers to prevent the agglomeration of graphene lamellae and prop up the graphene to form a three-dimensional structure to increase the conductive channels is beneficial for decreasing the resistance, as shown in Figure 9b. In the study, adding nano fillers to graphene sheets could effectively reduce the resistance value.

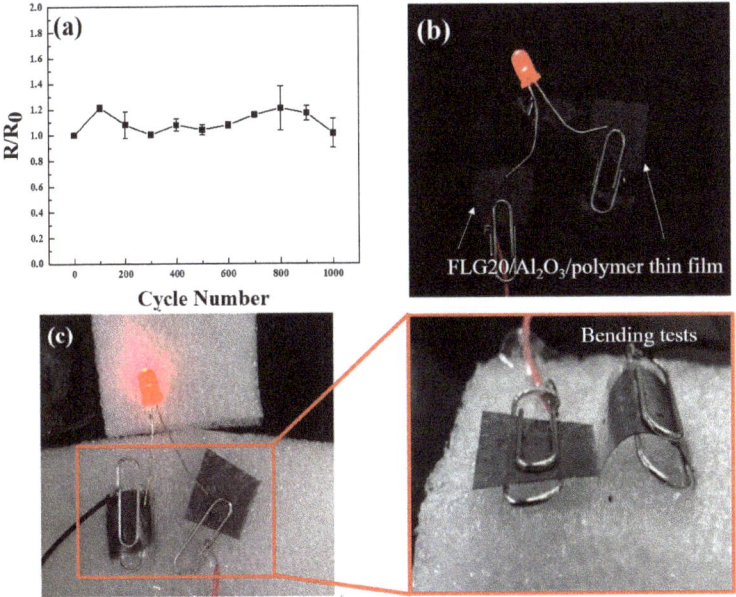

Figure 8. (a) The resistance increase ratio (R/R$_0$) of FLG20/Al$_2$O$_3$/polymer thin film on PET under a bending performance test. After connecting the FLG20/ Al$_2$O$_3$/polymer composite to an LED, the light remained bright at various degrees of bending: (**b**) flat and (**c**) bending.

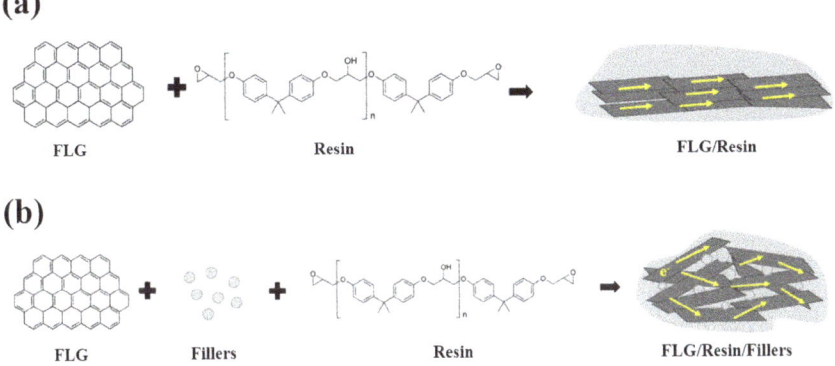

Figure 9. The before and after schematic diagram of the filler dispersion state. (**a**) Graphene composite with no filler, and (**b**) graphene composite with nano fillers.

4. Conclusions

We succeeded in delaminating artificial graphite and natural graphite by jet cavitation to prepare few-layered graphene (FLG3, FLG8, and FLG20). The structure and morphology of few-layered graphene had a two-dimensional structure and few-layered graphene was composed of thinner sheets and smaller sizes than graphite. We used G20 and FLG20 as the foremost material to apply to ECAs with different nano fillers (BN, TiO$_2$, and Al$_2$O$_3$) to prevent graphene stacking and generate a continuous conductive pathway. The results indicated that the solid content was 60% and the best condition was adding 2 wt.% Al$_2$O$_3$, for which the sheet resistance value was 51.8 Ω/■. The electrically conductive resin remained nearly the same after one thousand bending cycles. These results indicate that the formulated FLG/Al$_2$O$_3$/polymer composite adhesives have a great potential in conductive adhesives for flexible display applications.

Supplementary Materials: The following are available online at http://www.mdpi.com/2073-4360/11/10/1713/s1, Figure S1: SEM images of (a) KS-6, (b) 8 µm, (c) MoKS-6, (d) Mo8 µm, (e) BN and (f) TiO$_2$, Figure S2: Raman spectra of (a) KS-6, (b) 8 µm, (c) MoKS-6 and (d) Mo8 µm, Figure S3: Sheet resistance of MoKS-6, Mo8 µm and MoKS-44, Figure S4: KS-44 and MoKS-44 composite with various content (a) BN and (b) TiO$_2$, Table S1: KS-44 and MoKS-44 composite with various content BN and TiO$_2$.

Author Contributions: C.-H.Z. wrote the paper; W.-R.L. conceived and designed the experiments; C.-H.H. analyzed the data.

Funding: This research was funded by Ministry of Science Technology grant number [MOST 107-2622-E-033-015–CC3], [108-2911-I-033-502], [108-E-033-MY3], and [108-3116-F-006-002].

Acknowledgments: The authors gratefully acknowledge the Ministry of Science and Technology, Taiwan, project grant no's MOST 107-2622-E-033-015–CC3, 108-2911-I-033-502, 108-E-033-MY3, and 108-3116-F-006-002.

Conflicts of Interest: The authors declare no conflict of interest.

References

1. Ling, Q.-D.; Liaw, D.-J.; Zhu, C.; Chan, D.S.-H.; Kang, E.-T.; Neoh, K.-G. Polymer electronic memories: Materials, devices and mechanisms. *Prog. Polym. Sci.* **2008**, *33*, 917–978. [CrossRef]
2. Hsu, C.-W.; Hu, A.H. Green supply chain management in the electronic industry. *Int. J. Environ. Sci. Technol.* **2008**, *5*, 205–216. [CrossRef]
3. Li, Y.; Wong, C. Recent advances of conductive adhesives as a lead-free alternative in electronic packaging: Materials, processing, reliability and applications. *Mater. Sci. Eng. R Rep.* **2006**, *51*, 1–35. [CrossRef]
4. Yang, X.; He, W.; Wang, S.; Zhou, G.; Tang, Y. Preparation and properties of a novel electrically conductive adhesive using a composite of silver nanorods, silver nanoparticles, and modified epoxy resin. *J. Mater. Sci. Mater. Electron.* **2012**, *23*, 108–114. [CrossRef]
5. Lin, W.; Xi, X.; Yu, C. Research of silver plating nano-graphite filled conductive adhesive. *Synth. Met.* **2009**, *159*, 619–624. [CrossRef]
6. Wu, H.; Wu, X.; Liu, J.; Zhang, G.; Wang, Y.; Zeng, Y.; Jing, J. Development of a novel isotropic conductive adhesive filled with silver nanowires. *J. Compos. Mater.* **2006**, *40*, 1961–1969. [CrossRef]
7. Pu, N.-W.; Peng, Y.-Y.; Wang, P.-C.; Chen, C.-Y.; Shi, J.-N.; Liu, Y.-M.; Ger, M.-D.; Chang, C.-L. Application of nitrogen-doped graphene nanosheets in electrically conductive adhesives. *Carbon* **2014**, *67*, 449–456. [CrossRef]
8. Yim, M.J.; Li, Y.; Moon, K.-S.; Paik, K.W.; Wong, C. Review of recent advances in electrically conductive adhesive materials and technologies in electronic packaging. *J. Adhes. Sci. Technol.* **2008**, *22*, 1593–1630. [CrossRef]
9. Yang, C.; Wong, C.P.; Yuen, M.M. Printed electrically conductive composites: Conductive filler designs and surface engineering. *J. Mater. Chem. C* **2013**, *1*, 4052–4069. [CrossRef]
10. Ma, H.; Ma, M.; Zeng, J.; Guo, X.; Ma, Y. Hydrothermal synthesis of graphene nanosheets and its application in electrically conductive adhesives. *Mater. Lett.* **2016**, *178*, 181–184. [CrossRef]
11. Chen, D.; Qiao, X.; Qiu, X.; Tan, F.; Chen, J.; Jiang, R. Effect of silver nanostructures on the resistivity of electrically conductive adhesives composed of silver flakes. *J. Mater. Sci. Mater. Electron.* **2010**, *21*, 486–490. [CrossRef]
12. Amoli, B.M.; Marzbanrad, E.; Hu, A.; Zhou, Y.N.; Zhao, B. Electrical Conductive Adhesives Enhanced with High-A spect-R atio Silver Nanobelts. *Macromol. Mater. Eng.* **2014**, *299*, 739–747. [CrossRef]
13. Cui, H.-W.; Jiu, J.-T.; Nagao, S.; Sugahara, T.; Suganuma, K.; Uchida, H.; Schroder, K.A. Ultra-fast photonic curing of electrically conductive adhesives fabricated from vinyl ester resin and silver micro-flakes for printed electronics. *RSC Adv.* **2014**, *4*, 15914–15922. [CrossRef]
14. Nishikawa, H.; Mikami, S.; Miyake, K.; Aoki, A.; Takemoto, T. Effects of silver coating covered with copper filler on electrical resistivity of electrically conductive adhesives. *Mater. Trans.* **2010**, *51*, 1785–1789. [CrossRef]
15. Ho, L.-N.; Nishikawa, H. Influence of post-curing and coupling agents on polyurethane based copper filled electrically conductive adhesives. *J. Mater. Sci. Mater. Electron.* **2013**, *24*, 2077–2081. [CrossRef]

16. Santamaria, A.; Muñoz, M.E.; Fernández, M.; Landa, M. Electrically conductive adhesives with a focus on adhesives that contain carbon nanotubes. *J. Appl. Polym. Sci.* **2013**, *129*, 1643–1652. [CrossRef]
17. Czech, Z.; Kowalczyk, A.; Pełech, R.; Wróbel, R.; Shao, L.; Bai, Y.; Świderska, J. Using of carbon nanotubes and nano carbon black for electrical conductivity adjustment of pressure-sensitive adhesives. *Int. J. Adhes. Adhes.* **2012**, *36*, 20–24. [CrossRef]
18. Chou, T.-W.; Gao, L.; Thostenson, E.T.; Zhang, Z.; Byun, J.-H. An assessment of the science and technology of carbon nanotube-based fibers and composites. *Compos. Sci. Technol.* **2010**, *70*, 1–19. [CrossRef]
19. Janczak, D.; Słoma, M.; Wróblewski, G.; Młożniak, A.; Jakubowska, M. Thick Film Polymer Composites with Graphene Nanoplatelets for Use in Printed Electronics. In *Mechatronics 2013*; Březina, T., Jabloński, R., Eds.; Springer: Cham, Switzerland, 2014.
20. Jang, B.Z.; Zhamu, A. Nano Graphene Platelet-Based Conductive Inks. US Patent US20120007913A1, 7 January 2010.
21. Crain, J.M.; Lettow, J.S.; Aksay, I.A.; Korkut, S.A.; Chiang, K.S.; Chen, C.-H.; Prud'homme, R.K. Printed Electronics. US Patent 8,278,757, 2 October 2012.
22. Zhu, Y.; Murali, S.; Cai, W.; Li, X.; Suk, J.W.; Potts, J.R.; Ruoff, R.S. Graphene and graphene oxide: Synthesis, properties, and applications. *Adv. Mater.* **2010**, *22*, 3906–3924. [CrossRef]
23. Geim, A.K.; Novoselov, K.S. The rise of graphene. In *Nanoscience and Technology: A Collection of Reviews from Nature Journals*; World Scientific: Singapore, 2010; pp. 11–19.
24. Compton, O.C.; Nguyen, S.T. Graphene oxide, highly reduced graphene oxide, and graphene: Versatile building blocks for carbon-based materials. *Small* **2010**, *6*, 711–723. [CrossRef]
25. Dreyer, D.R.; Ruoff, R.S.; Bielawski, C.W. From conception to realization: An historial account of graphene and some perspectives for its future. *Angew. Chem. Int. Ed.* **2010**, *49*, 9336–9344. [CrossRef] [PubMed]
26. Geim, A.K. Graphene: Status and prospects. *Science* **2009**, *324*, 1530–1534. [CrossRef] [PubMed]
27. Gadipelli, S.; Guo, Z.X. Graphene-based materials: Synthesis and gas sorption, storage and separation. *Prog. Mater. Sci.* **2015**, *69*, 1–60. [CrossRef]
28. Stankovich, S.; Dikin, D.A.; Dommett, G.H.; Kohlhaas, K.M.; Zimney, E.J.; Stach, E.A.; Piner, R.D.; Nguyen, S.T.; Ruoff, R.S. Graphene-based composite materials. *Nature* **2006**, *442*, 282. [CrossRef]
29. Gao, J.; Liu, F.; Liu, Y.; Ma, N.; Wang, Z.; Zhang, X. Environment-friendly method to produce graphene that employs vitamin C and amino acid. *Chem. Mater.* **2010**, *22*, 2213–2218. [CrossRef]
30. Webb, M.J.; Palmgren, P.; Pal, P.; Karis, O.; Grennberg, H. A simple method to produce almost perfect graphene on highly oriented pyrolytic graphite. *Carbon* **2011**, *49*, 3242–3249. [CrossRef]
31. Chen, J.; Yao, B.; Li, C.; Shi, G. An improved Hummers method for eco-friendly synthesis of graphene oxide. *Carbon* **2013**, *64*, 225–229. [CrossRef]
32. Shen, Z.; Li, J.; Yi, M.; Zhang, X.; Ma, S. Preparation of graphene by jet cavitation. *Nanotechnology* **2011**, *22*, 365306. [CrossRef]
33. Yi, M.; Li, J.; Shen, Z.; Zhang, X.; Ma, S. Morphology and structure of mono-and few-layer graphene produced by jet cavitation. *Appl. Phys. Lett.* **2011**, *99*, 123112. [CrossRef]
34. He, Y.; Chen, W.; Li, X.; Zhang, Z.; Fu, J.; Zhao, C.; Xie, E. Freestanding three-dimensional graphene/MnO2 composite networks as ultralight and flexible supercapacitor electrodes. *ACS Nano* **2012**, *7*, 174–182. [CrossRef]
35. Peng, X.; Tan, F.; Wang, W.; Qiu, X.; Sun, F.; Qiao, X.; Chen, J. Conductivity improvement of silver flakes filled electrical conductive adhesives via introducing silver–graphene nanocomposites. *J. Mater. Sci. Mater. Electron.* **2014**, *25*, 1149–1155. [CrossRef]
36. Ghaleb, Z.; Mariatti, M.; Ariff, Z. Properties of graphene nanopowder and multi-walled carbon nanotube-filled epoxy thin-film nanocomposites for electronic applications: The effect of sonication time and filler loading. *Compos. Part A Appl. Sci. Manuf.* **2014**, *58*, 77–83. [CrossRef]
37. Fan, Y.; Wang, L.; Li, J.; Li, J.; Sun, S.; Chen, F.; Chen, L.; Jiang, W. Preparation and electrical properties of graphene nanosheet/Al2O3 composites. *Carbon* **2010**, *48*, 1743–1749. [CrossRef]
38. Breckenridge, R.G.; Hosler, W.R. Electrical properties of titanium dioxide semiconductors. *Phys. Rev.* **1953**, *91*, 793. [CrossRef]
39. Cui, H.-W.; Li, D.-S.; Fan, Q.; Lai, H.-X. Electrical and mechanical properties of electrically conductive adhesives from epoxy, micro-silver flakes, and nano-hexagonal boron nitride particles after humid and thermal aging. *Int. J. Adhes. Adhes.* **2013**, *44*, 232–236. [CrossRef]

40. Cui, H.-W.; Li, D.-S.; Fan, Q. Using nano hexagonal boron nitride particles and nano cubic silicon carbide particles to improve the thermal conductivity of electrically conductive adhesives. *Electron. Mater. Lett.* **2013**, *9*, 1–5. [CrossRef]
41. Tuinstra, F.; Koenig, J.L. Raman spectrum of graphite. *J. Chem. Phys.* **1970**, *53*, 1126–1130. [CrossRef]

© 2019 by the authors. Licensee MDPI, Basel, Switzerland. This article is an open access article distributed under the terms and conditions of the Creative Commons Attribution (CC BY) license (http://creativecommons.org/licenses/by/4.0/).

Article

Ultra-Sensitive Piezo-Resistive Sensors Constructed with Reduced Graphene Oxide/Polyolefin Elastomer (RGO/POE) Nanofiber Aerogels

Weibing Zhong [1], Haiqing Jiang [2], Liyan Yang [2], Ashish Yadav [2], Xincheng Ding [2], Yuanli Chen [2], Mufang Li [2], Gang Sun [1,3] and Dong Wang [1,2,*]

[1] College of Chemistry, Chemical Engineering and Biotechnology, Donghua University, Shanghai 201620, China; weibingzhong09@gmail.com (W.Z.); gysun@ucdavis.edu (G.S.)
[2] Hubei Key Laboratory of Advanced Textile Materials & Application, Wuhan Textile University, Wuhan 430200, China; hqjiang@wtu.edu.cn (H.J.); yangliyan0427@163.com (L.Y.); ashish84yadav@gmail.com (A.Y.); dxcooo@163.com (X.D.); chenyuanli2015@126.com (Y.C.); limufang223@126.com (M.L.)
[3] Division of Textiles and Clothing, University of California, Davis, CA 95616-8598, USA
* Correspondence: wangdon08@126.com

Received: 2 November 2019; Accepted: 12 November 2019; Published: 14 November 2019

Abstract: Flexible wearable pressure sensors have received extensive attention in recent years because of the promising application potentials in health management, humanoid robots, and human machine interfaces. Among the many sensory performances, the high sensitivity is an essential requirement for the practical use of flexible sensors. Therefore, numerous research studies are devoted to improving the sensitivity of the flexible pressure sensors. The fiber assemblies are recognized as an ideal substrate for a highly sensitive piezoresistive sensor because its three-dimensional porous structure can be easily compressed and can provide high interconnection possibilities of the conductive component. Moreover, it is expected to achieve high sensitivity by raising the porosity of the fiber assemblies. In this paper, the three-dimensional reduced graphene oxide/polyolefin elastomer (RGO/POE) nanofiber composite aerogels were prepared by chemical reducing the graphene oxide (GO)/POE nanofiber composite aerogels, which were obtained by freeze drying the mixture of the GO aqueous solution and the POE nanofiber suspension. It was found that the volumetric shrinkage of thermoplastic POE nanofibers during the reduction process enhanced the compression mechanical strength of the composite aerogel, while decreasing its sensitivity. Therefore, the composite aerogels with varying POE nanofiber usage were prepared to balance the sensitivity and working pressure range. The results indicated that the composite aerogel with POE nanofiber/RGO proportion of 3:3 was the optimal sample, which exhibits high sensitivity (ca. 223 kPa^{-1}) and working pressure ranging from 0 to 17.7 kPa. In addition, the composite aerogel showed strong stability when it is either compressed with different frequencies or reversibly compressed and released 5000 times.

Keywords: wearable pressure sensor; piezoresistive sensor; fiber assembly; nanofiber aerogel; reduced graphene oxide

1. Introduction

Flexible wearable pressure sensors have received extensive attention in recent years because of their benefits such as integratability, their lightweight nature, and their portability [1–7]. Compared to the capacitive, piezoelectric, and triboelectric sensors, piezoresistive sensors were widely applied in health management, humanoid robots, human machinery, and artificial intelligence due to their simple structure and easily collectable signal [8–14]. However, the low sensitivity still restricts the practical applications of the piezoresistive sensors. Recently, many structures were designed and constructed to improve the device performance including sensitivity and the working pressure range. Among them, fiber assemblies were considered as ideal substrates that can help improve the sensitivities due to their remarkable deformation ability [15–21]. When the external pressure was loaded, the porous structures constructed by the stacking fibers present larger deformation when compared to solid materials, which resulted in greater growth of the contacting areas of the conductive components. The increasing interconnection of the conductive components formed more effective conductive networks, which improves the sensitivity of the piezoresistive sensors. In addition, research indicates that a higher porosity of the porous substrates will further strengthen the sensitivity of the piezoresistive sensors [13,22].

Nanofiber aerogels, which were obtained by removing the solvent component from the nanofiber suspension in a supercritical state, possess the highest porosity (>80%) among the fiber assemblies. The randomly distributed POE nanofibers with a high length-diameter ratio would lead to hierarchical self-entanglement and would help form a three-dimensional nanofiber-based network [23,24]. Moreover, the ultra-high specific surface area (>500 m^2/g) of the nanofiber aerogel provides the structural basis for the interconnection of conductive components under compression. Therefore, a highly sensitive piezoresistive sensor can be expected by using the nanofiber aerogels as the flexible substrate. However, traditional aerogels always show narrow weak compression strength and low structural stability under revised external pressure [25–27]. Their internal structure will be easily destroyed under excessive external pressure. As a result, the working pressure range and the operational stability of the nanofiber aerogels-based piezoresistive sensor are generally difficult to meet the requirements of practical use. Literature [28] indicates that the aero carbon materials can enhance the mechanical property by forming the hierarchical three-dimensional structure. Therefore, designing and constructing a three-dimensional carbon material/nanofiber composite aerogel might be an effective solution to achieve ultra-high sensitivity, a wide working pressure range, and excellent cycle stability simultaneously for flexible piezo-resistive sensors.

In the present research, the three-dimensional RGO/POE nanofiber composite aerogels were prepared by chemically reducing the GO/POE nanofiber composite aerogels, which were obtained by freeze drying the mixture of the GO aqueous solution and the POE nanofiber suspension. The RGO/POE nanofiber proportions were adjusted to improve the sensing performance of the composite aerogels. It was found that the volumetric shrinkage of thermoplastic POE nanofibers during the reduction process enhances the compression strength of the composite aerogels while decreasing the sensitivity slightly. Therefore, the sensitivity and working pressure range of the composite aerogel were balanced by adjusting the additional amount of the POE nanofibers. The results indicate that the composite aerogel with POE nanofiber/RGO proportion of 3:3 was the optimal sample, which exhibits high sensitivity (ca. 223 kPa^{-1}), wide working pressure range (0–17.7 kPa), and strong stability either compressed with different frequencies or reversibly compressed and released for 5000 times.

2. Materials and Methods

Polyolefin elastomer (POE) and cellulose acetate butyrate (CAB. Butyrate content 35–39%) were purchased from Sigma-Aldrich (Saint Louis, MO, USA), Dow Chemical Company (Midland, MI, USA) and Acros Chemical Co. Ltd., Geel, Belgium, respectively. Tertiary butanol, acetone, concentrated sulfuric acid, potassium permanganate, and hydrochloric acid were from Sinopharm Chemical Reagent Co., Ltd., Shanghai, China. The dispersing agent was supplied by Lubrizol (Lake County, OH, USA). Deionized water is self-made in the laboratory. All the chemicals are used without further purification. The micro morphologies were observed by JSM-6510LV (JEOL, Tokyo, Japan). The chemical structures were measured via FTIR mechine of Tensor 27 (Bruck, Karlsruhe, Germany) The resistances of the aerogels were measured via the 15b+ multimeter (Fluke, Washington, USA). The I-t characteristics of the pressure sensors were collected through ST600L motorized dynamic resistance station (Shente, Shanghai, China)

The POE nanofibers were prepared by extraction removal of the CAB from the POE/CAB composite fiber obtained sea island method. Previous studies reported the detailed method [7,29–32]. Then, the POE nanofibers were dispersed in tertiary butanol-water and the dispersing agent under high speed (10,000 r/m) shearing. The dispersing agent marked with Lubrizol 27,000 was used to help the uniform dispersion process of the POE nanofiber with a 10% mass proportion. The suspension was purified via a filter with a diameter of 150 µm to remove the aggregations. The photographs of the obtained POE nanofiber were presented in Figure S1, which indicates the good uniformity.

The GO aqueous solution was prepared using a modified Hummer's method. The concentration using GO aqueous solution was 5 mg/mL. As shown in Figure 1, the prepared POE nanofiber suspension and the GO aqueous solution were directly mixed together by continuously stirring, according to a GO/POE nanofiber proportion of 6:0, 5:1, 4:2, 3:3, 2:4, and 0:6, respectively. After 5 min of ultrasonic treatment, the mixtures were transferred to a low temperature freezer (−38 °C) for 8 h. The GO/POE nanofiber aerogels were obtained after the mixtures were freeze dried for more than 24 h.

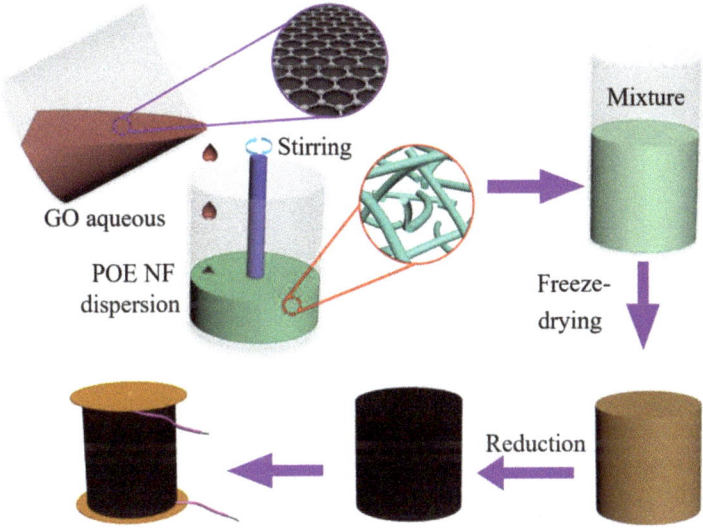

Figure 1. The illustration schematic of the preparation of the RGO/POE nanofiber composite aerogels.

A total of 10 mL of hydrazine hydrate aqueous solution was added to the bottom of the beaker. The obtained GO/POE nanofiber aerogels were placed on a suspension bracket hanging in the baker. Then the beaker was placed in an oven at 90 °C for 100 min after the beaker was sealed. The RGO/POE nanofiber aerogels were obtained after they were placed in a fume hood for more than 6 h.

3. Results and Discussions

The photographs of the prepared aerogels with a different proportion of the POE nanofiber/RGO were shown in Figure 2. As shown, the aerogel containing GO could favorably maintain the cylindrical shape after demolding, and it showed the same maple color as the pure GO aqueous solution. After air phase reduction by a hydrazine vapor, the reduced aerogels were all converted to a black color. The aerogel had a volume retraction during the reduction process. The shape retention rate of the aerogel during the reduction process was shown in Table S1. The aerogel with more POE content resulted in the lower shape retention rate. The lowest shape retention rate at only 14.91% appeared on the aerogel with a POE nanofiber/GO proportion of 5:1. This was caused by the softening and heat-induced shrinkage of the POE nanofibers under the high temperature (90 °C) during the reduction process. The shrinkage during this process may contribute to the physical crosslinking between the softening POE nanofibers, which helps improve the structural stability of the aerogels. The volume of the pure RGO aerogel reached 140.48% of the GO aerogel. This might be caused by the straightening of curled GO after undergoing a reduction. However, it can be easily observed that the pure GO aerogel had lamellar structures and the flake graphene is prone to chipping and slag. The lamellar structure easily collapsed under pressure.

Figure 2. The photographs of the composite aerogels with POE nanofiber/GO proportion of (**a**) 6:0, (**b**) 5:1, (**c**) 4:2, (**d**) 3:3, (**e**) 2:4, and (**f**) 0:6. The inserted images were the photographs of corresponding aerogels after being reduced.

The microstructure of the prepared aerogel was observed by scanning electron fiber microscopy (SEM). Figure 3 presented the SEM of pure POE nanofiber aerogel, pure RGO aerogel, GO/POE nanofibers aerogel, and RGO/POE nanofibers aerogel (POE: GO = 3:3). As observed, the pure POE nanofiber aerogel was composed of fluffy nanofibers, which were randomly distributed. In the pure RGO aerogel, the stacking structure of reduced RGO sheets with wrinkles on the surfaces can be easily observed. There is no nanofiber support between the sheets, which may result in the collapsing structure of the pure RGO aerogels. For the composite aerogels of GO/POE and RGO/POE nanofiber aerogels, the uniform porous structures were constructed since the graphene and the fibers were intertwined. As can be seen, the GO layer was extremely thin in GO/POE nanofiber aerogels, which made it look like a semi-transparent substance. After undergoing a reduction, the pore size decreased and the structure became compact, which may be induced by the volumetric shrinkage during the reduction process.

Figure 4 showed the density of the aerogel material before and after reduction and the compressive stress and strain curves of the aerogels after undergoing a reduction. It can be seen from Figure 4a that the density of the GO/POE nanofiber aerogel before reduction was relatively low, which distributed between 11.1–15.1 mg/cm^3. The density of the pure RGO aerogel decreased from 11.8 to 8.4 mg/cm^3 after undergoing a reduction. For the RGO/POE nanofiber composite aerogels, the density increased to some degree. The aerogels with more POE nanofibers content exhibited a greater density increase. Among them, the highest density reached 95.9 mg/cm^3. It is mainly related to the volumetric shrinkage during the reduction process. Since the aerogel before a reduction is easily collapsed and the compression test cannot be performed, Figure 4b only showed the compressive stress and strain diagram of the aerogel after undergoing a reduction. It can be seen that the compressive stress of aerogel increased very slowly in the low strain region but enhanced rapidly in the high strain region. This may be due to the high porosity of the aerogel, which requires a large amount of strain space to squeeze out the internal air. In addition, compared with the aerogels with different RGO/POE proportions, the aerogels with more POE nanofiber content have higher stress at the same strain, which represents stronger compression strength. This may be caused by the higher density and solid content that induced more force units during compression in aerogels with more POE nanofibers content, which makes it easier to generate greater stress.

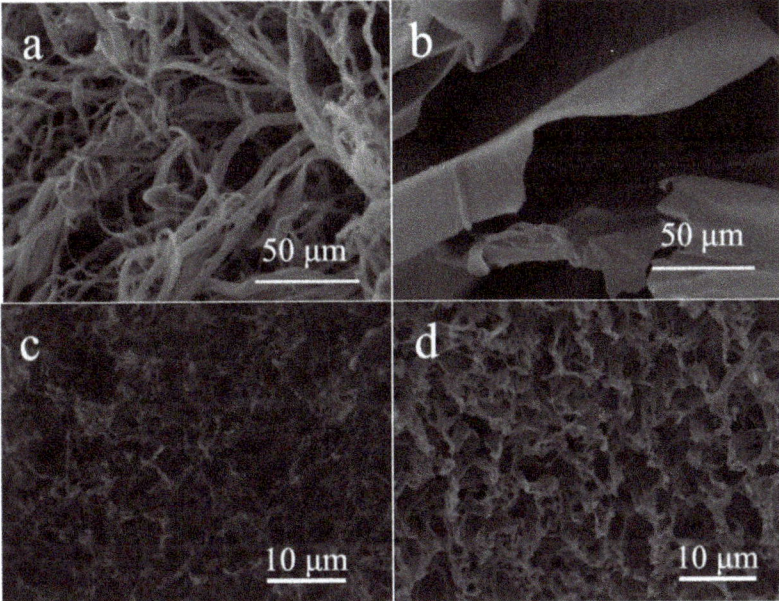

Figure 3. The SEM images of aerogels comprised of (**a**) pure POE nanofibers, (**b**) pure RGO, (**c**) GO/POE nanofibers, and (**d**) RGO/POE nanofibers.

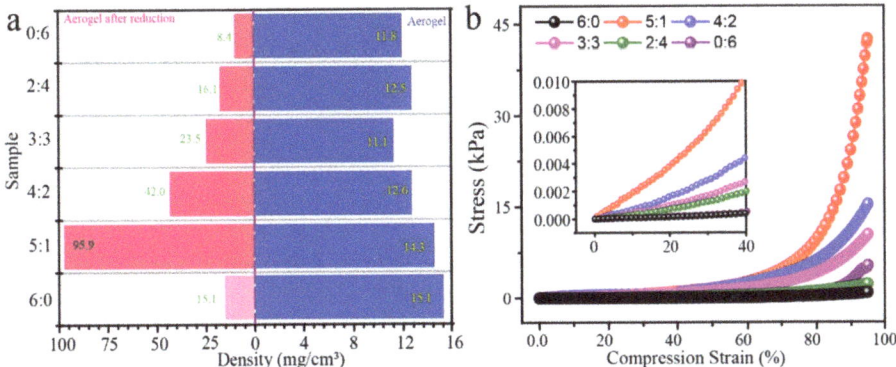

Figure 4. (**a**) The density of the prepared aerogels before and after a reduction, (**b**) the compression stress and strain curve of the prepared aerogels with different POE nanofiber/RGO proportions.

Figure 5 presented the XRD spectra of the prepared aerogel materials of pure GO and RGO. It can be seen that the peak of pure GO and RGO appeared at 9.92° and 24.10°, which means that the spacing of graphene sheets is 0.891 nm and 0.368 nm, respectively. Since the POE is partially crystallized, which acted as the physical crosslinking section, the POE exhibits several characteristic peaks. As shown, the peak position of pure POE appeared at 19.81°, 20.93°, 30.72°, and 41.64°. When the GO is added into the aerogel, the peak position of GO blue shifted (2θ = 8.50°), which suggested that the addition of POE leads to a larger spacing of the GO layer. The space reached 1.02 nm from 0.891 nm, which indicates that the POE nanofiber was uniformly mixed. In the RGO/POE nanofiber aerogel, the peak (2θ = 24.10°) is the same as pure RGO. Therefore, the mixing of POE and GO forms an impedance effect on the GO graphitization process, which results in undiminished RGO layer spacing.

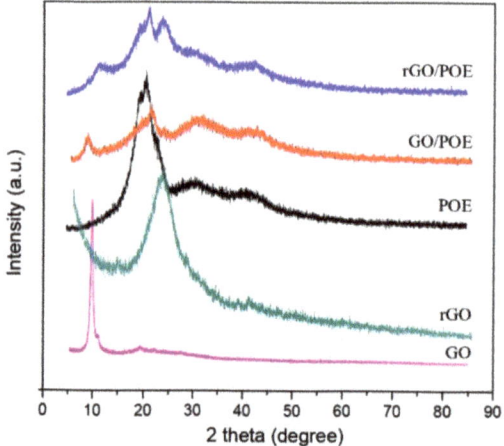

Figure 5. XRD spectrum of the prepared aerogels composed by GO, RGO, and POE nanofiber, GO/POE, and an RGO/POE nanofiber composite aerogel.

The FTIR of the prepared aerogel before and after a reduction were measured in Figure 6. The absorption peaks at 3300, 1718, 1625, 1230, and 1066 cm^{-1} represent the free hydroxyl bonds, carbon-oxygen double bonds, and epoxy groups on the carbonyl group, respectively. The absorptions were complex in the 1300 cm^{-1} to 400^{-1} region, which may include the carbon-oxygen stretching vibration, stretching resonance of carbon-sulfur bonds, and skeleton vibration. The absorption peaks of POE mainly at 2914, 2834, 1463, 1102, and 713 cm^{-1} represented the stretching resonance of the single bond in the methylene group and the in-plane and out-of-plane bending resonance absorption peaks of carbon-hydrogen bonds, respectively. It is noticeable that, with the decrease of GO content, the peak intensity of GO significantly decreased, while the absorption peak of POE gradually appeared. Moreover, even when the mixed ratio of POE and GO reached 5:1, the characteristic absorption peaks of POE were still weak. It can be inferred that the two-dimensional GO uniformly encapsulated the POE nanofibers, which makes it difficult to obtain the absorption peak of POE by using the measurement method of surface reflection. From Figure 6b, only extremely weak characteristic absorption peaks at 3729, 3630, 2328, and 1213 cm^{-1} remained after the reduction while most of the other absorption peaks disappeared. It suggested that there were rare free hydroxyl groups and a small amount of triple bonds or cumulative double bonds, which indicates a higher degree of reduction. Moreover, the few active groups mean the relatively stable chemical properties of the composite aerogels, which may make it stronger in the anti-interference ability in practical uses.

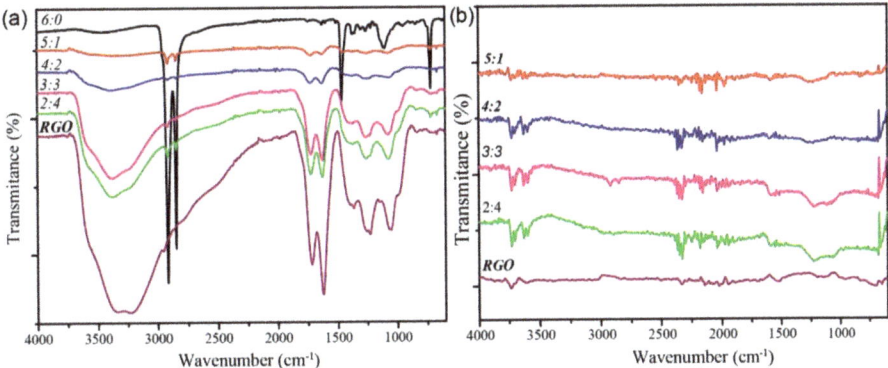

Figure 6. The FT-IR spectra of the prepared aerogels with different POE nanofiber/RGO proportions (a) before and (b) after undergoing a reduction.

The electrical resistance and relative current changes under different pressures of the RGO/POE nanofiber aerogels with a different RGO/POE proportion were measured and the results were shown in Figure 7. Because of the low solid content of the aerogels, the conductive RGO sheets were difficult to interconnect into an effective conductive network. Therefore, the aerogels exhibited very high electrical resistance of ca. 10^7–10^8 Ω (Figure 7a). The electrical resistance of the aerogel decreased when the amount of RGO increased. The aerogel with the POE/RGO proportion of 5:1 reached the highest electrical resistance of ca. 110.7 MΩ while the pure RGO aerogel shows the lowest resistance of ca. 16.2 MΩ. Figure 7b presented the $\Delta I/I_0$ versus pressure curves of aerogels with different RGO/POE nanofiber proportions. As shown, the obtained aerogel samples all exhibited an increasing relative current change ($\Delta I/I_0$) when exposed to raising external pressures. The more RGO usage led to a higher $\Delta I/I_0$ value but narrower working pressure ranges, which were determined by varying inner conductive networks and compression mechanical properties. Aerogels with more RGO usage present higher possibilities to form extra interconnections of conductive RGO sheets. Meanwhile, the aerogel with the higher RGO amount possess a lower density and smaller stress under the same compression deformation, according to the results in Figure 4.

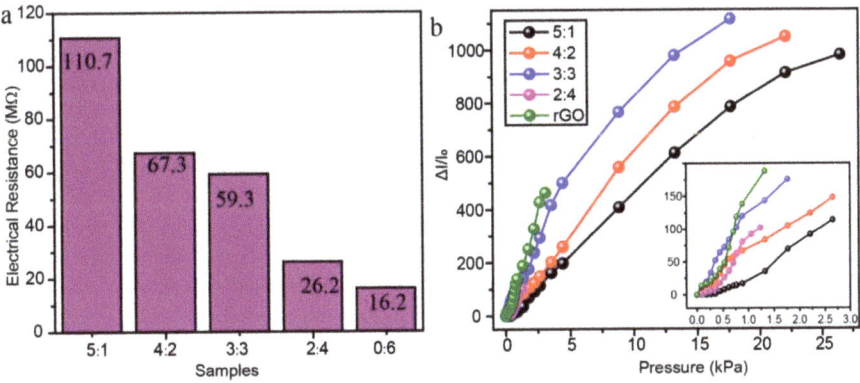

Figure 7. (a) The electrical resistance and (b) the relative current change curves under different pressures of the prepared composite aerogels with different POE nanofiber/RGO proportions.

The sensitivities of the aerogels were calculated via the definition equation of $S = \partial(\Delta I/I_0)/\partial p$. The results were shown in Figure 8a. As can be observed, the prepared aerogel exhibits high sensitivities in the low-pressure region (<5 kPa) and the peaks appeared at the pressure region of 0.5 to 1.0 kPa. The pure RGO aerogel showed the highest sensitivity of 391 kPa^{-1} at a pressure of ca. 0.8 kPa. The highest sensitivity of each aerogel decreased as the RGO usage reduced. The aerogel with POE/RGO proportion of 5:1 exhibited the lowest sensitivity of ca. 83 kPa^{-1}. The RGO/POE nanofiber composite aerogels exhibited different working pressure ranges, which was displayed in Figure 8b. As presented, the RGO/POE nanofiber composite aerogels with more POE content possess a larger working pressure range. The aerogel with POE/RGO proportion of 5:1 shows the widest working pressure range of 0–26.5 kPa while the narrowest was from 0–1.2 kPa when the POE/RGO proportion was 2:4. This correlated with the compression mechanical properties of the composite aerogels with different RGO/POE proportions. Taking the sensitivity and working pressure range into consideration, the composite aerogel with the RGO/POE nanofiber was selected as the optimal samples due to the high sensitivity of ca. 223 kPa^{-1} and a wide working pressure range of 0–17.7 kPa. The optimal sample was applied for the following stability testing.

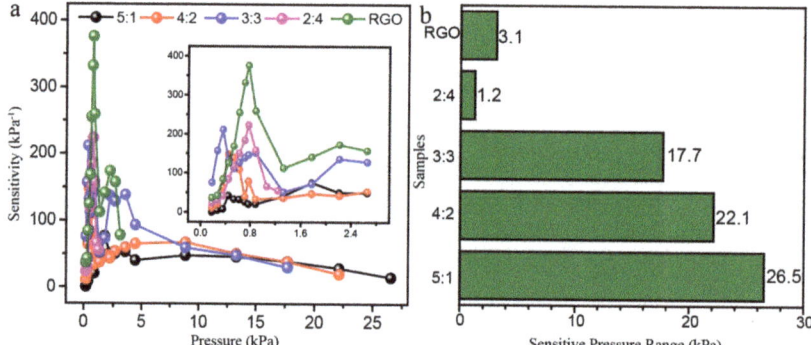

Figure 8. (a) The sensitivities and (b) the working pressure ranges of the prepared composite aerogels with different POE nanofiber/RGO proportions.

The stability of the composite aerogel with POE/RGO proportion of 3:3 was tested by reversibly applying and removing the compression strain with various frequencies. Due to the high moving speed of the indenter during the stability test, especially in the cycle stability test, the inertia causes a separation of the electrode and the aerogel while the indenter is on the highest point. This phenomenon leads to an inaccurate stability result. Therefore, the compression strain region of 10% to 50% was selected instead of a region between 0% to 50%. The results were presented in Figure 9a. The different compression and releasing periods of 3.2 s, 1.6 s, 1.2 s, and 0.8 s were implemented, respectively. As shown, the signal outputs were regularly arranged upward peaks. The shapes and the values of the signal outputs were almost the same under different compression frequencies, which indicated the good stability of the composite aerogel sensor. In addition, the composite aerogel sensor was repeatedly compressed and released under the strain from 10% to 50% for about 5000 times and the signal output was presented in Figure 9b. As observed, the shapes of the signal outputs remain the same from the beginning to the end of the test. During the 5000 times testing, only the values of the peak tops fluctuate slightly, which suggested good cycling stability of the composite aerogel. Moreover, the compression stress and strain curves under different compressing frequencies and different cycles during the 5000 times cycling tests were displayed in Figure 10a,b. It can be found that only a small amount of stress drawdowns occurred in different frequency compression tests and cycling tests. This further proves the compressive mechanical stability of the RGO/POE nanofiber composite aerogel, which may result in physical crosslinking by the softening of the POE nanofibers during the reduction process. Figure 11 presented several screenshots of the compression and releasing process from a cycling testing video, which suggested that there is no irreversible deformation and almost no reply hysteresis during the whole compressing and releasing process.

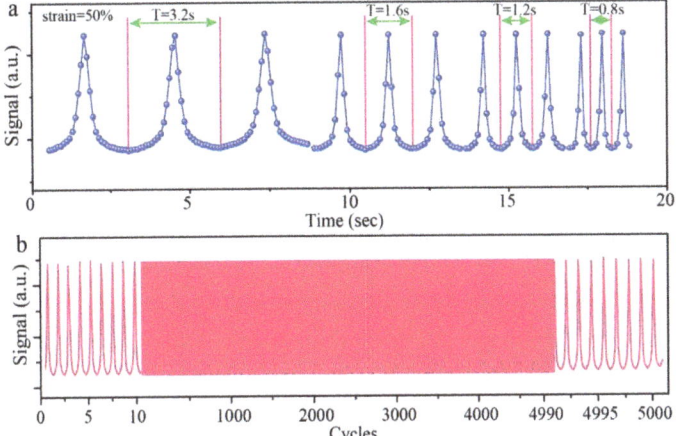

Figure 9. The stability of the prepared composite aerogel with POE nanofiber/RGO proportions of 3:3 under (**a**) dynamic pressure with a different frequency and (**b**) reversible compression and releasing for about 5000 times.

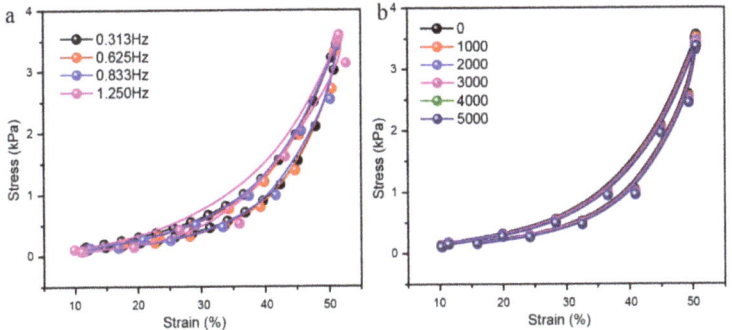

Figure 10. The relative compression stress and strain curve of the prepared sensor under (**a**) a different compression frequency and (**b**) a 5000 times cycling test.

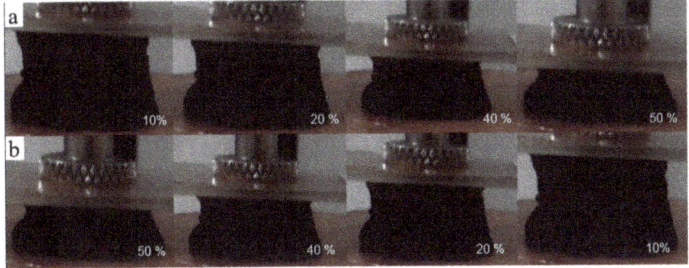

Figure 11. The digital camera photographs of the (**a**) compressing and (**b**) releasing process of the prepared aerogels.

4. Conclusions

In summary, the three-dimensional composite aerogels were prepared with one-dimensional POE nanofibers and two-dimensional RGO sheets. The RGO/POE nanofiber proportions were adjusted to improve the sensing performance of the composite aerogels. The results suggest that the proportions show great impact on volume shrinkage during reduction, compression mechanical behavior, sensitivity, and a working pressure range of the composite aerogels. The aerogels with more POE nanofiber content exhibit lower volume retention, larger stress under the same compression strain, lower sensitivity, and a wider working pressure range. Taking the sensitivity and the working pressure range into consideration, the composite aerogel with POE nanofiber/RGO proportion of 3:3 was selected as the optimal sample, which exhibits high sensitivity (ca. 223 kPa^{-1}) and a wide working pressure range (0–17.7 kPa). It also showed excellent operational stability either compressed with different frequencies or reversibly compressed and released for 5000 times.

Supplementary Materials: The following are available online at http://www.mdpi.com/2073-4360/11/11/1883/s1, Figure S1: The photographs of the uniform POE nanofiber suspension obtained from different observing angles, Table S1: The volume retention of the prepared GO/POE nanofiber aerogels after reduction.

Author Contributions: Conceptualization, D.W., G.S. and M.L.; methodology, W.Z.; software, X.D.; validation, H.J., L.Y. and Y.C.; formal analysis, W.Z.; investigation, W.Z. and X.D.; resources, D.W. and G.S.; data curation, X.D.; writing—original draft preparation, W.Z.; writing—review and editing, A.Y.; visualization, M.L.; supervision, D.W. and G.S.; project administration, W.Z.; funding acquisition, D.W., M.L., and Y.C.

Funding: We acknowledge funding support from the National Natural Science Foundation of China (Grant No. 51873165, 51873166, 51603155), Science and Technology Innovation Projects of Hubei Province (2017AHB065), and central guidance for local science and technology development projects (2018ZYYD057).

Conflicts of Interest: The authors declare no conflict of interest.

References

1. Chang, H.; Kim, S.; Jin, S.; Lee, S.W.; Yang, G.T.; Lee, K.Y.; Yi, H. Ultrasensitive and Highly Stable Resistive Pressure Sensors with Biomaterial-Incorporated Interfacial Layers for Wearable Health-Monitoring and Human-Machine Interfaces. *ACS Appl. Mater. Interfaces* **2017**, *10*, 1067–1076. [CrossRef] [PubMed]
2. Chen, Z.; Wang, Z.; Li, X.; Lin, Y.; Luo, N.; Long, M.; Zhao, N.; Xu, J.B. Flexible Piezoelectric-Induced Pressure Sensors for Static Measurements Based on Nanowires/Graphene Heterostructures. *ACS Nano* **2017**, *11*, 4507–4513. [CrossRef] [PubMed]
3. Hong-Bin, Y.; Jin, G.; Chang-Feng, W.; Xu, W.; Wei, H.; Zhi-Jun, Z.; Yong, N.; Shu-Hong, Y. Pressure sensors: A flexible and highly pressure-sensitive graphene-polyurethane sponge based on fractured microstructure design. *Adv. Mater.* **2013**, *25*, 6691.
4. Lee, J.H.; Yoon, H.J.; Kim, T.Y.; Gupta, M.K.; Lee, J.H.; Seung, W.; Ryu, H.; Kim, S.W. Energy Harvesting: Micropatterned P(VDF-TrFE) Film-Based Piezoelectric Nanogenerators for Highly Sensitive Self-Powered Pressure Sensors. *Adv. Funct. Mater.* **2015**, *25*, 3276. [CrossRef]
5. Shu, W.; Bi, H.; Zhou, Y.; Xiao, X.; Shi, S.; Yin, K.; Sun, L. Graphene oxide as high-performance dielectric materials for capacitive pressure sensors. *Carbon* **2017**, *114*, 209–216.
6. Zang, Y.; Zhang, F.; Di, C.A.; Zhu, D. Advances of flexible pressure sensors toward artificial intelligence and health care applications. *Mater. Horiz.* **2015**, *2*, 25–59. [CrossRef]
7. Zhong, W.; Liu, C.; Liu, Q.; Piao, L.; Jiang, H.; Wang, W.; Liu, K.; Li, M.; Sun, G.; Wang, D. Ultrasensitive Wearable Pressure Sensors Assembled by Surface-Patterned Polyolefin Elastomer Nanofiber Membrane Interpenetrated with Silver Nanowires. *ACS Appl. Mater. Interfaces* **2018**, *10*, 42706–42714. [CrossRef]
8. Atalay, O.; Atalay, A.; Gafford, J.; Walsh, C. A Highly Sensitive Capacitive-Based Soft Pressure Sensor Based on a Conductive Fabric and a Microporous Dielectric Layer. *Adv. Mater. Technol.* **2018**, *3*, 1700237. [CrossRef]
9. Bandari, N.M.; Ahmadi, R.; Hooshiar, A.; Dargahi, J.; Packirisamy, M. Hybrid piezoresistive-optical tactile sensor for simultaneous measurement of tissue stiffness and detection of tissue discontinuity in robot-assisted minimally invasive surgery. *J. Biomed. Opt.* **2017**, *22*, 077002. [CrossRef]

10. Dong, N.; Jiang, W.; Ye, G.; Wang, K.; Lei, Y.; Shi, Y.; Chen, B.; Luo, F.; Liu, H. Graphene-elastomer nanocomposites based flexible piezoresistive sensors for strain and pressure detection. *Mater. Res. Bull.* **2018**, *102*, 92–99.
11. He, X.; Liu, Q.; Zhong, W.; Chen, J.; Sun, D.; Jiang, H.; Liu, K.; Wang, W.; Wang, Y.; Lu, Z. Strategy of Constructing Light-Weight and Highly Compressible Graphene-Based Aerogels with an Ordered Unique Configuration for Wearable Piezoresistive Sensors. *ACS Appl. Mater. Interfaces* **2019**, *11*, 19350–19362. [CrossRef] [PubMed]
12. Liu, W.; Liu, N.; Yue, Y.; Rao, J.; Cheng, F.; Su, J.; Liu, Z.; Gao, Y. Piezoresistive Pressure Sensor Based on Synergistical Innerconnect Polyvinyl Alcohol Nanowires/Wrinkled Graphene Film. *Small* **2018**, *14*, 1704149. [CrossRef] [PubMed]
13. Zhao, T.; Li, T.; Chen, L.; Yuan, L.; Li, X.; Zhang, J. Highly Sensitive Flexible Piezoresistive Pressure Sensor Developed Using Biomimetically Textured Porous Materials. *ACS Appl. Mater. Interfaces* **2019**, *11*, 29466–29473. [CrossRef] [PubMed]
14. Zhong, W.; Ding, X.; Li, W.; Shen, C.; Yadav, A.; Chen, Y.; Bao, M.; Jiang, H.; Wang, D. Facile Fabrication of Conductive Graphene/Polyurethane Foam Composite and Its Application on Flexible Piezo-Resistive Sensors. *Polymers* **2019**, *11*, 1289. [CrossRef]
15. Atalay, O.; Kennon, W.R.; Husain, M.D. Textile-Based Weft Knitted Strain Sensors: Effect of Fabric Parameters on Sensor Properties. *Sensors* **2013**, *13*, 11114–11127. [CrossRef]
16. Chang, K.; Li, M.; Zhong, W.; Wu, Y.; Luo, M.; Chen, Y.; Liu, Q.; Liu, K.; Wang, Y.; Lu, Z. A novel, stretchable, silver-coated polyolefin elastomer nanofiber membrane for strain sensor applications. *J. Appl. Polym. Sci.* **2019**, *136*, 47928. [CrossRef]
17. Cheng, Y.; Wang, R.; Zhai, H.; Sun, J. Stretchable electronic skin based on silver nanowire composite fiber electrodes for sensing pressure, proximity, and multidirectional strain. *Nanoscale* **2017**, *9*, 3834–3842. [CrossRef]
18. Gao, R.D.; Yang, J.Z.; Fa-Zhou, L.I. Compressibility of variation cashmere fiber assemblies. *Wool Text. J.* **2011**, *39*, 39–41.
19. Gao, J.; Pan, N.; Yu, W. Compression behavior evaluation of single down fiber and down fiber assemblies. *J. Text. Inst. Proc. Abstr.* **2010**, *101*, 253–260. [CrossRef]
20. Wei, Z.; Lin, S.; Qiao, L.; Song, C.; Fei, W.; Xiao, T. Fiber-based wearable electronics: A review of materials, fabrication, devices, and applications. *Adv. Mater.* **2014**, *26*, 5310.
21. Liu, K.; Zhou, Z.; Yan, X.; Meng, X.; Tang, H.; Qu, K.; Gao, Y.; Li, Y.; Yu, J.; Li, L. Polyaniline Nanofiber Wrapped Fabric for High Performance Flexible Pressure Sensors. *Polymers* **2019**, *11*, 1120. [CrossRef] [PubMed]
22. Zhao, Z.; Li, B.; Xu, L.; Qiao, Y.; Wang, F.; Xia, Q.; Lu, Z. A sandwich-structured piezoresistive sensor with electrospun nanofiber mats as supporting, sensing, and packaging layers. *Polymers* **2018**, *10*, 575. [CrossRef] [PubMed]
23. Schütt, F.; Signetti, S.; Krüger, H.; Röder, S.; Smazna, D.; Kaps, S.; Gorb, S.; Mishra, Y.K.; Pugno, N.; Adelung, R. Hierarchical self-entangled carbon nanotube tube networks. *Nat. Commun.* **2017**, *8*, 1215. [CrossRef] [PubMed]
24. Rasch, F.; Schütt, F.; Saure, L.M.; Kaps, S.; Strobel, J.; Polonskyi, O.; Nia, A.; Lohe, M.; Mishra, Y.K.; Faupel, F.; et al. Wet-Chemical Assembly of 2D Nanomaterials into Lightweight, Microtube-Shaped, and Macroscopic 3D Networks. *ACS Appl. Mater. Interfaces* **2019**. [CrossRef] [PubMed]
25. Maleki, H.; Durães, L.; Portugal, A. An overview on silica aerogels synthesis and different mechanical reinforcing strategies. *J. Non-Cryst. Solids* **2014**, *385*, 55–74. [CrossRef]
26. Madyan, O.A.; Fan, M.; Feo, L.; Hui, D. Enhancing mechanical properties of clay aerogel composites: An overview. *Compos. Part B Eng.* **2016**, *98*, 314–329. [CrossRef]
27. Meador, M.A.B.; Fabrizio, E.F.; Ilhan, F.; Dass, A.; Zhang, G.; Vassilaras, P.; Johnston, J.C.; Leventis, N. Cross-linking amine-modified silica aerogels with epoxies: Mechanically strong lightweight porous materials. *Chem. Mater.* **2005**, *17*, 1085–1098. [CrossRef]
28. Mecklenburg, M.; Schuchardt, A.; Mishra, Y.K.; Kaps, S.; Adelung, R.; Lotnyk, A.; Kienle, L.; Schulte, K. Aerographite: Ultra lightweight, flexible nanowall, carbon microtube material with outstanding mechanical performance. *Adv. Mater.* **2012**, *24*, 3486–3490. [CrossRef]

29. Li, F.; Wang, D.; Liu, Q.; Wang, B.; Zhong, W.; Li, M.; Liu, K.; Lu, Z.; Jiang, H.; Zhao, Q. The construction of rod-like polypyrrole network on hard magnetic porous textile anodes for microbial fuel cells with ultra-high output power density. *J. Power Sources* **2019**, *412*, 514–519. [CrossRef]
30. Wang, Y.; Zhou, Z.; Qing, X.; Zhong, W.; Liu, Q.; Wang, W.; Li, M.; Liu, K.; Wang, D. Ion sensors based on novel fiber organic electrochemical transistors for lead ion detection. *Anal. Bioanal. Chem.* **2016**, *408*, 5779–5787. [CrossRef]
31. Zhong, W.; Liu, C.; Xiang, C.; Jin, Y.; Li, M.; Liu, K.; Liu, Q.; Wang, Y.; Sun, G.; Wang, D. Continuously producible ultrasensitive wearable strain sensor assembled with three-dimensional interpenetrating Ag nanowires/polyolefin elastomer nanofibrous composite yarn. *ACS Appl. Mater. Interfaces* **2017**, *9*, 42058–42066. [CrossRef] [PubMed]
32. Zhong, W.; Liu, Q.; Wu, Y.; Wang, Y.; Qing, X.; Li, M.; Liu, K.; Wang, W.; Wang, D. A nanofiber based artificial electronic skin with high pressure sensitivity and 3D conformability. *Nanoscale* **2016**, *8*, 12105–12112. [CrossRef] [PubMed]

© 2019 by the authors. Licensee MDPI, Basel, Switzerland. This article is an open access article distributed under the terms and conditions of the Creative Commons Attribution (CC BY) license (http://creativecommons.org/licenses/by/4.0/).

Article

The Electrochemical Oxidation of Hydroquinone and Catechol through a Novel Poly-geminal Dicationic Ionic Liquid (PGDIL)–TiO$_2$ Composite Film Electrode

Yanni Guo [1], Deliang He [1,*], Aomei Xie [2], Wei Qu [1], Yining Tang [1], Lei Zhou [1] and Rilong Zhu [1,*]

[1] State key Laboratory of Chemo/Biosensing and Chemometrics, College of Chemistry and Chemical Engineering, Hunan University, Applied Chemistry, Changsha 410082, China; gyn0303@hnu.edu.cn (Y.G.); qw779470790@163.com (W.Q.); yiningt@163.com (Y.T.); zl3139552204@hnu.edu.cn (L.Z.)
[2] College of Marxism, Jishou University, Jishou 416000, China; xieaomei@163.com
* Correspondence: delianghe@163.com (D.H.); zrlden@hnu.edu.cn (R.Z.)

Received: 17 October 2019; Accepted: 15 November 2019; Published: 19 November 2019

Abstract: A novel poly-geminal dicationic ionic liquid (PGDIL)-TiO$_2$/Au composite film electrode was successfully prepared by electrochemical polymerization of 1,4-bis(3-(m-aminobenzyl) imidazol-1-yl)butane bis(hexafluorinephosphate) containing polymerizable anilino groups in the electrolyte containing nano-TiO$_2$. The basic properties of PGDIL–TiO$_2$/Au composite films were studied by SEM, cyclic voltammetry, electrochemical impedance spectroscopy, and differential pulse voltammetry. The SEM results revealed that the PGDIL–TiO$_2$ powder has a more uniform and smaller particle size than the PGDIL. The cyclic voltammetry results showed that the catalytic effect on electrochemical oxidation of hydroquinone and catechol of the PGDIL–TiO$_2$ electrode is the best, yet the R_{ct} of PGDIL–TiO$_2$ electrode is higher than that of PGDIL and TiO$_2$ electrode, which is caused by the synergistic effect between TiO$_2$ and PGDIL. The PGDIL–TiO$_2$/Au composite electrode presents a good enhancement effect on the reversible electrochemical oxidation of hydroquinone and catechol, and differential pulse voltammetry tests of the hydroquinone and catechol in a certain concentration range revealed that the PGDIL–TiO$_2$/Au electrode enables a high sensitivity to the differentiation and detection of hydroquinone and catechol. Furthermore, the electrochemical catalytic mechanism of the PGDIL–TiO$_2$/Au electrode was studied. It was found that the recombination of TiO$_2$ improved the reversibility and activity of the PGDIL–TiO$_2$/Au electrode for the electrocatalytic reaction of HQ and CC. The PGDIL–TiO$_2$/Au electrode is also expected to be used for catalytic oxidation and detection of other organic pollutants containing –OH groups.

Keywords: nano-TiO$_2$; poly-geminal dicationic ionic liquid; hydroquinone; catechol

1. Introduction

Poly ionic liquids (PILs) are a kind of functional polymer material that contain at least one ion center in a polymer chain and a repeating unit similar to a common ionic liquid (IL) structure; they combine the properties of polymers and ionic liquids, and show are in the foreground of applications for ionic conductors, adsorption and separation, dispersants, and catalysts [1–4]. Therefore, the research on the preparation and performance of PILs has aroused wide interest and concern in recent years [5,6].

The dicationic ionic liquids (DILs) are considered to be a combination of three structural moieties: (1) cationic head groups; (2) a linkage chain (also called a spacer); and (3) counter anions. DILs can be classified as either homoanionic or heteroanionic, which can further be categorized as symmetrical or asymmetrical. Homoanionic dicationic ionic liquids are typical DILs that consist of a dication and two identical anions. Symmetrical or geminal dicationic ILs (GDILs) can be synthesized by joining two of the same cations, such as imidazolium or pyrrolidinium, and may contain a cyclic or aliphatic chain via either a rigid or a flexible spacer. A common spacer is an alkyl chain [7,8]. Armstrong et al. [9] studied the relationship between the structure and properties of DILs by synthesizing 39 imidazolium-based and pyrrolidinium-based DILs. The head groups were linked with an alkyl chain (from three to 12 carbons long), and hence reacted with four different traditional anions. The thermal stability of these ILs was found to be in the range of −4 to > 400 °C, which is greater than that of most traditional monocationic ILs. GDILs have more unique physical and chemical properties and solvation characteristics, and can be used as separation materials [10], surfactants [11], and catalyst candidates [12,13], although they have not yet been studied heavily [14,15]. The same is true of poly-GDILs (PGDILs).

As is known, nano-TiO_2 has better chemical properties and photon characteristics due to its good absorbability and lower electron/hole recombining rate, and can be used as a new kind of electrical catalyst material [16–18]. There have been many reports on the application of nano-TiO_2 composite polymers in the electrocatalysis of organic materials [19], including a study by the present research group [20]. Aniline is a kind of monomer that is easy to polymerize. Polyaniline is an important conductive polymer [21–23] that can be used as anode or cathode material with electrocatalytic function due to its excellent electrical and electrochemical properties [24]. The combination of polyaniline and nano-TiO_2 can not only effectively inhibit the agglomeration of TiO_2 nanoparticles, but also improve the physical and chemical properties of composites, and may be widely applied in the electrochemical catalysis field [20,25,26].

Hydroquinone (HQ) and catechol (CC) are important phenolic compounds that are widely used as basic raw materials in the organic chemical, agriculture, and medicine industries, among others [27]. They are typical and important electro-active molecules in fundamental electrochemical research. Therefore, it is of great significance to establish a high-sensitivity detection method for HQ and CC in the fields of environmental monitoring and food inspection. Moreover, the chemical structures and the physicochemical properties of HQ and CC are very similar, and they are, therefore, difficult to distinguish [28]. Current methods for the determination of HQ and CC include fluorescence [29], the electrochemical method [30], the photometric method [31], and high-performance liquid chromatography [32]. The electrochemical method is highly valued because of its versatility, simple operation, easy automation, and environmental compatibility. In electro-catalytic reactions, in which an electrode is an electrical catalyst, different electrode materials can change the electrochemical reaction rate by different magnitudes, so new and efficient catalytic electrode materials have always been a focus of related research.

The present research group previously successfully synthesized 1,4-bis(3-(m-aminobenzyl) imidazol-1-yl)butane bis(hexafluorinephosphate) ([C_4(m-ABIM)$_2$][PF_6]$_2$), which is a novel GDIL containing anilino groups, and poly-[C_4(m-ABIM)$_2$][PF_6]$_2$ (PGDIL) with a polyaniline-like structure was prepared by electro-polymerization [33]. In this study, a novel PGDIL–TiO_2 composite was prepared via electrodeposition on an Au electrode surface, and the electrochemical redox behaviors of HQ and CC were respectively investigated via the cyclic voltammetry (CV) and differential pulse voltammetry (DPV) methods. It was expected that the PGDIL–TiO_2 composite film would have the advantages of both PGDIL and TiO_2.

2. Experimental

2.1. Main Reagents and Instrumentation

The primary reagents used in the experiments, [C_4(m-ABIM)$_2$][PF_6]$_2$ ([33]), sodium dihydrogen phosphate, disodium hydrogen phosphate, potassium chloride, hydroquinone, and anatase nano-TiO_2 powder (10–25 nm), were of analytical grade, and were used as received. The water used in the experiment was double distilled water.

The instruments used in the experiments were an automatic double distilled water distillation apparatus D1810C (Shanghai Asia-Pacific Glass instrument Company, Shanghai, China); a heat-collecting, constant temperature magnetic heating stirrer CJJ 78-1 (Zhengzhou Greatwall Scientific Industrial and Trade Co., Ltd., Zhengzhou, China); a rotary evaporator R.1002 (Zhengzhou Greatwall Scientific Industrial and Trade Co., Ltd., Zhengzhou, China); an ultrasonic cleaner SK5200HP (Shanghai Kudos Ultrasonic instrument Co., Ltd., Shanghai, China); an infrared spectrophotometer TJ270-30 (A) (Tianjin Jinwei Electronic Instrument Co., Ltd., Tianjin, China); and an electrochemical workstation IviumStat (Ivium Technologies BV, Eindhoven, Netherlands).

2.2. Synthesis of Poly-[C_4(m-ABIM)$_2$][PF_6]$_2$–TiO_2 Composite Film

An electrochemical tri-electrode system was introduced for electropolymerization. An Au electrode, platinum wire, and Ag/AgCl electrode were used as the working electrode, counter electrode, and reference electrode, respectively. Prior to the surface modification, Au electrodes were separately polished by alumina particles with diameters of 1, 0.3, and 0.05 μm to obtain mirror-like surfaces. After successive sonication in ethanol and double distilled water successively for 10 minutes (5 minutes each), the polished electrodes were rinsed with water and then dried with an air blower. All electrochemical experiments were carried out at room temperature under a nitrogen atmosphere.

A 0.5 mg/mL TiO_2 uniform suspension solution (prepared via the ultrasonic dispersion of 10 mg nano-TiO_2 power in 20 mL anhydrous ethanol containing 5 wt % nafion) was mixed with an equal volume of 0.05 mol/L acetonitrile solution of GDIL (containing 0.1 mol/L sodium perchlorate) via ultrasonic agitation. Then, the Au electrode as the working electrode was placed in this mixture to prepare the PGDIL–TiO_2 composite film by means of chronoamperometric polymerization at a constant potential of 1.1 V for 300 s. As control subjects, PGDIL film without TiO_2 was also prepared in the same chronoamperometric polymerization conditions, and TiO_2 film was prepared by dropping 10 μL 0.5 mg/mL TiO_2 uniform suspension solution onto the bare gold electrode.

Chronoamperometry is an important diagnostic technique for the initial stage of electro-crystallization [34]. Figure 1 sows a current transient (CTT) recorded during the polymerization process of GDIL-TiO_2 at 1.1 V. The electropolymerization process should be hardly affected by other reactions. The CTT can be divided into three regions, which is consistent with the aggregation polymerization process of GDIL [33]. In the first region ($t < 2$ s), the decrease in oxidation current is related to charging of the double layer due to the specific adsorption of GDIL on the Au electrode. The second region (2 s $< t <$ 21 s) corresponds to the increase in the oxidation current up to a maximum, which is typical of nucleation and growth processes. The third region ($t > 21$ s) corresponds to the decrease in the oxidation current, which is typical of a diffusion-controlled process. An analysis of the CTT was performed by fitting the experimental data to a dimensionless theoretical curve for crystal nucleation and diffusion-controlled growth in three dimensions (3D), as proposed by Scharifker and Hills [35]. The instantaneous and progressive theoretical transients are given by Equations (1) and (2), respectively:

$$\frac{i^2}{i_m^2} = \frac{1.9542}{t/t_m}\{1 - \exp[-1.2564(t/t_m)]\}^2 \quad (1)$$

$$\frac{i^2}{i_m^2} = \frac{1.2254}{t/t_m}\{1 - \exp[-2.3367(t/t_m)^2]\}^2 \quad (2)$$

where i_m and t_m represent the maximum current density and its corresponding time, respectively.

The experimental data were fitted as shown in the inset of Figure 1. The instantaneous nucleation is dominant at the oxidant peak potential in the electropolymerization of GDIL-TiO_2, because the experimental curve conforms to the theoretical curve of the instantaneous nucleation model.

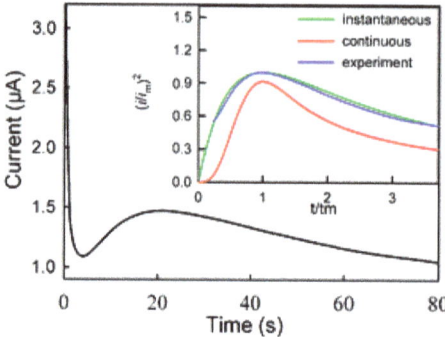

Figure 1. Current versus time transient during the potentiostatic electropolymerization of geminal dicationic ionic liquid (GDIL)–TiO$_2$ at a potential of 1.1 V. Inset: comparison of the experiment curve with theoretical curves from the Scharifker–Hills model.

2.3. Electrochemical Performance Tests

All electrochemical treatments were conducted at room temperature. in 0.1 mol/L KCl containing K$_3$[Fe(CN)$_6$]/K$_4$[Fe(CN)$_6$] (5.0 × 10^{-3} mol/L each). Electrochemical impedance spectra (EIS) were tested in the frequency range between 100,000 and 0.01 Hz at an alternating current voltage amplitude of 5 mV. Cyclic voltammetry (CV) measurements were performed at a scan rate of 0.05 V/s in the scanning range of −0.2–0.6 V.

The electrochemical behaviors of HQ and CC on the PGDIL–TiO$_2$/Au, PGDIL/Au, TiO$_2$/Au, and Au electrodes were preliminarily investigated by CV and differential pulse voltammetry (DPV) methods, respectively. Platinum wire and Ag/AgCl electrode were used as the counter electrode and reference electrode, respectively. The buffer solutions used in all electrochemical experiments were pH = 7 NaH$_2$PO$_4$–Na$_2$HPO$_4$ (PBS). For continuous determination, the HQ and CC adsorbed on the electrode surface could be removed by CV for 10 cycles at a 0.05 V/s scan rate in the buffer solution.

3. Results and Discussion

3.1. Polymerization Mechanism of PGDIL–TiO$_2$ Electrode

The CV curves of GDIL–TiO$_2$ on the Au electrode are presented in Figure 2. In the first scan process, the apparent oxidation peak g could be found at the anode 1.0 V, and was induced by oxidation of the anilino-groups of GDIL to a cation radical. The reduction peak g' of the cation radical appeared in the process of the back sweep cathode, and decreased with the scanning. This indicates that with the continuation of the reaction, the cation radical generated was continuously consumed, and the polymerization generated PGDIL. The current of the redox peak f–f', which is the redox peak pair for the formation of a chain-type polymer reaction, increased with the polymerization, and finally stabilized. The CV curves of GDIL on the Au electrodes also have the appearance of this redox peak pair, which proves that f–f' peak is the redox peak pair of polymerization. However, the polymerization process of GDIL on the Au electrode did not contain this reduction peak of the cation radical. It is possible that the cationic base generated was more active and reacted immediately; this is also illustrated by the peak current response of GDOL on the Au electrode, which is significantly greater than the peak current response of GDIL–TiO$_2$. It can be seen that the electropolymerization mechanism of GDIL–TiO$_2$ and GDIL were the same, but the polymerization rate of GDIL–TiO$_2$ was slower than that of GDIL [33].

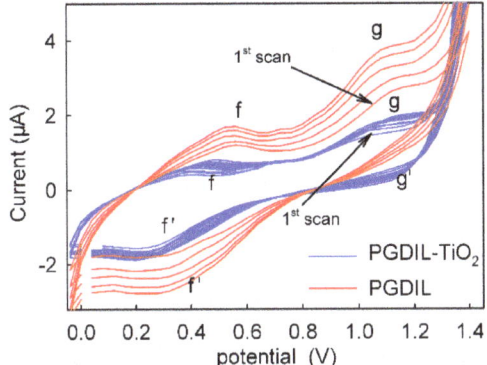

Figure 2. The cyclic voltammetry curve of GDIL. The arrows were the 1st scan of the CV curves of GDIL and GDIL–TiO$_2$ on the Au electrode. g and g′ were the oxidation and reduction peak of the monomer, respectively. f and f′ were the oxidation and reduction peak of the polymer molecular chain, respectively.

3.2. SEM Characterization of PGDIL–TiO$_2$ and PGDIL

The PGDIL–TiO$_2$ film formed on the surface of the Au electrode was scraped down and dissolved in anhydrous ethanol. The ultrasonic shock made it disperse evenly and then remain stationary. The clear droplets were absorbed into the copper sheet after treatment.

Figure 3 shows the scanning electron micrograph of the PGDIL–TiO$_2$ powder. As shown in the images, the PGDIL–TiO$_2$ powder was composed of homogeneous, stacked, spherical particles, which is similar in shape to the PGDIL powder. Additionally, the PGDIL–TiO$_2$ powder had a more uniform and smaller particle size than the PGDIL [33]. This is because, at the beginning of electropolymerization, TiO$_2$ nanoparticles adsorbed on the surface of Au electrode had a good promoting effect on the electropolymerization of GDIL with anilino groups because its O-vacancy can adsorb GDIL molecules, which forms the core of polymerization. The growth process of PGDIL–TiO$_2$ particles was slow and uniform, and the resulting particles were smaller, and the film layer was thicker, than that of PGDIL. This is consistent with the CV results.

Figure 3. SEM image of PGDIL–TiO$_2$ powder.

3.3. EIS Characterization of PGIL–TiO$_2$/Au, TiO$_2$/Au, and Bare Au Electrode

The conductivity of the electrode can be judged by the redox reaction of potassium ferricyanide solution on the electrode. The stronger the conductivity of the electrode, the stronger the redox reaction of potassium ferricyanide, and the greater the redox current value, and vice versa. Figure 4a presents the cyclic voltammetry curves of the PGDIL–TiO$_2$/Au, PGDIL/Au, TiO$_2$/Au, and Au electrodes in 5 mmol/L Fe(CN)$_6^{3-/4-}$ solution containing 0.1 mmol/L KCl. From largest to smallest, the redox peak current values were Au > PGDIL > TiO$_2$ > PGDIL–TiO$_2$ [36].

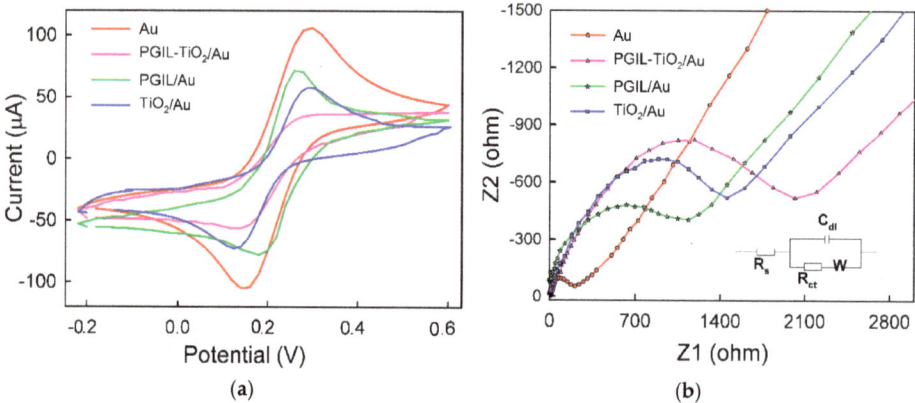

Figure 4. (a) The cyclic voltammetry curves and (b) the EIS plots of PGDIL–TiO$_2$/Au, PGDIL/Au, TiO$_2$/Au, and Au in the presence of 5.0 mmol L^{-1} Fe(CN)$_6^{3-/4-}$ with 0.1 mol L^{-1} KCl as the electrolyte. The illustration equivalent circuit.

Electrochemical impedance spectroscopy (EIS) can be used to characterize the surface electron transfer of different modified electrodes. In the resulting graph of the data (Figure 4b), the linear part indicates a diffusion-controlled process at low frequency, and the semicircle part is consistent with the electron transfer resistance at a high frequency; together, they make up the EIS results in a typical Nyquist plot. The diameter of the arc represents the charge-transfer resistance (R_{ct}) on the surface of the electrode. The impedance is mainly composed of film resistance, ion transmission resistance in the film, and double layer capacitance at the interface between the film and the solution, which formed a large capacitive arc. It can be seen from Figure 4b that the conductivity of each electrode from smallest to largest was PGDIL–TiO$_2$ < TiO$_2$ < PGDIL < Au. Although the conductivity of PGDIL was inferior to that of the metal Au, it was superior to that of TiO$_2$, which may be related to the low electrical conductivity of the polyaniline-like material itself. In addition, it is also possible that the PGIL film with a positive charge adsorbed part of the Fe(CN)$_6^{3-/4-}$, resulting in the generation of an electrostatic repulsion between the surface of the electrode material and the ions in the bulk solution; thus, there was not a significant difference in concentration between the electrode surface and the bulk solution, which prevented the reaction of Fe(CN)$_6^{3-/4}$ on the electrode surface [37]. The conductivity of PGDIL–TiO$_2$ was worse than that of TiO$_2$ and PGDIL because it included not only the resistance of TiO$_2$, but the resistance of PGDIL. In the low frequency region, the PGDIL–TiO$_2$ composites presented a straight line close to 45°, and the length of the Warburg curve was smaller than that of the PGDIL, which demonstrates that the electrolyte ions had a better rapid diffusion and transfer ability, and lower interfacial resistance.

3.4. CVs of HQ and CC on Different Modified Electrodes

The CV results of 0.1 mmol/L HQ and 0.05 mmol/L CC on different modified electrodes are shown in Figure 5. The related electrochemical parameters are listed in Table 1. HQ and CC are composed of one benzene ring and two hydroxyl groups, but the CVs are slightly different due to the different hydroxyl positions. The charge density of *p*-hydroxyl group in benzene ring is higher than that in *ortho*-position, and is the lowest in *meso*-position. Because of the higher charge density, the anodic peak potential (E_{pa}) of HQ is lower than that of CC. On the contrary, the lower the charge density, the easier it can be reduced, so the reduction peak potential (E_{pc}) of CC is higher than that of HQ [38].

Table 1. Electrochemical parameters of CC and HQ on different modified electrodes.

Electrode	PGIL-TiO$_2$/Au			PGIL/Au			TiO$_2$/Au			Au		
E/V vs Ag/AgCl	E_{pa}	E_{pc}	ΔE_p	E_{pa}	E_{pc}	ΔE_p	E_{pa}	E_{pc}	ΔE_p	E_{pa}	E_{pc}	ΔE_p
HQ	0.070	0.030	0.040	0.165	0.065	0.100	>0.225	−0.005	>0.23	>0.175	<−0.05	>0.23
CC	0.235	0.205	0.03	0.249	0.175	0.074	0.234	0.140	0.094	0.244	0.195	0.049

It is demonstrated by the CV results that there was no significant redox peak current response to either HQ (Figure 5a) or CC (Figure 5b) on the Au electrode, whereas a pair of obvious redox peaks appeared on the CVs of the PGDIL–TiO$_2$/Au electrode of both HQ (Figure 5a) and CC (Figure 5b) solutions. The peak potential difference (ΔE_p) between the anodic and cathodic peaks on the PGDIL–TiO$_2$/Au electrode was smaller than that for the PGDIL/Au and TiO$_2$/Au electrodes (Table 1), indicating that the reversibility of the electrochemical reaction was improved. We propose that the electrochemical processes of HQ and CC on the PGDIL–TiO$_2$/Au electrode were quasi-reversible because the cathodic current (I_{pc}) and anodic peak current (I_{pa}) were near equal. On the other hand, the peak currents were much higher compared with those on the PGDIL/Au and TiO$_2$/Au electrodes. HQ exhibited a peak current of 1.533 µA on the PGDIL–TiO$_2$/Au electrode, which is 1.66 times higher than that on the PGDIL/Au electrode (0.926 µA). In contrast, the peak current of CC on the PGDIL–TiO$_2$/Au electrode was 1.771 µA, exhibiting an increase of 1.62 times that on the PGDIL/Au electrode (1.090 µA). The increasing current signals and minimization of over potentials confirms that the PGDIL–TiO$_2$/Au electrode has high electrocatalytic activity for electrochemical oxidation of CC and HQ.

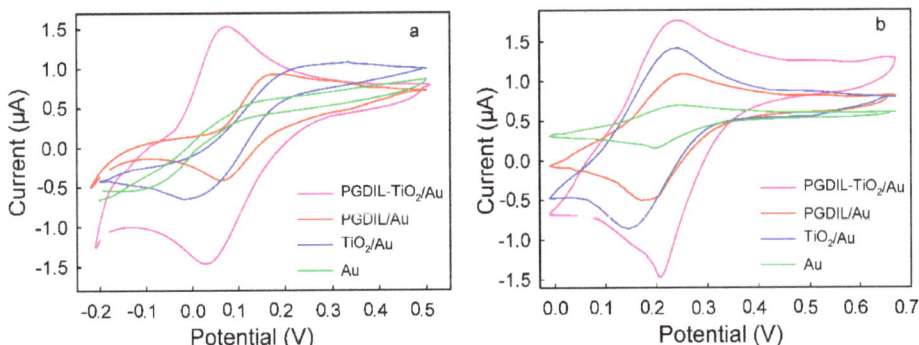

Figure 5. The CV results of (**a**) 0.1 mmol/L HQ and (**b**) 0.05 mmol/L CC at different modified electrodes. The scan rate was 0.1 V/s.

According to the characteristics of a reversible electrode reaction [39], at 25 °C,

$$\Delta E_p(\text{mV}) = 22.5RT/nF = 59/n(\text{mV}) \tag{3}$$

The electron transfer number *n* of the electrochemical reaction can be obtained from the formula. It can be determined that the reaction processes of HQ and CC on the PGDIL–TiO$_2$/Au electrode are two electron-involved quasi-reversible reactions. According to the experimental results, the possible electrocatalytic redox mechanism of HQ and CC can be inferred from the relevant literature [40].

As shown in Scheme 1, the structure of PGDIL is a mixture of benzenoid and quinoid units [33]. In the electrocatalytic oxidation system, due to the polyaniline-like structure in the structure of the PGDIL–TiO_2 electrode, the imino nitrogen atoms (–N=) in the conjugated quinoid moieties of polyaniline-like structure will interact with the hydroxyl groups (–OH) in HQ or CC through hydrogen bonds, accelerating the oxidation of HQ or CC [41,42]. First, CC in the solution was absorbed on the surface of the PGDIL–TiO_2 electrode and formed two hydrogen bonds with the two hydrogen atoms of imino on the polyaniline-like structure. Then, two single electron transition processes occurred. As shown in Scheme 1, the reacting center itself is based on the hydrogen bonds, or, better yet, a hydrogen bridge that regulates the redox state of quinone (catechol or intermediate product) on the PGDIL–TiO_2 electrode. The phenolic hydroxyl groups of HQ or CC can be easily adsorbed on the O-vacancy of the TiO_2-anatase surface to form an adsorption structure, and, the H atoms on the phenolic hydroxyl group can be easily moved to the O atom surface at the adjacent position via hydrogen transfer. Finally, the corresponding benzodiquinone was formed. This may also explain why, although its catalytic ability was better, the R_{ct} value of the TiO_2/Au electrode was higher than that of PGDIL (as shown in Figures 4 and 5) [43,44]. In addition, there is a certain synergistic effect between TiO_2 and PGDIL. Due to the electrocatalytic reaction, the quinoid units of PGDIL became benzenoid units and lost their oxidative activity, while the lone pair electrons on the N atom (–N̈H–) in benzenoid units occupied the O-vacancy of TiO_2 to be restored by dehydrogenation to form quinoid units again. Therefore, the PGDIL–TiO_2/Au electrode has higher electrocatalytic activity for electrochemical oxidation of CC and HQ than the PGDIL/Au and TiO_2/Au electrode.

Cyclic voltammetry at different scan rates (0.01–0.09 V/s) was employed to further study the conduction characteristics of HQ and CC on the PGDIL–TiO_2/Au composite film modified electrode in 0.1 mmol/L HQ and 0.05 mmol/L CC. As shown in Figure 6a, with the increase of scan rate, the anodic peak currents of HQ and CC increased, and as the anodic peak potential shifted positively, the reduction peak currents increased, and the reduction peak potential shifted negatively; these findings further prove that the redox reactions of HQ and CC are quasi-reversible processes. Both the anodic peak current value (I_{pa}) and the cathodic peak current value (I_{pc}) are linearly related to $v^{1/2}$ (Figure 6b,c) under the given conditions. These results indicated that the electrode reactions of CC and HQ on PGDIL–TiO_2/Au were typical diffusion-controlled process.

As presented in Figure 6d, the anodic peak potential (E_{pa}) and the cathodic peak potential (E_{pc}) of HQ and CC are linearly related to $\ln v$ when the ΔE_p value of HQ is greater than 32 mV, and when the ΔE_p of CC is greater than 30 mV, it is in accordance with the Laviron equation [45]:

$$E_{pc} = E^\theta + \frac{RT}{\alpha nF}\ln\frac{RT\kappa_s}{\alpha nF} - \frac{RT}{\alpha nF}\ln v \qquad (4)$$

$$E_{pa} = E^\theta + \frac{RT}{(1-\alpha)nF}\ln\frac{RT\kappa_s}{(1-\alpha)nF} + \frac{RT}{(1-\alpha)nF}\ln v \qquad (5)$$

where E_{pc} is the cathodic peak potential, E_{pa} is the anodic peak potential, E^θ is the standard electrode potential, α is the electron transfer coefficient, n is the electron transfer number, k_s is the standard heterogeneous reaction rate constant, R is the gas constant, T is the thermodynamic temperature, and F is the faraday constant.

By the slope of the E_{pa}–$\ln v$ curve, $\alpha_{HQ} = 0.5826$ and $\alpha_{CC} = 0.5909$ can be obtained.

According to the Randles–Sevcik formula [46] for a quasi-reversible reaction controlled by diffusion:

$$I_{pa} = 2.69 \times 10^5 \, n^{3/2} \, A \, D^{1/2} \, C_0 \, v^{1/2} \qquad (6)$$

where D is the reactant diffusion coefficient (cm^2/s), A is the electroactive surface area (cm^2), n is the electron transfer number, v is the scan rate (V/s), C_0 is the reactant concentration (mol/cm^3), and I_{pa} is the anodic peak current (A).

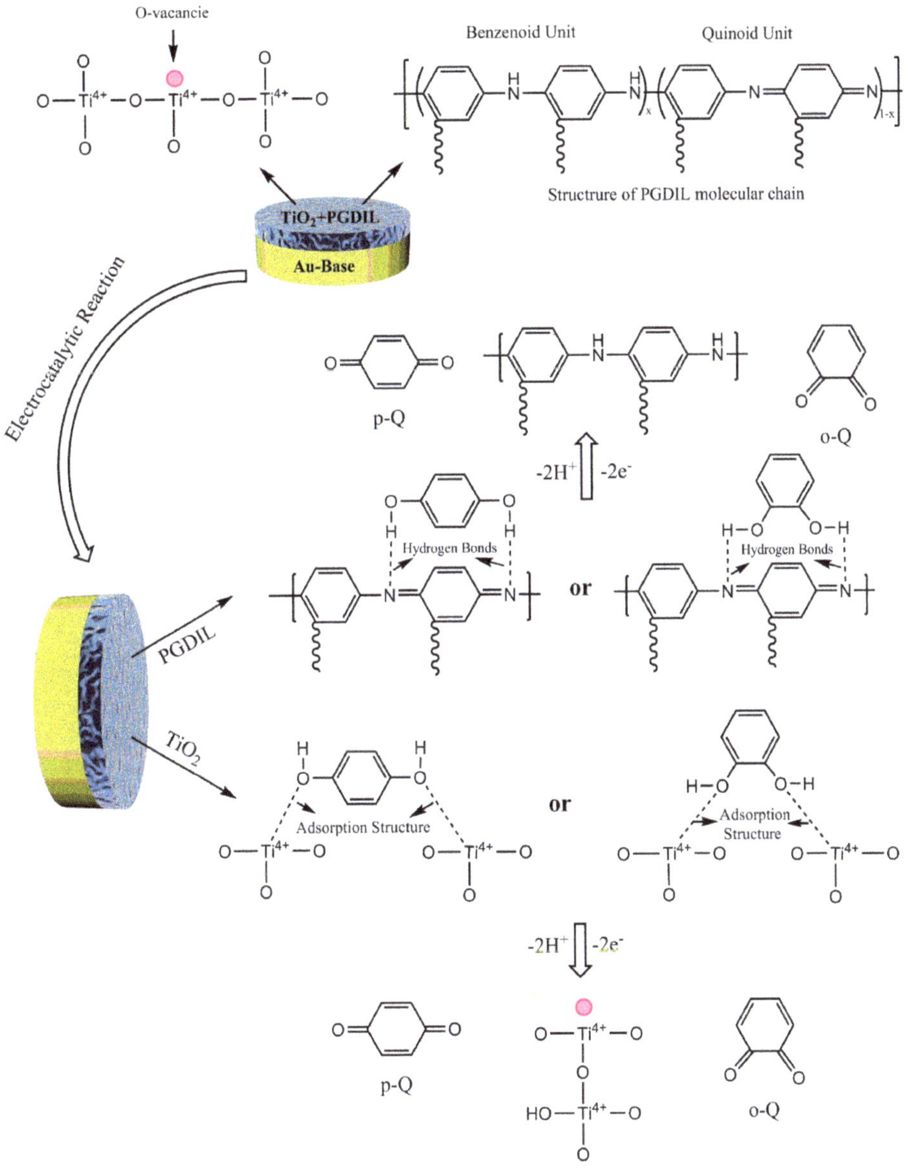

Scheme 1. Structure of PGDIL–TiO$_2$ and the oxidation processes of HQ and CC.

With a concentration of 1 mmol/L K$_3$Fe(CN)$_6$ as the model material (diffusion coefficient $D = 7.6 \times 10^{-6}$ cm^2/s, $n = 1$), the relationship between the anodic peak current (I_{pa}) and scan rates (v) is $I_{pa} = 1.09 \times 10^{-5} + 0.7959 \times 10^{-4} v^{1/2}$, and the slope is $0.7959 \times 10^{-4} = 2.69 \times 10^5 n^{3/2} AD^{1/2} C_0$. The electroactive surface area of modified electrode can be concluded to be $A = 0.1073$ cm^2. By the linear relationship of the anodic peak current (I_{pa}) and scan rates (v) on the modified electrode $I_{pa,HQ}$ (μA) = $0.4469 + 4.892\ v^{1/2}$ and according to the Randles–Sevcik formula, there is a straight slope $4.892 \times 10^{-6} = 2.69 \times 10^5\ n^{3/2}AD^{1/2}C_0$ ($n = 2$, $A = 0.1073$ cm^2, $C_0 = 0.1$ mmol/L). The diffusivity of HQ in pH = 7 phosphate buffer solution can be obtained as $D = 3.591 \times 10^{-7}$ cm^2/s. Similarly, $I_{pa,CC}$ (μA) = $0.5529 + 5.646\ v^{1/2}$, and the diffusion coefficient of CC in pH = 7 phosphate buffer solution can be determined as 1.913×10^{-6} cm^2/s.

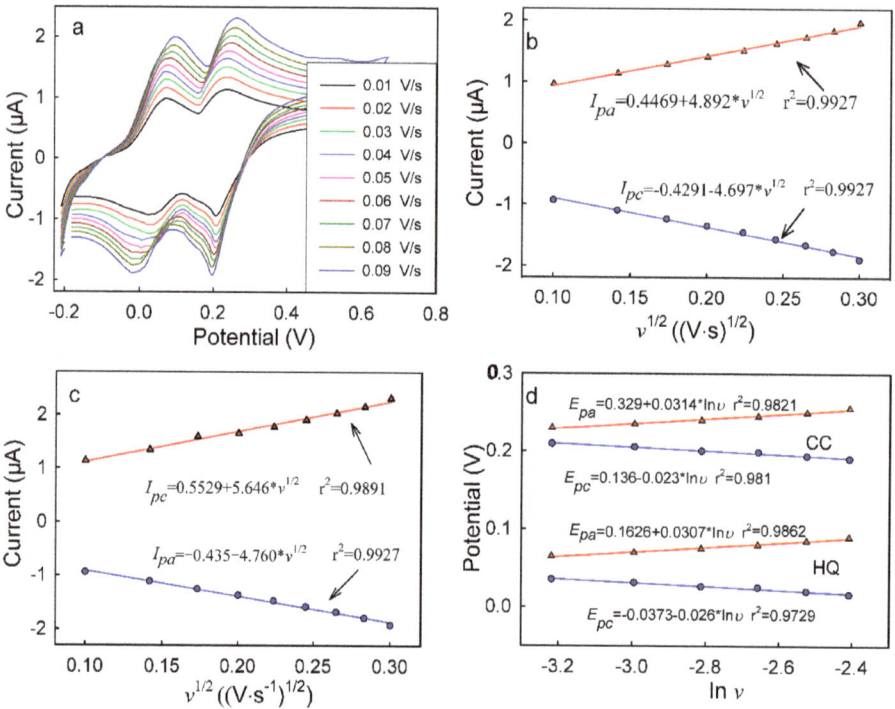

Figure 6. (a) The CV curves of HQ and CC on the PGIL–TiO$_2$/Au electrode at different scan rates (0.01, 0.02, 0.03, 0.04, 0.05, 0.06, 0.07, 0.08, and 0.09 V/s). (b) The relationship between the redox peak current (I_p) of HQ and scan rates (v). (c) The relationship between redox peak current (I_p) of CC and scan rates (v). (d) The relationship between E_p and scan rates (v).

The CV curves of 0.1 mmol/L HQ and 0.05 mmol/L CC on the PGDIL–TiO$_2$/Au composite film electrode at different temperatures were tested experimentally, and the relationship between the obtained oxidation peak current (I_{pa}) and the square root of scan rate ($v^{1/2}$) is shown in Figure 7. The values of the diffusion coefficients of HQ and CC (D_{HQ} and D_{CC}, respectively) at different temperatures were calculated according to the diffusion coefficient formula (Table 2). It can be found from the table that both the D_{HQ} and D_{CC} increased with the increase of temperature, possibly because the increase of temperature led to the increase of kinetic energy of HQ and CC and the decrease of the viscosity of the solution medium, which is beneficial to diffusion.

According to Arrhenius equation [47], the relationship between temperature and diffusion coefficient is in accordance with Equation (7),

$$D = D_0 e^{-E_D/RT} \tag{7}$$

where D is the diffusion coefficient when the temperature is T (cm^2/s), and D_0 is the empirical parameter. The formula can be obtained by logarithmic transformation of the formula; i.e., there is a linear relationship between $\ln D$ and $1/T$.

$$\ln D = \ln D_0 - E_D/RT \tag{8}$$

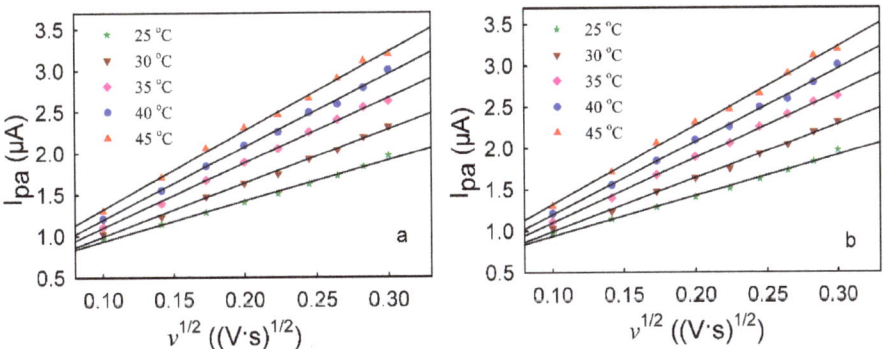

Figure 7. The relationship between I_{pa} and $v^{1/2}$ of (a) HQ and (b) CC at different temperatures.

Table 2. Diffusion coefficients of HQ and CC at different temperatures.

	T (°C)	25	30	35	40	45
HQ	$k \times 10^{-6}$	4.892	6.467	7.840	8.794	9.521
	r	0.9964	0.9982	0.9992	0.9987	0.9984
	$D \times 10^{-7}$ (cm^2/s)	3.591	6.171	8.923	11.05	12.75
CC	$k \times 10^{-6}$	5.646	6.674	7.245	8.272	9.172
	r	0.9945	0.9964	0.9955	0.9986	0.9990
	$D \times 10^{-6}$ (cm^2/s)	1.913	2.629	3.569	3.910	4.731

According to the data of D_{HQ} and temperature in Table 2, the fitting equation is $\ln D_{HQ} = 5.2187 - 5941/T$ (As shown in Figure 8a). As can be seen from Equation (8), the slope of the fitting line is $-E_D/R$. Therefore, the diffusion activation energy of HQ is $E_{D,HQ} = 49393$ J/mol = 49.39 kJ/mol. Similarly, the diffusion activation energy of CC is $E_{D,CC} = 34935$ J/mol = 34.935 kJ/mol (as shown in Figure 8b).

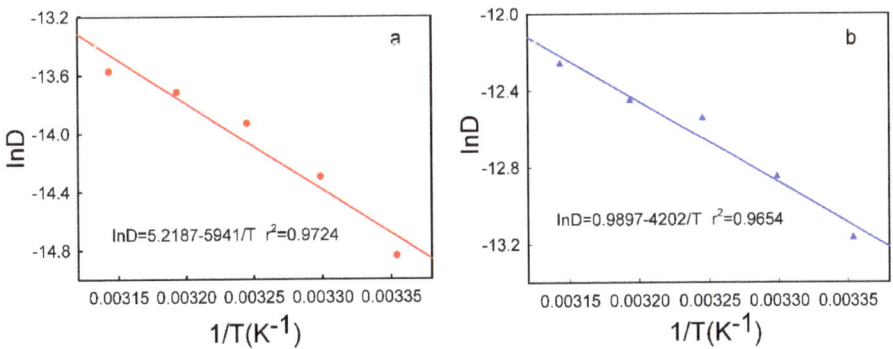

Figure 8. The relationship between $\ln D$ and $(1/T)$ of HQ (a) and CC (b).

3.5. DPV Analyses of HQ and CC on Different Modified Electrodes

The electrochemical behaviors of the PGDIL–TiO$_2$/Au, PGDIL/Au, TiO$_2$/Au, and Au electrodes in the buffer solution containing 0.1 mmol/L HQ and 0.05 mmol/L CC were studied by differential pulse voltammetry (DPV) to investigate the practicability of the PGDIL–TiO$_2$/Au electrode. As shown in Figure 9, there was no peak current response in the blank buffer solution in the range of −0.1–0.4 V, which indicates that the buffer solution can be used as the base solution for the electrocatalytic reaction of HQ and CC without impure peak interferences. There was no difference between the anodic peaks of HQ and CC on the bare gold electrode; only a small and wide anodic peak could be observed. The PGDIL/Au and TiO$_2$/Au electrodes exhibited respective anodic peaks, but their response sensitivity was low, their current was small, and the two peaks were indistinct, so it was difficult to distinguish them. However, there were two completely separated and responsive anodic peaks of HQ and CC on the PGDIL–TiO$_2$/Au electrode. The anodic peak potentials were 0.18 and 0.3 V, which correspond to the oxidation reactions of HQ and CC, respectively. The anodic peak currents of HQ and CC on the PGDIL–TiO$_2$/Au modified electrode were 3.22 and 3.42 µA, respectively, and are much higher than those of the PGDIL/Au and TiO$_2$/Au electrodes. The results demonstrate that the PGDIL–TiO$_2$/Au electrode can effectively improve the electrochemical behaviors of HQ and CC. The anodic peaks of HQ and CC can be completely separated into two sensitive anodic peaks. The peak-to-peak potential difference was 0.12 V, which is slightly higher than the potential reported in the work by [48]. It, therefore, provides a new electrode material for the simultaneous determination of HQ and CC.

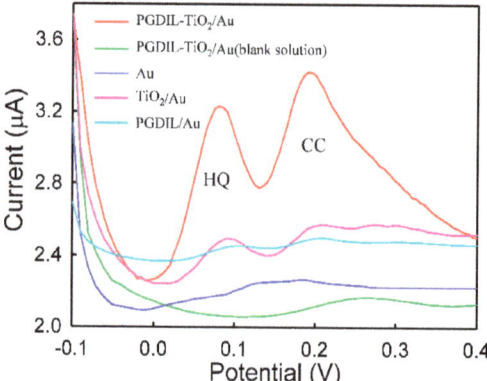

Figure 9. DPV results of 0.1 mmol/L HQ and 0.05 mol/L CC on the PGDIL–TiO$_2$/Au, PGDIL/Au, TiO$_2$/Au, and Au in pH = 7 PBS at a scan rate of 0.1 V/s.

The effects of pH on the electrochemical behaviors of HQ and CC were studied by DPVs. As shown in Figure 10, in the range of pH = 5–9, the anodic peaks of HQ and CC were completely separated. The anodic peak currents increased with the decrease of pH value, and reached the maximum value at pH = 7.0.

The oxidation peak potentials (E_p) of HQ and CC were negatively shifted with the increase of pH values (Figure 10c), and presented good linear relationship with slopes of 0.0556 and 0.059, respectively ($E_{p,HQ}$ = 0.4678−0.0556 pH, r^2 = 0.9985; $E_{p,CC}$ = 0.605−0.059 pH, r^2 = 0.9924). These values are close to the Nernst theoretical value, indicating that H$^+$ entered into the electrode reaction, and the number of protons was equal to the number of electrons in the reaction [49]. According to the above conclusion, the number of electron transfers for both HQ and CC during electrochemical oxidation was two, so the redox reactions of HQ and CC were two-electron and two-proton transfer processes, which is consistent with the existing literature [40]. In total, 0.1 mol/L PBS (pH = 7.0) was selected as the support electrolyte for the detection of HQ and CC to obtain high sensitivity and selectivity.

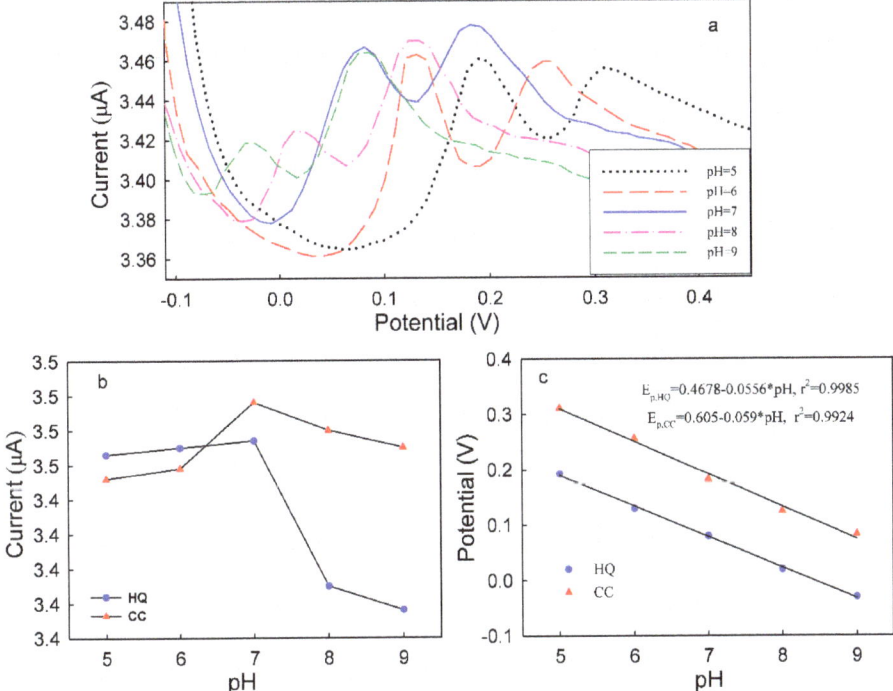

Figure 10. (a) DPV results of 0.1 mmol/L HQ and 0.05 mmol/L CC on the PGDIL–TiO$_2$/Au electrode under different pHs at a scan rate of 0.1 V/s. (b) The relationship between I_p and pH. (c) The relationship between E_p and pH.

The selective and simultaneous determination of HQ and CC were performed on the PGDIL–TiO$_2$/Au electrode at pH = 7 using DPV. The selective determinations of HQ and CC were carried out by changing the concentration of each isomer separately. Figure 11a,b presents the variations of HQ and CC concentrations in the range of 1–100 μmol/L and 2–100 μmol/L. The results show that the oxidation peak currents are related to the increase of HQ concentration, while the peak currents of CC remain basically unchanged (as shown in Figure 11a). Similarly, the peak currents of CC increased with the increase of the concentration, whereas the peak currents of HQ remained basically unchanged (as shown in Figure 10b). The peak current also had a good linear relationship. The regression equations were $I_{p,HQ}$ (μA) = 2.380 + 0.009804 C_{HQ} (μmol/L) and $I_{p,CC}$ (μA) = 2.380 + 0.009804 C_{CC} (μmol/L), respectively. The detection limits (LODs) for HQ and CC were estimated to be 0.23 μmol/L and 0.41 μmol/L (S/N = 3), respectively.

Figure 12a displays the DPV results of the binary mixture of CC and HQ at various concentrations. The composition of the binary mixture of CC and HQ is shown in Table 3. The results show two well-defined and separated oxidation peaks for CC and HQ with their respective concentrations as displayed in calibration plots (insets in Figure 12b,c). The regression equations for HQ, and CC are $I_{p,HQ}$ (μA) = 3.413 + 0.008152 C_{HQ} (μmol/L) (r^2 = 0.9929), and $I_{p,CC}$ (μA) = 2.368 + 0.02179 C_{CC} (μmol/L) (r^2 = 0.9952), respectively. The LODs for HQ and CC were estimated to be 0.17 μmol/L and 0.25 μmol/L, respectively. Therefore, the sensitive and simultaneous determination of HQ and CC are favored without valid interference between them. Thus, the PGDIL–TiO$_2$/Au modified electrode is a competitive candidate in simultaneous determination of dihydroxybenzene isomers compared with other modified electrodes [50].

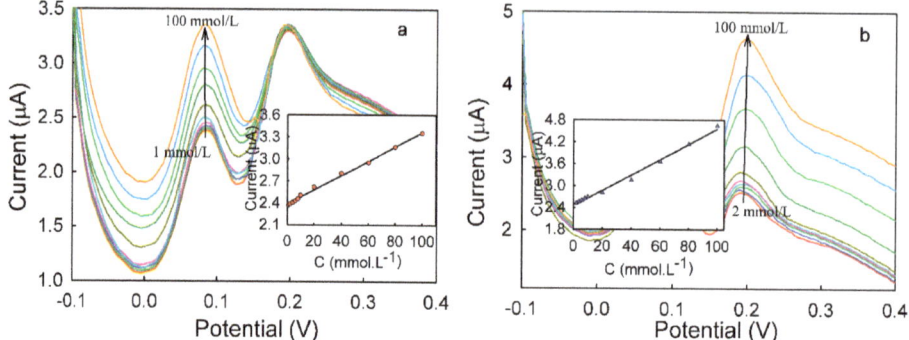

Figure 11. (a) DPV results of the PGDIL–TiO$_2$/Au in the presence of 0.1 mmol/L HQ with CC at different concentrations (the inset is the calibration curve of the peak current and target concentration). (b) DPV results of the PGDIL–TiO$_2$/Au in the presence of 0.05 mmol/L CC with HQ at different concentrations (the inset is the calibration curve of the peak current and target concentration). The scan rate is 0.1 V/s.

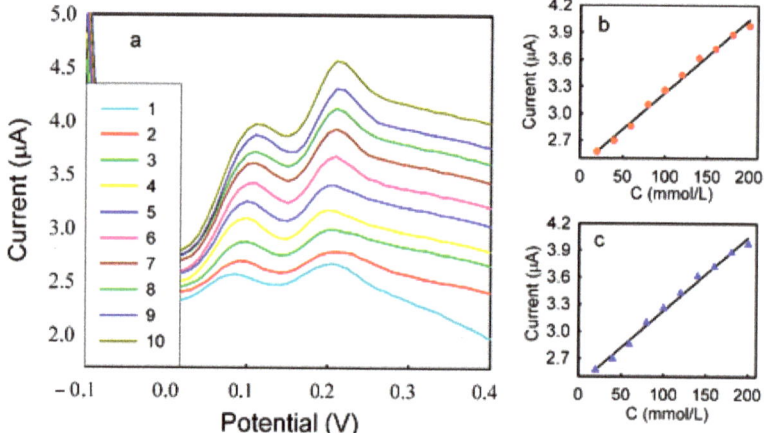

Figure 12. (a) DPV results of the binary mixture of CC and HQ at various concentrations on the PGDIL–TiO$_2$/Au in pH = 7 PBS. (b) The relationship between I_{pHQ} and C_{HQ}. (c) The relationship between I_{pCC} and C_{CC}.

Table 3. The composition of the binary mixture of CC and HQ.

Binary Mixture	Concentration of HQ (mmol/L)	Concentration of CC (mmol/L)
1	0.02	0.01
2	0.04	0.02
3	0.06	0.03
4	0.08	0.04
5	0.10	0.05
6	0.12	0.06
7	0.14	0.07
8	0.16	0.08
9	0.18	0.09
10	0.20	0.10

4. Conclusions

A novel poly-geminal dicationic ionic liquid-TiO$_2$ (PGDIL)-TiO$_2$/Au composite electrode was synthesized via the electrochemical polymerization of 1,4-bis(3-(m-aminobenzyl)imidazol-1-yl)butane bis(hexafluorinephosphate) in an electrolyte containing nano-TiO$_2$. The SEM results showed that the morphologies of the PGDIL and PGDIL–TiO$_2$ powder are similar. Additionally, the PGDIL–TiO$_2$ powder has a more uniform and smaller particle size than the PGDIL because the O-vacancies of TiO$_2$ can adsorb GDIL molecules to form the core of polymerization, and thus, the polymerization process can be carried out smoothly and slowly. EIS results revealed that the R_{ct} value of the PGDIL–TiO$_2$ electrode is higher than those of the PGDIL and TiO$_2$ electrodes, but its catalytic effect on HQ and CC is the best due to the synergistic effect between TiO$_2$ and PGDIL. Furthermore, the DPV method was used to study the simultaneous qualitative and quantitative determination of HQ and CC at the PGDIL–TiO$_2$/Au electrode. The results showed that the peak current of each component in the mixture had almost non-interference during electrochemical selectivity determination, and the peak current values were linearly related to their concentrations. The equation for HQ is $I_{p, HQ}$ (μA) = 3.413 + 0.008152 C_{HQ} (μmol/L), and that for CC is $I_{p, CC}$ (μA) = 2.368 + 0.02179 C_{CC} (μmol/L). The PGDIL–TiO$_2$/Au electrode can be used for simultaneous qualitative and quantitative electrochemical determination of HQ and CC in wastewater by the DPV tests. The PGDIL–TiO$_2$/Au electrode is also expected to be used for catalytic oxidation and detection of other organic pollutants containing –OH groups. Further experiments and studies are required to determine whether there are other synergistic effects between TiO$_2$ and PGDIL.

Author Contributions: Y.G., D.H., and R.Z. conceived and designed the experiments; A.X. performed the experiments; Y.G. and W.Q. analyzed the data; Y.G. wrote the paper. Y.T. and L.Z. revised the manuscript.

Funding: This research received no external funding.

Conflicts of Interest: The authors declare no conflict of interest.

References

1. Zheng, C.; Dong, Y.; Liu, Y.; Zhao, X.; Yin, J. Enhanced stimuli-responsive electrorheological property of poly(ionic liquid) s-capsulated polyaniline particles. *Polymers* **2017**, *9*, 385. [CrossRef] [PubMed]
2. Vollas, A.; Chouliaras, T.; Deimede, V.; Ioannides, T.; Kallitsis, J. New pyridinium type poly(Ionic Liquids) as membranes for CO$_2$ separation. *Polymers* **2018**, *10*, 912. [CrossRef] [PubMed]
3. Ohno, H. Molten salt type polymer electrolytes. *Electroehimiea Acta* **2001**, *46*, 1407–1411. [CrossRef]
4. Men, Y.; Kuzmicz, D.; Yuan, J. Poly(ionic liquid) colloidal particles. *Curr. Opin. Colloid Interface Sci.* **2014**, *19*, 76–83. [CrossRef]
5. He, D.; Li, F.; Xia, S.; Liu, F.; Xiong, Y.; Zhang, Q. Synthesis and characterization of a novel ionic liquid polymer: Poly(1-(3-aminobenzyl)-3-methylimidazolium chloride). *Chin. J. Polym. Sci.* **2014**, *32*, 163–168. [CrossRef]
6. He, D.; Guo, Y.; Zhou, Z.; Xia, S.; Xie, X.; Yang, R. Electropolymerization of ionic liquid substituted polyphenylene as supercapacitors materials. *Electrochem. Commun.* **2009**, *11*, 1671–1674. [CrossRef]
7. Masri, A.N.; MI, A.M.; Leveque, J. A review on dicationic ionic liquids: Classification and application. *Ind. Eng. Manag.* **2016**, *5*, 197–204. [CrossRef]
8. Zhang, H.; Liu, J.; Li, M.; Yang, B. Functional groups in geminal imidazolium ionic compounds and their influence on thermo-physical properties. *J. M. Liq.* **2018**, *269*, 738–745. [CrossRef]
9. Anderson, J.L.; Ding, R.F.; Ellern, A.; Armstrong, D.W. Structure and properties of high stability geminal dicationic ionic liquids. *J. Am. Chem. Soc.* **2005**, *127*, 593–604. [CrossRef]
10. Shahkaramipour, N.; Adibi, M.; Seifkordi, A.A.; Fazli, Y. Separation of CO$_2$/CH$_4$ through alumina-supported geminal ionic liquid membranes. *J. Membr. Sci.* **2014**, *455*, 229–235. [CrossRef]
11. Ding, Y.S.; Zha, M.; Zhang, J.; Wang, S.S. Synthesis, characterization and properties of geminal imidazolium ionic liquids. *Colloids and Surfaces A: Physicochem. Eng. Aspects* **2007**, *298*, 201–205. [CrossRef]

12. Chinnappan, A.; Kim, H. Environmentally benign catalyst: Synthesis, characterization, and properties of pyridinium dicationic molten salts (ionic liquids) and use of application in esterification. *Chem. Eng. J.* **2012**, *187*, 283–288. [CrossRef]
13. Cheng, J.; Xiang, C.; Zou, Y.; Chu, H.; Qiu, S.; Zhang, H.; Sun, L.; Xu, F. Highly active nanoporous Co-B-TiO$_2$ framework for hydrolysis of NaBH$_4$. *Ceram. Int.* **2015**, *41*, 899–905. [CrossRef]
14. Zafer, C.; Ocakoglu, K.; Ozsoy, C.; Icli, S. Dicationic bis-imidazolium molten salts for efficient dye sensitized solar cells: Synthesis and photovoltaic properties. *Electrochim. Acta* **2009**, *54*, 5709–5714. [CrossRef]
15. Li, X.J.; Bruce, D.W.; Shreeve, J.M. Dicationic imidazolium-based ionic liquids and ionic liquid crystals with variously positioned fluoro substituents. *J. Mater. Chem.* **2009**, *19*, 8232–8238. [CrossRef]
16. Fatichi, A.Z.; Mello, M.G.; Caram, R.; Cremasco, A. Self-organized TiO$_2$ nanotube layer on Ti–Nb–Zr alloys: Growth, characterization, and effect on corrosion behavior. *J. Appl. Electrochem.* **2019**, in press. [CrossRef]
17. Wang, X.; Zhao, J.; Xiao, T.; Li, Z.; Wang, X. Preparation and properties of Co$_3$O$_4$-doped TiO$_2$ nanotube array electrodes. *J. Appl. Electrochem.* **2019**, *49*, 305–314. [CrossRef]
18. Bashiri, R.; Mohamed, N.M.; Kait, C.F.; Sufian, S. Optimization of hydrogen production over TiO$_2$ supported copper and nickel oxides: Effect of photoelectrochemical features. *J. Appl. Electrochem.* **2019**, *49*, 27–38. [CrossRef]
19. Kannan, R.; Kim, A.R.; Nahm, K.S.; Yoo, D.J. Manganese-titanium-oxide-hydroxide-supported palladium nanostructures–A facile electrocatalysts for the methanol, ethylene glycol and xylitol electrooxidation. *Int. J. Hydrog. Energy* **2016**, *41*, 6787–6797. [CrossRef]
20. Guo, Y.; He, D.; Xia, S.; Xin, X.; Gao, X.; Zhang, Q. Preparation of a novel nanocomposite of polyaniline core decorated with anatase-TiO$_2$ nanoparticles in ionic liquid/water microemulsion. *J. Nanomater.* **2011**, *2012*, 1–8. [CrossRef]
21. Zhou, Z.; He, D.; Guo, Y.; Cui, Z.; Wang, M.; Li, G.; Yang, R. Fabrication of polyaniline-silver nanocomposites by chronopotentiometry in different ionic liquid microemulsion systems. *Thin Solid Film.* **2009**, *517*, 6767–6771. [CrossRef]
22. Abdul-Manaf, N.A.; Yusoff, W.Y.W.; Demon, S.Z.N.; Shaari, N.A.; Shamshuddin, A.; Mohamed, N.S. Anodic and cathodic deposition of polyaniline films: A comparison between the two methods. *Mater. Res. Express* **2019**, *6*, 096453. [CrossRef]
23. Wang, X.; Wei, H.; Liu, X.; Du, W.; Zhao, X.; Wang, X. Novel three-dimensional polyaniline nanothorns vertically grown on buckypaper as high-performance supercapacitor electrode. *Nanotechnology* **2019**, *30*, 325401. [CrossRef] [PubMed]
24. Arukula, R.; Vinothkannan, M.; Kim, A.R.; Yoo, D.J. Cumulative effect of bimetallic alloy, conductive polymer and graphene toward electrooxidation of methanol: An efficient anode catalyst for direct methanol fuel cells. *J. Alloy. Compd.* **2019**, *771*, 477–488. [CrossRef]
25. Asiltürk, M.; Sayılkan, F.; Arpaç, E. Effect of Fe^{3+} ion doping to TiO$_2$ on the photocatalytic degradation of Malachite Green dye under UV and vis-irradiation. *J. Photochem. Photobiol. A: Chem.* **2009**, *203*, 64–71. [CrossRef]
26. Ramakrishnan, S.; Karuppannan, M.; Vinothkannan, M.; Ramachandran, K.; Kwon, O.J.; Yoo, D.J. Ultrafine Pt nanoparticles stabilized by MoS$_2$/N-doped reduced graphene oxide as a durable electrocatalyst for alcohol oxidation and oxygen reduction reactions. *ACS Appl. Mater. Interfaces* **2019**, *11*, 12504–12515. [CrossRef]
27. Wang, J.; Park, J.N.; Wei, X.Y.; Lee, C.W. Room-temperature heterogeneous hydroxylation of phenol with hydrogen peroxide over Fe^{2+}, Co^{2+} ion-exchanged Naβ zeolite. *Chem. Commun.* **2003**, *5*, 628–629. [CrossRef]
28. Hong, Z.; Zhou, L.; Li, J.; Tang, J. A sensor based on graphitic mesoporous carbon/ionic liquids composite film for simultaneous determination of hydroquinone and catechol. *Electrochim. Acta* **2013**, *109*, 671–677. [CrossRef]
29. Wang, Y.; Zhang, S.; Dong, Y.; Qu, J. Research progress of methods for detecting catechol and hydroquinone in water. *Chem. Res.* **2015**, *26*, 100–104. [CrossRef]
30. Feng, X.; Gao, W.; Zhou, S.; Shi, H.; Huang, H.; Song, W. Discrimination and simultaneous determination of hydroquinone and catechol by tunable polymerization of imidazolium-based ionic liquid on multi-walled carbon nanotube surfaces. *Anal. Chim. Acta* **2013**, *805*, 36–44. [CrossRef]
31. Kai, X.; Shen, Y.; Zhang, G.; Xie, J. Spectrophotometric simultaneous determination of pyrocatechol, resorcinol and hydroquinone by LM-BP neural network. *Spectrosc. Spectr. Anal.* **2005**, *25*, 2070–2072. [CrossRef]

32. Penner, N.A.; Nesterenko, P.N. Simultaneous determination of dihydroxybenzenes, aminophenols and phenylenediamines in hair dyes by high-performance liquid chromatography on hypercross-linked polystyrene. *Analyst* **2000**, *125*, 1249–1254. [CrossRef] [PubMed]
33. Guo, Y.; He, D.; Xie, A.; Qu, W.; Tang, Y.; Shang, J.; Zhu, R. Preparation and characterization of a novel poly-geminal dicationic ionic liquid (PGDIL). *J. Mol. Liq.*. In press. [CrossRef]
34. Ivanova, Y.A.; Ivanou, D.K.; Streltsov, E.A. Electrodeposition of Te onto monocrystalline n-and p-Si (1 0 0) wafers. *Electrochim. Acta* **2007**, *52*, 5213–5218. [CrossRef]
35. Gunawardena, G.; Hills, G.; Montenegro, I.; Scharifker, B. Electrochemical nucleation: Part, I. General considerations. *J. Electroanal. Chem. Interfacial Electrochem.* **1982**, *138*, 225–239. [CrossRef]
36. Weidlich, C.; Mangold, K.-M.; Jüttner, K. EQCM study of the ion exchange behaviour of polypyrrole with different counterions in different electrolytes. *Electrochim. Acta* **2005**, *50*, 1547–1552. [CrossRef]
37. Stoller, M.D.; Park, S.; Zhu, Y.; An, J.; Ruoff, R.S. Graphene-based ultracapacitors. *Nano Lett.* **2008**, *8*, 3498–3502. [CrossRef]
38. Zhang, Y.; Xiao, S.; Xie, J.; Yang, Z.; Pang, P.; Gao, Y. Simultaneous electrochemical determination of catechol andhydroquinone based on grapheme-TiO$_2$nanocomposite modifiedglassy carbon electrode. *Sens. Actuators B* **2014**, *204*, 102–108. [CrossRef]
39. Bard, A.J.; Faulkner, L.R. *Electrochemical Methods Fundamentals and Applications*, 2nd ed.; Shao, Y.; Zhu, G.; Dong, X.; Zhang, B., Translators; Chemical Industry Press: Beijing, China, 2005; pp. 124–125. ISBN 978-7-5025-96704-0.
40. He, D.; Mho, S. Electrocatalytic reactions of phenolic compounds at ferric ion co-doped SnO$_2$: Sb^{5+} electrodes. *J. Electroanal. Chem.* **2004**, *568*, 19–27. [CrossRef]
41. Feng, X.; Shi, Y.; Hu, Z. Polyaniline/polysulfone composite film electrode for simultaneous determination of hydroquinone and catechol. *Mater. Chem. Phys.* **2011**, *131*, 72–76. [CrossRef]
42. Lee, C.W.; Jin, S.H.; Jeong, H.M.; Chi, K.-W. Efficient oxidation of hydroquinone and alcohols by tailor-made solid polyaniline catalyst. *Tetrahedron Lett.* **2009**, *50*, 559–561. [CrossRef]
43. Mao, J.; Zhao, B.; Zhou, J.; Zhang, L.; Yang, F.; Guo, X.; Zhang, Z.C. Identification and characteristics of catalytic quad-functions on Au/Anatase TiO$_2$. *ACS Catal.* **2019**, *9*, 7900–7911. [CrossRef]
44. Wan, W.; Nie, X.; Janik, M.J.; Song, C.; Guo, X. Adsorption, dissociation, and spillover of hydrogen over Au/TiO$_2$ catalysts: The effects of cluster size and meta-support interaction from DFT. *J. Phys. Chem. C* **2018**, *122*, 17895–17916. [CrossRef]
45. Laviron, E. General expression of the linear potential sweep voltammogram in the case of diffusionless electrochemical systems. *J. Electroanal. Chem. Interfacial Electrochem.* **1979**, *101*, 19–28. [CrossRef]
46. Zanello, P. *Inorganic Electrochemistry: Theory, Practice and Application*; The Royal Society of Chemistry: Cambridge, UK, 2003; p. 38. ISBN 0-85404-661-5.
47. Rodríguez-Aragón, L.J.; López-Fidalgo, J. Optimal designs for the Arrhenius equation. *Chemom. Intell. Lab. Syst.* **2005**, *77*, 131–138. [CrossRef]
48. Luo, Q.; Wang, H.; Liu, D. Simultaneous Determination of Catechol and Hydroquinone Based on Gold Electrode Modified with Carbon Nanotubes-graphene Nanosheet Hybrid Films. *Chin. J. Appl. Chem.* **2014**, *31*, 983–989. [CrossRef]
49. Qi, H.; Zhang, C. Simultaneous determination of hydroquinone and catechol at a glassy carbon electrode modified with multiwall carbon nanotubes. *Electroanalysis* **2005**, *17*, 832–838. [CrossRef]
50. Kumar, A.A.; Swamy, B.E.K.; Rani, T.S.; Ganesh, P.S.; Raj, Y.P. Voltmetric determination of catechol and hydroquinone at poly(murexide) modified glassy carbon electrode. *Mater. Sci. Eng. C* **2019**, *98*, 746–752. [CrossRef]

© 2019 by the authors. Licensee MDPI, Basel, Switzerland. This article is an open access article distributed under the terms and conditions of the Creative Commons Attribution (CC BY) license (http://creativecommons.org/licenses/by/4.0/).

Article

Preparation and Application of Organic-Inorganic Nanocomposite Materials in Stretched Organic Thin Film Transistors

Yang-Yen Yu [1,2,*] and Cheng-Huai Yang [1]

1 Department of Materials Engineering, Ming Chi University of Technology, New Taipei City 243, Taiwan; harry830308@gmail.com
2 Department of Chemical and Materials Engineering, Chang Gung University, Taoyuan City 33302, Taiwan
* Correspondence: yyyu@mail.mcut.edu.tw

Received: 15 March 2020; Accepted: 2 May 2020; Published: 5 May 2020

Abstract: High-transparency soluble polyimide with COOH and fluorine functional groups and TiO_2-SiO_2 composite inorganic nanoparticles with high dielectric constants were synthesized in this study. The polyimide and inorganic composite nanoparticles were further applied in the preparation of organic-inorganic hybrid high dielectric materials as the gate dielectric for a stretchable transistor. The optimal ratio of organic and inorganic components in the hybrid films was investigated. In addition, Jeffamine D2000 and polyurethane were added to the gate dielectric to improve the tensile properties of the organic thin film transistor (OTFT) device. PffBT4T-2OD was used as the semiconductor layer material and indium gallium liquid alloy as the upper electrode. Electrical property analysis demonstrated that the mobility could reach 0.242 $cm^2 \cdot V^{-1} \cdot s^{-1}$ at an inorganic content of 30 wt.%, and the switching current ratio was 9.04 × 10^3. After Jeffamine D2000 and polyurethane additives were added, the mobility and switching current could be increased to 0.817 $cm^2 \cdot V^{-1} \cdot s^{-1}$ and 4.27 × 10^5 for Jeffamine D2000 and 0.562 $cm^2 \cdot V^{-1} \cdot s^{-1}$ and 2.04 × 10^5 for polyurethane, respectively. Additives also improved the respective mechanical properties. The stretching test indicated that the addition of polyurethane allowed the OTFT device to be stretched to 50%, and the electrical properties could be maintained after stretching 150 cycles.

Keywords: soluble polyimide; polyurethane; Jeffamine; organic-inorganic hybrid film; Stretchable transistor

---

1. Introduction

Stretchable electronic components have attracted much research interest due to their considerable potential in biomedical instruments, smart skins, displays, and battery devices [1–3]. From 2010 to 2020, thin film transistors have been made predominantly from inorganic materials. The main reason for this is that the carrier mobility values of organic materials are too low compared with those of inorganic materials [4]. Because the performance of an organic thin film transistor (OTFT) [5,6] has not been able to reach the same performance of inorganic transistor, researchers have continued to study the use of various semiconductor materials [7–11] to improve their carrier mobility. In addition, the plastic soft board-based OTFTs can also be used on flexible substrates [12–15]. The rise of plastic substrates [16–18] has necessitated some flexural quality measurements and novel processing methods, such as stretching and coating, to increase the flexibility and mobility of components [19,20]. A roll-to-roll process that can be fabricated on a flexible substrate in a low-temperature environment could support future commercial development [21].

Hybrid materials [22–24] are organic-inorganic polymer blends that are molecularly mixed and blended through van der Waals forces, hydrogen bonds, ionic bonds, or covalent bonds, thus overcoming the phase separation that can usually be observed in traditional materials. These hybrid materials have the advantages of organic and inorganic materials, providing excellent material properties, including thermal, mechanical, optical, and electrical properties. To achieve a good nanoscale dispersion of organic-inorganic materials, the sol-gel method is the most commonly used method because it is flexible and materials prepared with the sol-gel method have high thermal stability and optical transparency.

This study used spin coating to replace the traditional vaporization for the fabrication of thin film transistors. Polyimide [25–27] was used for the preparation of OTFT due to its good thermal stability, chemical resistance, and mechanical properties. As practical applications continue to advance, the requirements for thermal and mechanical properties are becoming more and more demanding, so inorganic materials are often used to enhance the relevant properties. The most common inorganic materials are SiO_2 and TiO_2, which can be prepared using tetraethoxysilane (TEOS) and titanium butoxide, respectively. The use of inorganic composite material TiO_2-SiO_2 has also been featured in the literature [28,29]. Such an organic-inorganic hybrid film [30] was applied in OTFT as a dielectric film. The donor material, PffBT4T-2OD [31], was also used to replace the traditional pentacene [32] as a semiconductor layer in organic photovoltaic devices. However, PffBT4T-2OD has not been applied to OTFTs in other research.

Electronic products increasingly require the properties of flexibility and stretchability [33–37]. Therefore, some suitable polymers have been added to these advanced electronic products [38,39], such as Jeffamine D2000 [40] and polyurethane [41]. Another approach to enhance flexibility and stretchability is to connect the sidechain of the semiconductor layer material with an elastic polymer, such as poly(butyl acrylate)(PBA) or 2,6-pyridine dicarboxamide (PDCA). After this modification, the researchers expected that the device could retain its original performance after being subjected to stretching many cycles. The chemical structures of the polyimide-TiO_2-SiO_2, Jeffamine D2000, and polyurethane as well as the structural diagrams for the OTFT device and the experimental stretching directions are shown in Figure 1. Tensile properties depend on the ratio of TiO_2-SiO_2 and whether Jeffamine D2000 or polyurethane is added. The addition ratio of TiO_2-SiO_2 ranges from A0–A40 in the order of 0 wt.% to 40 wt.%, B0–B40 when Jeffamine D2000 is added, and C0–C40 when polyurethane is added.

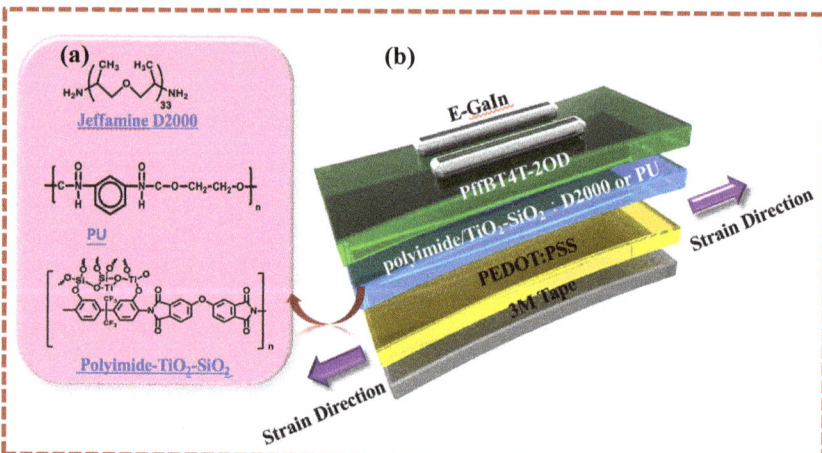

Figure 1. (a) Chemical structures of polyimide-TiO_2-SiO_2, Jeffamine D2000, and PU. (b) Device structure with illustration of each layer and strain direction in an organic thin film transistor.

2. Experimental Section

In this study, a stretchable OTFT was fabricated using Elastomer Tape 3M tape as the stretchable substrate and poly(3,4-ethylenedioxythiophene) polystyrene sulfonate (PEDOT:PSS, Sigma Aldrich, Darmstadt, Germany) as the lower electrode. TiO_2-SiO_2 inorganic nanoparticles and a soluble polyimide with COOH and a fluorine atom functional group were used to prepare the dielectric layer. The COOH on polyimide could be hydrolyzed and condensed with TiO_2-SiO_2 to form a dense network structure, and the size of the CF group in PI molecule is quite big, which can cause an increase of free volume and a reduction of the interaction between the molecular chains, so as to increase the solubility and transparency for the prepared polyimide-TiO_2-SiO_2 hybrid films. The film was used as an OTFT gate dielectric. In addition, soluble polyimide overcame the problem of the high temperature dehydration cyclization of thermal polymerization and was applicable to a stretchable OTFT device. In addition, Jeffamine D2000 and polyurethane could be used as additives to increase the tensile properties without the original electrical properties being affected.

2.1. Preparation of Dielectric Gate Dielectric

Briefly, the 4,4-oxydiphthalic anhydride (97%, Sigma Aldrich, Darmstadt, Germany) and the 2,2-*bis*(3-amino-4-hydroxyphenyl) hexafluoropropane (98%, Matrix Scientific, Columbia, SC, USA) in a three-necked flask were dissolved in the *n*-methyl-2-pyrrolidone (NMP, 99.9%, TEDIA, USA) with 1:1 molar ratio and mixed uniformly. After the further addition of isoquinoline (95%, Tokyo Chemical Industry) in a nitrogen atmosphere for 5 h, a yellow-brown solution was obtained, which was poly (amic acid) (PAA). The PAA was placed in an oil bath at 150 °C for 18 h. The polyimide solution obtained was placed in a water: methanol (98%, Mallinckrodt Baker, Phillipsburg, KS, USA) (1:3) mixed solvent to produce the precipitate. The filtrated precipitate was placed in a vacuum oven and dried at 60 °C for 2 days to obtain a soluble polyimide powder containing COOH and a fluorine functional group. Tetraethyl orthosilicate (TEOS, Sigma Aldrich, Darmstadt, Germany) was dissolved in ethanol (99.5%, Acros Organics, NJ, USA) added to an aqueous solution of nitric acid, and stirred for 30 min. Simultaneously, titanium(IV) butoxide (Ti(OBu)$_4$, Sigma Aldrich, Darmstadt, Germany) was dissolved in 2-methyl-2,4-pentanediol (98%, Alfa Aesar, MA, USA) solvent, stirred for 30 min. The two aforementioned solutions were mixed and stirred for sol-gel reaction for 30 min, and the solvent was then removed with a rotary evaporator and finally placed in an oven to obtain the TiO_2-SiO_2 inorganic nanoparticles. The polyimide dissolved in N,N-Dimethylacetamide (DMAc, 99.8%, TEDIA, USA) and the TiO_2-SiO_2 nanoparticles dispersed in butanol solvent were mixed and stirred for 30 min to prepare three series of hybrid materials, namely polyimide-TiO_2-SiO_2, polyimide-TiO_2-SiO_2:D2000, and polyimide-$TiO.SiO_2$:PU. To prepare the polyimide-TiO_2-SiO_2 hybrid material, the different ratios of SiO_2-TiO_2 (0, 10, 20, 30, and 40 wt.%) were mixed with polyimide and stirred for 1 h to obtain the PI-TiO_2-SiO_2 precursor solution represented by AX (X = weight percentage of SiO_2-TiO_2 in hybrid material). For preparation of the polyimide-TiO_2-SiO_2:D2000 and polyimide-TiO_2-SiO_2: PU hybrid material, the preparation procedure was the same as for polyimide-SiO_2-TiO_2. The only difference was that the polymer (Jeffamine D2000, Alfa Aesar, Massachusetts, USA) or polyurethane, (Sigma Aldrich, Darmstadt, Germany) was dropped gradually into the polyimide solution before the mixing with TiO_2-SiO_2 inorganic nanoparticles. The polyimide-TiO_2-SiO_2:D2000 and polyimide-TiO_2-SiO_2: PU hybrid materials were represented by BX and CX, respectively, where X was the weight proportion of SiO_2-TiO_2 in the hybrid material.

2.2. OTFT Device Preparation

First, the elastomer tape was attached to the glass and subjected to plasma treatment for 3 min to clean the tape surface. PEDOT:PSS was then spin coated on the elastomer tape and annealed at 100 °C for 30 min. The solution of polyimide-SiO_2-TiO_2 (or polyimide-SiO_2-TiO_2:Jeffamine D2000 or polyimide-TiO_2-SiO_2:polyurethane) was spin coated onto PEDOT:PSS elastomer tape at 2000 rpm/20 s.

The coated wafer was placed on a hot plate and thermally polymerized through stepwise heating. The baking process was performed at 60, 80, and 100 °C for 10 min and then, finally, at a temperature of 120 °C for 10 min. Three series of hybrid dielectric films, namely AX, BX, and CX, were obtained. The poly[(5,6-difluoro-2,1,3-benzothiadiazol-4,7-diyl)-alt-(3,3'''-di(2-octyldodecyl)-2,2',5',2'',5'',2'''-quaterthiophen-5,5'''-diyl) (PffBT4T-2OD, Sigma Aldrich) as the active layer was then spin coated onto the dielectric layer on a hot plate and heated at 90 °C for 5 min as an annealing process. The upper electrode (source and drain) EGaIn (99.99%, Alfa Aesar, MA, USA) was dropped onto the lower electrode and the PffBT4T-2OD surface, respectively, to fabricate the OTFT device. The device structure is shown in Figure 1.

2.3. Characterization

The thermal properties of the prepared hybrids were assessed using a thermogravimetric analysis (TGA, TA Instruments, Q50) and differential scanning calorimeter analysis (DSC, TA Instruments, Q20/RSC90) at heating rates of 20 °C and 10 °C/min, respectively. The transmittances of the hybrid films coated on the quartz substrates were collected using an ultraviolet-visible spectrum (UV-Vis, Jasco, V-650). The morphologies of the thin films were observed with a high-resolution transmission electron microscope (HR-TEM, JEOL, JEM-2100), a scanning electron microscope (SEM, Hitachi, H-2400), and an atomic force microscope (AFM, Veeco, DI 3100). The thicknesses of the hybrid thin films were analyzed with a microfigure measuring instrument (Surface Profiler, α-step, ET-4000, Kosaka Laboratory Ltd.). For the metal-insulator-metal (MIM) structure analysis, 0.6-mm diameter Al electrodes were deposited directly onto the gate dielectric films through shadow masking. MIM direct current measurements and OTFT measurements were performed in ambient conditions using a probe station interface with an Agilent E4980A precision LCR meter (10 kHz to 1 MHz) and an Agilent B1500A semiconductor device parameter analyzer.

3. Results and Discussion

Figure 1 shows the chemical structures of the polyimide-TiO_2-SiO_2 composite dielectric material, Jeffamine D2000, and polyurethane additives and the schematic for the OTFT device structure and the tensile direction. The OTFT devices exhibit tensile properties that depend on the addition ratio of TiO_2-SiO_2 inorganic nanoparticles and the presence or absence of Jeffamine D2000 or polyurethane additives. The addition ratio of TiO_2-SiO_2 inorganic nanoparticles ranges from 0 wt.% to 40 wt.%; the cases with those ratio values are denoted by A0–A40, B0–B40, and C0–C40, respectively, indicate the addition of Jeffamine D2000 and polyurethane additives in the order of 0 wt.% to 40 wt.%.

3.1. Analysis of Optical and Thermal Properties

All of the prepared hybrid films have optical transmittances greater than 90% with the film thickness about 200 nm. Figure 2a shows the UV-vis spectra of the optical transmittance of A0, A30, B30, and C30 hybrid thin films in the visible light region of 400–700 nm, the optical transmittance is greater than 90%. This result shows that the composite dielectric film has good transparency (as listed in Table 1). Supplementary Figure S1 shows the TEM image of inorganic TiO_2-SiO_2 nanoparticles. It shows that the size of the particles of the as-prepared TiO_2-SiO_2 is about 30–40 nm with a spherical morphology. When the particle size is less than 50 nm, light scattering can be negligible [42]. Moreover, Supplementary Figure S2 shows the optical transmittance of A0, A30, B30, and C30 films as the dielectric layer of OTFTs device in the visible light region of 400–700 nm. This indicates that the optical transmittances of all samples are greater than 75%. The thermal properties of the prepared polyimide-TiO_2-SiO_2 composite dielectric films were analyzed by thermogravimetric analysis (TGA). Figure 2b shows the TGA curves undertaken in a nitrogen atmosphere. It reveals that the decomposition temperature (T_d) of A0, A30, B30, and C30 hybrid thin films are 418, 450, 461, and 443 °C, respectively. The relative parameters for thermal properties are listed in Table 1. This indicates that the thermal decomposition temperature increases with the content of TiO_2-SiO_2 nanoparticles due to the formation of chemical bonding between polyimide and TiO_2-SiO_2, which can restrict the polyimide chain reaction, and the T_d and thermal stability for the hybrid films thus increases as TiO_2-SiO_2 content increases [22]. In addition, the addition of Jeffamine D2000 and polyurethane also increase the T_d from 426 °C to 477 °C for B0–B40 and 405 °C to 454 °C for C0–C40. The increase in T_d for B0–B40 is due to the hydrogen bonding between the N atom in the Jeffamine D2000 and the composite dielectric material. However, the polyurethane is a softer polymer, so the T_d of B0–B40 is expected to be lower than that for the other two series of hybrid films. However, the T_d for all of hybrid films nonetheless exceed 400 °C, indicating good thermal stability. In addition, none of the hybrid films exhibit weight loss at temperatures lower than 300 °C, and the residual quantity of A0–A40 increased with increasing quantities of TiO_2-SiO_2 added when the temperature increased to 900 °C. At 900 °C most of the polyimide has completely decomposed, and the remaining residual quantity is an inorganic oxide forming a cross-linked stable network. This result demonstrates that inorganic TiO_2-SiO_2 nanoparticles have been successfully incorporated into organic materials. Figure 2c shows the differential scanning calorimeter analysis (DSC) measured in a nitrogen atmosphere. It reveals that the glass transition temperatures (T_g) of A0 (PI), B0 (PI:D2000), and C0 (PI:PU) are 266 °C, 286 °C, and 270 °C, respectively. In addition, no T_g point of all samples can be observed in Figure 2c in the temperature range of 25–350 °C, showing the T_g of all hybrid materials (PI/TiO_2-SiO_2) prepared in this study exceeds 350 °C (as listed in Table 1). It is known that the inorganic TiO_2-SiO_2 nanoparticles can be uniformly distributed in the polyimide matrix, and form a crosslinking structure between the polyimide and nanoparticles, which restricts the chain motion and strengthens the polyimide strength, thereby causing an increase in T_g and T_d. The results of thermal analysis suggest that all of the hybrid films prepared in this study exhibit good heat resistance and no phase separation between the polyimide and TiO_2-SiO_2 nanoparticles [43].

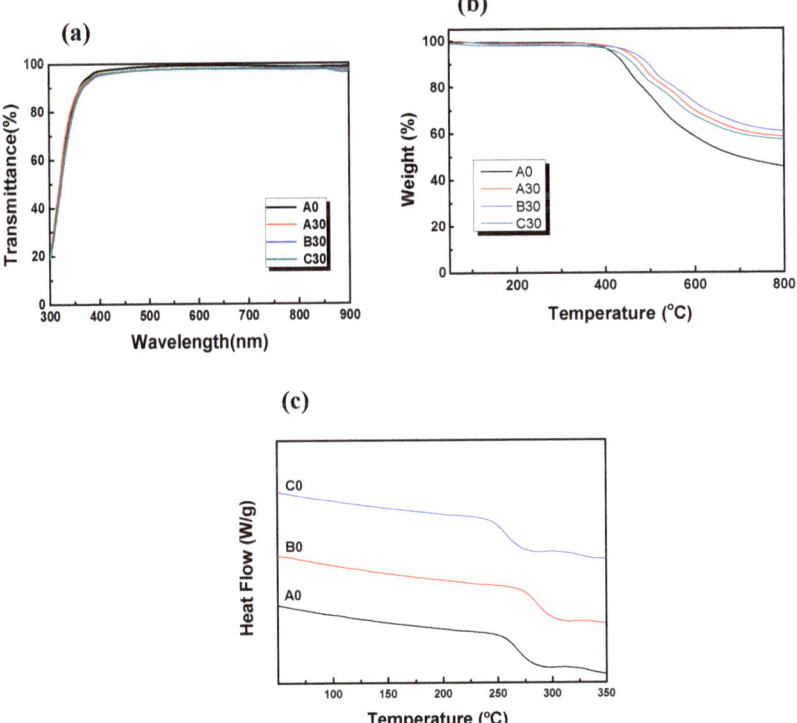

Figure 2. (a) UV-vis spectra of the optical transmittance, (b) TGA curves, and (c) DSC curves of hybrid thin films.

Table 1. Summary of properties of the prepared dielectric hybrid film.

No.	H (nm) [a]	Ra (nm) [b]	Ra/h (%)	T_d (°C)	T_g (°C)	T (%)
A0	215	0.44	0.20	418	266	
A10	200	0.56	0.28	427	-	
A20	215	0.61	0.28	431	-	>90%
A30	220	0.65	0.29	450	-	
A40	210	0.75	0.35	461	-	
B0	200	0.31	0.15	426	286	
B10	210	0.35	0.17	431	-	
B20	220	0.41	0.19	440	-	>90%
B30	215	0.57	0.27	461	-	
B40	215	0.64	0.29	477	-	
C0	220	0.43	0.19	405	270	
C10	210	0.49	0.23	419	-	
C20	220	0.53	0.24	432	-	>90%
C30	215	0.60	0.28	443	-	
C40	210	0.69	0.33	454	-	

[a] Thickness of the prepared thin film. [b] Ra is the average roughness of the prepared thin films, respectively.

3.2. Analysis of Stretching Properties

Figure 3 shows the optical microscopy images of (a) A0, B0, and C0 and of (b) A30, B30, and C30 thin films subject to various strain levels. Figure 3a shows that bare polyimide film generates cracks when subjected to a stretching ratio of 30%, and numerous cracks and wrinkles are produced when the stretching ratio reaches 50%. The B0 film exhibits only wrinkles under the stretching ratio of 30%, but numerous cracks are also present as the stretching ratio increases to 50%. Compared with the result for A0, the addition of Jeffamine D2000 (B0) is seen to improve the film's stretchability. For the C0, the film shows no cracks or wrinkles subject to stretching ratios of 30% and 50%. The results reveal that the addition of polyurethane is more effective than the addition of Jeffamine D2000 for improving the tensile properties of the thin films in the absence of inorganic particles in the polyimide matrix. Moreover, the effect of nanoparticles TiO_2-SiO_2 on the tensile properties is seen in Figure 3b. Adding TiO_2-SiO_2 is seen to cause the tensile properties of the A30 films to decrease obviously because more cracks are observed for a stretching ratio of 30%. For the case of B30 films, wrinkles continue to be generated at a stretching ratio of 30%, but no cracks are observed. The C30 films' stretchability is optimal at the stretching ratios of 30% and 50%. No cracks or wrinkles are observed on the C30 films. Therefore, from the optical microscopy diagram of these dielectric hybrid films, the addition of Jeffamine D2000 and polyurethane polymers is seen to increase the film stretchability and the addition of polyurethane produces tensile properties superior to those obtained from the addition of Jeffamine D2000.

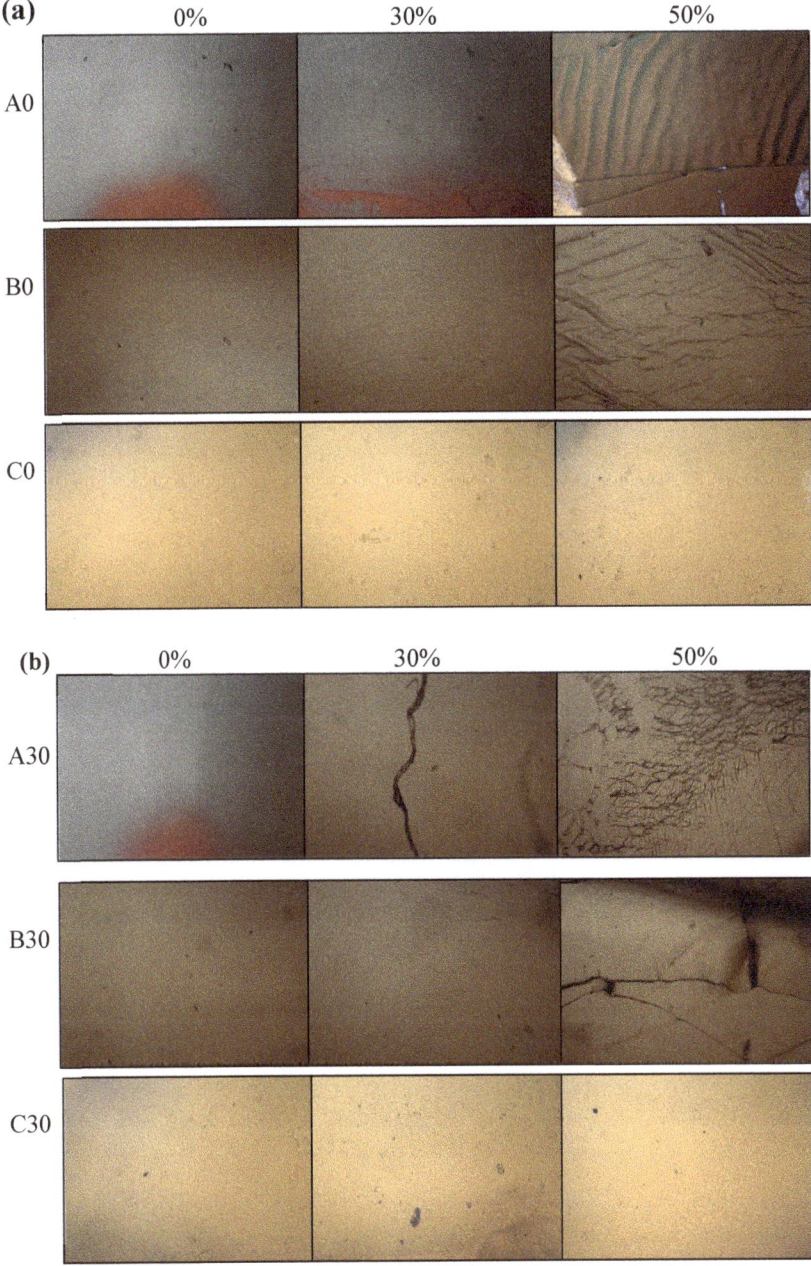

Figure 3. Optical microscopy images of (**a**) A0, B0, and C0 and of (**b**) A30, B30, and C30 thin films subject to various strain levels.

3.3. Analysis of Surface Morphology and Surface Energy

The surface flatness of the hybrid dielectric film s was measured using an atomic force microscope (AFM). The AFM result demonstrates that the surface roughness (Ra) of the three series of A0–A40, B0–B40, and C0–C40 are 0.44–0.75, 0.31–0.64, and 0.43–0.69 nm, respectively. Ra increases with the increase in TiO_2-SiO_2 content. However, all hybrid dielectric films were produced without pinholes, and the surface flatness (the ratio of Ra to film thickness, Ra/h) for all hybrid films was less than 0.35% (Table 1), indicating that all prepared hybrid dielectric films in this study had a good surface flatness. In the previous studies, it has been confirmed that, when the ratio of surface roughness to thickness is less than 0.5%, the material has a good flatness. The prepared dielectric thin film in this study has a better surface flatness than those of studies in literature [39,42,43]. According to the aforementioned results, the three series of hybrid dielectric films have good light transmission, thermal stability, and surface flatness, and no phase separation was observed. Therefore, the prepared hybrid dielectric films can be effectively applied as the gate dielectric materials for the OTFT application.

Typically, thin films show a light scattering behavior due to the surfaces roughness. An innovative hybrid thin film with a lower-than-usual surface roughness can reduce the light loss on the waveguide surface. This result also confirms the potential for the use of polyimide-TiO_2-SiO_2 hybrid material in OTFT as the dielectric film. The surface flatness of the dielectric layer greatly influences the characteristics of the OTFT. When the dielectric layer has a low surface roughness, it effectively reduces the leakage current of OTFT and promotes the order growth of crystals in the active layer. Figure 4 shows the AFM images of the semiconductor layer (BffBT4T-2OD) coated on the various dielectric composite films, namely A0, B0, C0, A30, B30, and C30. This indicates that island-like aggregation was produced from the BffBT4T-2OD on the semiconductor layer (A30) after the addition of the inorganic nanoparticles, and the degree of aggregation became more obvious after the addition of the Jeffamine D2000 (B30) and polyurethane (C30). These dense aggregates of BffBT4T-2OD can help to increase the tensile properties of these composite films and the stretchability of OTFT [41]. BffBT4T-2OD is hydrophobic in nature, and we observed B30 to have lower surface energy (42.07 mJ·m^{-2}) than A30 and C30 did, which enables the growth of BffBT4T-2OD. As the adding ratio of the TiO_2-SiO_2 nanoparticles increases, the size of BffBT4T-2OD crystal grains in the semiconductor layer also increases, resulting in better characteristics for the OTFT. This may be related to the affinity of the dielectric surface for BffBT4T-2OD. As the TiO_2-SiO_2 content increases from 0% (A0) to 30 wt.% (A30), the surface energy of the dielectric layer decreases, which causes the grain size of BffBT4T-2OD in semiconductor layer to increase. The well-connected domain of the A30 provides an efficient channel for charge transport and increases the charge carrier density at the interface between dielectric and semiconductor, which can prevent the charge defects from occurring at the interface and improve the performance of the OTFTs [38–40].

The contact angles of the gate dielectric films were measured using deionized water and diiodomethane as the test drops, respectively, due to their different polarities. The surface energies of the polyimide-TiO_2-SiO_2 hybrid films could then be calculated from the values of the contact angles obtained from water and diiodomethane drops, respectively. The results are listed in Table 2. If the surface of a hybrid film has a small water contact angle, this indicates that the surface is hydrophilic and has a large surface energy. Conversely, a large water contact angle indicates that the surface is hydrophobic and has a low surface energy. As shown in Table 2, whether water or diiodomethane is used as the test drop, the contact angles for the hybrid films (A, B, and C) increase first and then decrease with the addition of inorganic particles. For A0–A40 hybrid films, the water contact angles increase from 79.78 for A0 to 83.99 for A30 and then decrease to 78.30 for A40.

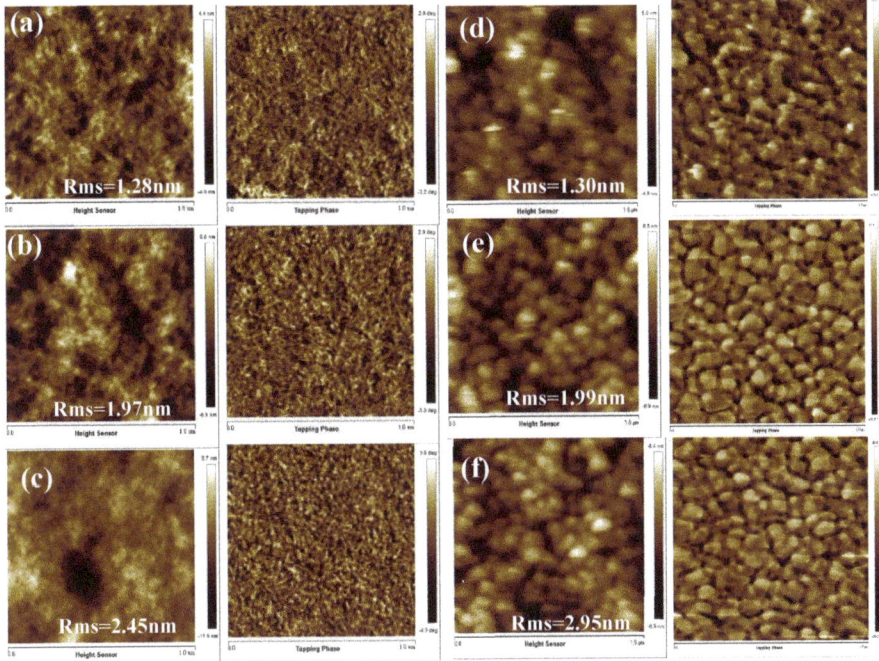

Figure 4. Tapping-mode AFM (1 × 1 μm) images (left: topographic images, right: phase images) of blend films deposited with various TiO_2-SiO_2 ratios and addition of Jeffamine D2000 or polyurethane; (**a**) A0, (**b**) B0, (**c**) C0, (**d**) A30, (**e**) B30, and (**f**) C30.

Table 2. Summary of dielectric constant and surface energy for various hybrid dielectric films.

No.	Dielectric Constant [-]				Water Contact Angle [°]	Diidomethane Contact Angle [°]	Surface Energy [mJ·m^{-2}]
	1 kHz	10 kHz	100 kHz	1 MHz			
A0	4.53	4.49	4.43	4.10	79.78	35.39	51.02
A10	5.78	5.20	4.86	4.27	80.26	42.65	47.66
A20	6.93	6.30	6.11	5.02	80.76	43.09	45.73
A30	7.24	7.08	6.75	6.35	83.99	50.53	44.17
A40	9.51	8.88	7.47	6.53	78.30	29.23	52.33
B0	4.47	4.44	4.41	4.07	78.46	39.10	48.74
B10	5.57	5.11	4.72	4.03	80.44	45.49	45.97
B20	6.03	6.02	5.98	5.00	81.17	51.38	44.68
B30	7.14	7.01	6.72	5.98	84.91	51.83	42.07
B40	8.58	7.96	7.20	6.31	76.12	37.04	51.37
C0	4.22	4.10	4.01	3.98	77.38	39.24	49.47
C10	5.44	5.07	4.69	4.00	78.60	46.55	47.20
C20	5.99	5.79	5.69	5.12	78.80	51.27	44.70
C30	6.91	6.83	6.59	5.45	82.00	51.38	43.34
C40	7.87	7.78	7.01	6.29	74.49	38.22	51.02

The increase in water contact angle is mainly due to the change in surface roughness of hybrid films. In addition, the PI/SiO_2-TiO_2 hybrid films in this study have a polarizable and weakly hydrophobic surface, resulting in this dielectric layer having a low surface energy. This is a very important property for wetting of the latter deposited organic semiconductor layer, and can improve the performance of the device [44]. Therefore, when 30 wt.% of TiO_2-SiO_2 was added, the surface energy of the hybrid film was lowered from 51.02 mJ·m^{-2} (A0) to 44.17 mJ·m^{-2} (A30), indicating that the addition of high dielectric TiO$_2$-SiO$_2$ in a low dielectric PI matrix can change the surface roughness of the PI/TiO$_2$-SiO$_2$ hybrid film, which in turn affects the surface energy of the hybrid film. In addition, due to the inherent hydrophobic nature of the polymer, the addition of Jeffamine D2000 and polyurethane additives can produce a highly hydrophobic surface, which further reduces the surface energy. The results show that the lowest surface energy obtained from B30 is 42.07 mJ·m^{-2} [31]. Typically, dielectric surfaces with low surface energy can provide a venue for the growth of organic semiconductor chains.

3.4. Analysis of Electrical Properties

The result of volumetric capacity measurement shows that the capacitance (at 1 kHz–1 MHz) increases as the TiO$_2$-SiO$_2$ ratio increases, and the relationship is linear. At lower frequencies, the capacitance may increase slightly due to the increased response time available for polarization. The dielectric constant (k) is evaluated using the following equation:

$$C = \frac{k\varepsilon_0 A}{d} \quad (1)$$

where C is the measured capacitance, ε_0 is the vacuum dielectric constant, A is the area of the capacitor, and d is the thickness of the dielectric layer. As shown in Table 2, the dielectric constants obtained at 1 kHz were 4.53 for A0, 9.51 for A40, 4.47 for B0, 8.58 for B40, 4.22 for A0, and 7.87 for A40. When a higher concentration of TiO$_2$-SiO$_2$ was used in a film's fabrication, its dielectric constant was higher. Moreover, the electrical data of OTFTs fabricated by various hybrid dielectrics are shown in Table 3. The results show that the values of mobility (μ) and switch current ratio (on-off current ratios, I_{on}-I_{off}) increase with increasing the TiO$_2$-SiO$_2$ content. Moreover, the leakage current density (LCD) measured at -2 MV·cm^{-1} also increases as the content of TiO$_2$-SiO$_2$ nanoparticles increases. Figure S3. shows the transfer curves of OTFTs prepared by different dielectric materials, A0, A30, B30, and C30. This indicates that the mobility and I_{on}/I_{off} of the device prepared by the polyimide dielectric layer without TiO$_2$-SiO$_2$ nanoparticles (A0) are 0.0181 cm^2·V^{-1}·s^{-1} and 1.13 × 10^3, respectively. When the PI/TiO$_2$-SiO$_2$ hybrid material (A30) was used as a dielectric layer, the mobility and I_{on}/I_{off} of device increased to 0.242 cm^2·V^{-1}·s^{-1} and 9.04 × 10^3. When adding Jeffamine D2000 (B30) and polyurethane (C30) additives into the dielectric layer, the mobility and I_{on}/I_{off} were further increased to 0.817 cm^2·V^{-1}·s^{-1} and 4.27 × 10^5 for B30 and 0.562 cm^2·V^{-1}·s^{-1} and 2.04 × 10^5 for C30, respectively, which shows that the proper amount of TiO$_2$-SiO$_2$ nanoparticles and additives can effectively improve the device performance. The larger the I_{on}/I_{off} ratio, the better the switch characteristics of OTFTs. It is known that high-k dielectric layer could cause a low operation voltage and low power consumption. The LCD value is related to the thickness of the hybrid dielectric layer and the pinholes density, because the chemical bonding between the inorganic and polyimide can make the dielectric layer structure more dense, and the high-k dielectric can improve the capacitive coupling effect between the gate and active channel layer, which can increase the driving current and reduce the operating voltage. In addition, the surface morphology of the dielectric layer affects the structure of the deposited organic semiconductor layer, which in turn affects the performance of the device. A smooth and pinhole-free surface of dielectric layer is important for the interfacial connection during the deposition of organic semiconductor layers, because a smooth interface reduces the charge scattering sites in the channel. Reducing the current leakage from the dielectric interface has a great influence on the electrical performance of the OTFTs. For high-performance OTFTs, the gate dielectric should have a low LCD and a high-k, which can provide greater surface charge accumulation and simultaneously reduce the operating voltage. It should be noted that the thickness of the dielectric layer must be carefully controlled to minimize the current leakage without greatly reducing the capacitance [44]. An increase in the LCD value means that the effect of the insulating layer is reduced. As shown in Table 3, the LCD values of all dielectric layers were less than 10^{-8} A·cm^{-2}. Therefore, these hybrid dielectric layers are suitable for OTFT applications. The gate-current behavior is usually similar to the capacitor leakage current. In this work, we used a metal-insulator-metal capacitor to study the dielectric leakage current. Moreover, as the surface energy decreases, the particle size and alignment of the semiconductor layer become denser, and the carrier mobility of the OTFT increases. Namely, more hydrophobic material increases the carrier mobility of the OTFT. Therefore, the increase in inorganic content helps to form a good organic polymer film, reducing the structural defects in the film and increasing compactness, thus improving carrier mobility. However, the device mobility decreases when the TiO_2-SiO_2 content is more than 30 wt.%, which may be attributed to the coarser surface and the aggregation of the TiO_2-SiO_2 particles, disturbing the formation of BffBT4T-2OD crystal structure in the semiconductor layer. Supplementary Figure S4 shows the output characteristics of the OTFTs using (a) (A0), (b) A30, (c) B30 and (d) C30 as the dielectric layer, respectively. The threshold voltages (V_t) of OTFTs based on hybrid films are small, so only a small gate voltage is needed to turn on the gate. Surface polarization may result in smaller threshold voltages, which can lead to the filling defects of local carrier. Most of the V_t displacement is affected by three factors, which are the charge defect trapping, surface polarization, and ions. In this study, the variation of V_t displacement might be attributed to the addition of different proportions of inorganic nanoparticles at the interface between the dielectric and the semiconductor layers [45,46].

Table 3. The electrical data of OTFTs fabricated by various hybrid dielectrics.

No	LCD [A·cm^{-2}] (at −2 MV·cm^{-1})	V_t [V]	µ [cm^2·V^{-1}·s^{-1}]	I_{ON}/I_{OFF} [-]
A0	7.5×10^{-10}	−2.1	1.81×10^{-2}	1.13×10^3
A10	1.5×10^{-9}	4.1	7.01×10^{-2}	3.24×10^3
A20	2.7×10^{-9}	−2.3	1.21×10^{-1}	5.57×10^3
A30	4.8×10^{-9}	−7.3	2.42×10^{-1}	9.04×10^3
A40	7.7×10^{-9}	3.2	1.07×10^{-2}	1.16×10^3
B0	6.3×10^{-10}	3.3	5.04×10^{-2}	1.51×10^4
B10	8.6×10^{-10}	−1.5	2.09×10^{-1}	2.24×10^4
B20	1.7×10^{-9}	3.9	4.23×10^{-1}	3.01×10^4
B30	2.5×10^{-9}	−8.1	8.17×10^{-1}	4.27×10^5
B40	5.8×10^{-9}	2.2	5.51×10^{-1}	8.50×10^4
C0	5.9×10^{-10}	4.6	4.81×10^{-2}	8.13×10^3
C10	7.7×10^{-10}	2.4	1.89×10^{-1}	1.24×10^4
C20	1.6×10^{-9}	−2.6	3.21×10^{-1}	2.57×10^4
C30	3.4×10^{-9}	3.8	5.62×10^{-1}	2.04×10^5
C40	8.5×10^{-9}	2.2	2.07×10^{-1}	2.16×10^4

Figure 5a shows the mobility values of A0, B0, C0, A30, B30, and C30 at various strain values. These results prove that devices with Jeffamine D2000 (B0, B30) and polyurethane (C0, C30) as additives can be stretched to 20% and 50%, respectively. Figure 5b shows the mobility of A0, B0, C0, A30, B30, and C30 at various stretch cycles, indicating that the devices with Jeffamine D2000 (B0, B30) and polyurethane (C0, C30) as additives can be stretched up to 150. The aforementioned results show that the devices with polyurethane additive can achieve superior stretching properties of 50% for stretching 150 cycles. The mobility has almost no change, which is because the polyurethane additive is a softer chain polymer. AFM analysis revealed that the hybrid films with polyurethane always exhibit a denser and more concentrated film structure that is advantageous for the stretching propertes of the stretchable devices. In addition, the devices with Jeffamine D2000 can also achieve a good stretching properties of 50% for stretching 150 cycles. Although the mobility is reduced by approximately 10%. However, the A0 and A30 samples without added any additives have a significant problem in that the mobility decreases sharply and does not have stretchability.

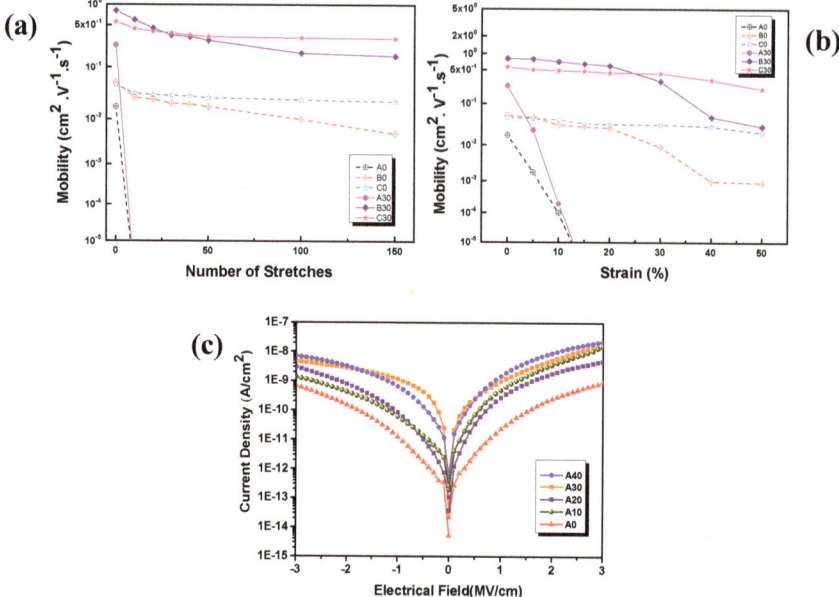

Figure 5. (a) mobility of the strained percentage, (b) the strained cycles and (c) leakage characteristics of hybrid thin films.

Finally, we applied three series of dielectric materials (A, B, C) in the OTFT device as gate materials. Table 3 summarizes the electrical characteristics of these OTFTs, including the LCD, mobility, and I_{on}-I_{off}. As mentioned, the electrical properties of the fabricated OTFT devices are identical to those obtained using AFM and surface energy. In the upstretched case, the LCD becomes larger with the increase in the content of TiO_2-SiO_2 inorganic nanoparticles, but the LCD value is less than 10^{-8} A·cm^{-2} (−2 MV·cm^{-1}) (Figure 5c). The addition of Jeffamine D2000 and polyurethane additives reduces the LCD values. The mobility and I_{on}-I_{off} of the dielectric layers of pure polyimide without inorganic nanoparticles and polymer additives (A0, B0, and C0) are 1.81×10^{-2} cm^2·V^{-1}·s^{-1} and 1.13×10^3 for A0, 5.04×10^{-2} cm^2·V^{-1}·s^{-1} and 1.51×10^4 for B0, and 4.81×10^{-2} cm^2·V^{-1}·s^{-1} and 8.13×10^3 for C0, respectively. The mobility and I_{on}-I_{off} increase obviously with the content of TiO_2-SiO_2 inorganic nanoparticles. The mobility and I_{on}-I_{off} of A30 hybrid film reach 2.42×10^{-1} cm^2·V^{-1}·s^{-1} and 9.04×10^3, respectively. Moreover, after the addition of Jeffamine D2000 (B30) and polyurethane (C30), the mobility and I_{on}-I_{off} improve to 8.17×10^{-1} cm^2·V^{-1}·s^{-1} and 4.27×10^5, respectively, for B30 and to 5.62×10^{-1} cm^2·V^{-1}·s^{-1} and 2.04×10^5, respectively, for C30. However, when the content of SiO_2-TiO_2 increases to 40 wt.% (A40, B40, and C40), the mobility and switching current ratio decrease to 1.07×10^{-2} cm^2·V^{-1}·s^{-1} and 1.16×10^3, respectively, for A40 and to 5.51×10^{-1} cm^2·V^{-1}·s^{-1}, and 8.50×10^4, respectively, for B40, and to 2.07×10^{-1} cm^2·V^{-1}·s^{-1} and 2.16×10^3, respectively, for C40.

The decreases in mobility and switching current ratio are attributed to the phase separation of the mixed solution when the content of inorganic particles of TiO_2-SiO_2 is too high. The results show that the use of TiO_2-SiO_2 inorganic nanoparticles and Jeffamine D2000 and polyurethane additives can improve the mobility and I_{on}-I_{off}. However, when the content of TiO_2-SiO_2 inorganic nanoparticles reaches 40 wt.%, precipitation and subsequent phase separation occurs in the precursor solution, resulting in poor film properties and thus poor electrical properties for A40, B40, and C40 samples. Therefore, the optimal mass ratio of polyimide and TiO_2-SiO_2 inorganic nanoparticles is 70:30 wt.%. The aforementioned results demonstrate that the content of TiO_2-SiO_2 nanoparticles exert an obvious influence on the electrical performance of OTFTs. In summary, the mobility and current-switching ratio of B30- and C30-based OTFT after being stretched 150 cycles at 50% of strain are 0.29 cm^2·V^{-1}·s^{-1} and 8.17 × 10^4 for B30-based OTFT and 0.38 cm^2·V^{-1}·s^{-1} and 1.34 × 10^5 for C30-based OTFT, respectively, which is higher than those obtained from other hybrid dielectrics-based devices. These results show that Jeffamine D2000 additives (B30) and polyurethane additives (C30) improve the properties of stretchable OTFE devices, of which C30 exerts the better effect. This result is consistent with the optical microscopy result of the tensile test for the hybrid dielectric films.

4. Conclusion

In this study, we successfully synthesized a series of hybrid dielectric films using polyimide and SiO$_2$-TiO$_2$ nanoparticles without polymer additives, namely polyimide-TiO$_2$-SiO$_2$, polyimide-TiO$_2$-SiO$_2$:D2000, and polyimide-TiO$_2$-SiO$_2$:PU. Pffbt4t-2OD was used in the semiconductor layer. The addition of Jeffamine D2000 (D2000) and polyurethane (PU) as additives was observed to increase the tensile properties without affecting the original electrical properties. The results suggest that the C30-based OTFT achieves the best tensile effect of 50% train after 150 cycles subject to the 10% mobility reduction because the polyurethane polymers are softer and can provide a denser and more concentrated film structure, which facilitates the stretching of the device. Through the adjustment of the ratios of various TiO$_2$-SiO$_2$ inorganic nanoparticles, the dielectric constant of the hybrid material can be adjusted, thereby significantly improving the dielectric properties of the dielectric layer. The device properties (mobility and threshold voltage) and film properties (dielectric constant, surface morphology, and hydrophilic hydrophobicity) exhibit a strong correlation to the proportion of TiO$_2$-SiO$_2$ inorganic nanoparticles. This study shows that the prepared hybrid films can be customized according to the requirements for practical applications. In addition, our PI-hybrid material has the advantage of transparency, high thermal stability, and environmental safety. The addition of Jeffamine D2000 and polyurethane can increase tensile properties without affecting the original electrical properties and widen the applicability of OTFT devices.

Supplementary Materials: The following are available online at http://www.mdpi.com/2073-4360/12/5/1058/s1, Figure S1: The TEM image of inorganic TiO$_2$-SiO$_2$ nanoparticles; Figure S2: The optical transmittance of A0, A30, B30, and C30 films as the dielectric layer of OTFTs device; Figure S3: The transfer curves of OTFTs prepared by different dielectric layer materials, A0, A30, B30, and C30; Figure S4: The output characteristics of the OTFTs prepared by different dielectric layer materials: (a) A0, (b) A30, (c) B30, and (d) C30.

Author Contributions: Conceptualization, Y.-Y.Y.; Data curation, C.-H.Y.; Funding acquisition, Y.-Y.Y.; Investigation, Y.-Y.Y. and C.-H.Y.; Methodology, C.-H.Y.; Project administration, Y.-Y.Y.; Supervision, Y.-Y.Y.; Writing—original draft, Y.-Y.Y.; Writing—review & editing, Y.-Y.Y. All authors have read and agreed to the published version of the manuscript.

Acknowledgments: We thank the Ministry of Science and Technology of Taiwan (MOST 108-2221-E-131-003) for providing financial support.

Conflicts of Interest: The authors declare no conflict of interest.

References

1. Qian, Y.; Zhang, X.W.; Xie, L.H.; Qi, D.P.; Chandran, B.K.; Chen, X.D.; Huang, W. Stretchable Organic Semiconductor Devices. *Adv. Mater.* **2016**, *28*, 9243–9265. [CrossRef] [PubMed]
2. Forrest, S.R. The path to ubiquitous and low-cost organic electronic appliances on plastic. *Nature* **2004**, *428*, 911–918. [CrossRef] [PubMed]
3. Hsieh, Y.T.; Chen, J.Y.; Fukuta, S.; Lin, P.C.; Higashihara, T.; Chueh, C.C.; Chen, W.C. Realization of Intrinsically Stretchable Organic Solar Cells Enabled by Charge-Extraction Layer and Photoactive Material Engineering. *ACS Appl. Mater. Interfaces* **2018**, *10*, 21712–21720. [CrossRef]
4. Shekar, B.C.; Lee, J.Y.; Rhee, S.W. Organic thin film transistors: Materials, process and devices. *Korean J. Chem. Eng.* **2004**, *21*, 267–285. [CrossRef]
5. Kumar, B.; Kaushik, B.K.; Negi, Y.S. Organic Thin Film Transistors: Structures, Models, Materials, Fabrication, and Applications: A Review. *Polym. Rev.* **2014**, *54*, 33–111. [CrossRef]

6. Golmar, F.; Gobbi, M.; Llopis, R.; Stoliar, P.; Casanova, F.; Hueso, L.E. Non-conventional metallic electrodes for organic field-effect transistors. *Org. Electron.* **2012**, *13*, 2301–2306. [CrossRef]
7. Wen, H.F.; Wu, H.C.; Aimi, J.; Hung, C.C.; Chiang, Y.C.; Kuo, C.C.; Chen, W.C. Soft Poly(butyl acrylate) Side Chains toward Intrinsically Stretchable Polymeric Semiconductors for Field-Effect Transistor Applications. *Macromolecules* **2017**, *50*, 4982–4992. [CrossRef]
8. Wang, J.T.; Takshima, S.; Wu, H.C.; Shih, C.C.; Isono, T.; Kakuchi, T.; Satoh, T.; Chen, W.C. Stretchable Conjugated Rod Coil Poly(3-hexylthiophene)-block-poly(butyl acrylate) Thin Films for Field Effect Transistor Applications. *Macromolecules* **2017**, *50*, 1442–1452. [CrossRef]
9. Lee, W.Y.; Wu, H.C.; Lu, C.; Naab, B.D.; Chen, W.C.; Bao, Z.N. n-Type Doped Conjugated Polymer for Nonvolatile Memory. *Adv. Mater.* **2017**, *29*. [CrossRef]
10. Oh, J.Y.; Rondeau-Gagne, S.; Chiu, Y.C.; Chortos, A.; Lissel, F.; Wang, G.J.N.; Schroeder, B.C.; Kurosawa, T.; Lopez, J.; Katsumata, T.; et al. Intrinsically stretchable and healable semiconducting polymer for organic transistors. *Nature* **2016**, *539*, 411–415. [CrossRef]
11. Kumar, B.; Kaushik, B.K.; Negi, Y.S. Perspectives and challenges for organic thin film transistors: Materials, devices, processes and applications. *J. Mater. Sci. Mater. Electron.* **2014**, *25*, 1–30. [CrossRef]
12. Wang, B.H.; Huang, W.; Chi, L.F.; Al-Hashimi, M.; Marks, T.J.; Facchetti, A. High-k Gate Dielectrics for Emerging Flexible and Stretchable Electronics. *Chem. Rev.* **2018**, *118*, 5690–5754. [CrossRef] [PubMed]
13. Han, G.Q.; Wang, X.M.; Zhang, J.; Zhang, G.C.; Yang, H.H.; Hu, D.B.; Sun, D.W.; Wu, X.M.; Ye, Y.; Chen, H.P.; et al. Interface engineering with double-network dielectric structure for flexible organic thin film transistors. *Org. Electron.* **2018**, *52*, 213–221. [CrossRef]
14. Hung, C.C.; Wu, H.C.; Chiu, Y.C.; Tung, S.H.; Chen, W.C. Crosslinkable high dielectric constant polymer dielectrics for low voltage organic field-effect transistor memory devices. *J. Polym. Sci. Part A Polym. Chem.* **2016**, *54*, 3224–3236. [CrossRef]
15. McCoul, D.; Hu, W.L.; Gao, M.M.; Mehta, V.; Pei, Q.B. Recent Advances in Stretchable and Transparent Electronic Materials. *Adv. Electron. Mater.* **2016**, *2*. [CrossRef]
16. Han, D.D.; Chen, Z.F.; Cong, Y.Y.; Yu, W.; Zhang, X.; Wang, Y. High-Performance Flexible Tin-Zinc-Oxide Thin-Film Transistors Fabricated on Plastic Substrates. *IEEE Trans. Electron Devices* **2016**, *63*, 3360–3363. [CrossRef]
17. Sekine, T.; Fukuda, K.; Kumaki, D.; Tokito, S. Highly stable flexible printed organic thin-film transistor devices under high strain conditions using semiconducting polymers. *Jpn. J. Appl. Phys.* **2015**, *54*. [CrossRef]
18. Lim, J.W.; Koo, J.B.; Yun, S.J.; Kim, H.T. Characteristics of pentacene thin film transistor with Al_2O_3 gate dielectrics on plastic substrate. *Electrochem. Solid-State Lett.* **2007**, *10*, J136–J138. [CrossRef]
19. Shih, C.C.; Lee, W.Y.; Lu, C.; Wu, H.C.; Chen, W.C. Enhancing the Mechanical Durability of an Organic Field Effect Transistor through a Fluoroelastomer Substrate with a Crosslinking-Induced Self-Wrinkled Structure. *Adv. Electron. Mater.* **2017**, *3*. [CrossRef]
20. Song, L.; Wang, Y.; Gao, Q.; Guo, Y.; Wang, Q.J.; Qian, J.; Jiang, S.; Wu, B.; Wang, X.R.; Shi, Y.; et al. Speed up Ferroelectric Organic Transistor Memories by Using Two-Dimensional Molecular Crystalline Semiconductors. *ACS Appl. Mater. Interfaces* **2017**, *9*, 18127–18133. [CrossRef]
21. Hsu, H.H.; Chang, C.Y.; Cheng, C.H. Room-temperature flexible thin film transistor with high mobility. *Curr. Appl. Phys.* **2013**, *13*, 1459–1462. [CrossRef]
22. Yu, Y.Y.; Liu, C.L.; Chen, Y.C.; Chiu, Y.C.; Chen, W.C. Tunable dielectric constant of polyimide-barium titanate nanocomposite materials as the gate dielectrics for organic thin film transistor applications. *RSC Adv.* **2014**, *4*, 62132–62139. [CrossRef]
23. Zhang, C.Y.; Wang, H.; Shi, Z.S.; Cui, Z.C.; Yan, D.H. UV-directly patternable organic-inorganic hybrid composite dielectrics for organic thin-film transistors. *Org. Electron.* **2012**, *13*, 3302–3309. [CrossRef]
24. Lee, W.H.; Wang, C.C.; Ho, J.C. Influence of nano-composite gate dielectrics on OTFT characteristics. *Thin Solid Films* **2009**, *517*, 5305–5310. [CrossRef]
25. Jung, S.W.; Koo, J.B.; Park, C.W.; Na, B.S.; Park, N.M.; Oh, J.Y.; Moon, Y.G.; Lee, S.S.; Koo, K.W. Non-volatile organic ferroelectric memory transistors fabricated using rigid polyimide islands on an elastomer substrate. *J. Mater. Chem. C* **2016**, *4*, 4485–4490. [CrossRef]
26. Damaceanu, M.D.; Constantin, C.P.; Bruma, M.; Belomoina, N.M. Highly fluorinated polyimide blends—Insights into physico-chemical characterization. *Polymer* **2014**, *55*, 4488–4497. [CrossRef]

27. Ahn, T.; Kim, J.W.; Choi, Y.; Yi, M.H. Hybridization of a low-temperature processable polyimide gate insulator for high performance pentacene thin-film transistors. *Org. Electron.* **2008**, *9*, 711–720. [CrossRef]
28. Kim, Y.J.; Kim, J.; Kim, Y.S.; Lee, J.K. TiO$_2$-poly(4-vinylphenol) nanocomposite dielectrics for organic thin film transistors. *Org. Electron.* **2013**, *14*, 3406–3414. [CrossRef]
29. Chu, C.W.; Li, S.H.; Chen, C.W.; Shrotriya, V.; Yang, Y. High-performance organic thin-film transistors with metal oxide/metal bilayer electrode. *Appl. Phys. Lett.* **2005**, *87*. [CrossRef]
30. Tsai, C.L.; Chen, C.J.; Wang, P.H.; Lin, J.J.; Liou, G.S. Novel solution-processable fluorene-based polyimide/TiO$_2$ hybrids with tunable memory properties. *Polym. Chem.* **2013**, *4*, 4570–4573. [CrossRef]
31. Zhao, J.B.; Li, Y.K.; Yang, G.F.; Jiang, K.; Lin, H.R.; Ade, H.; Ma, W.; Yan, H. Efficient organic solar cells processed from hydrocarbon solvents. *Nat. Energy* **2016**, *1*. [CrossRef]
32. Scenev, V.; Cosseddu, P.; Bonfiglio, A.; Salzmann, I.; Severin, N.; Oehzelt, M.; Koch, N.; Rabe, J.P. Origin of mechanical strain sensitivity of pentacene thin-film transistors. *Org. Electron.* **2013**, *14*, 1323–1329. [CrossRef]
33. Tsai, M.H.; Huang, Y.C.; Tseng, I.H.; Yu, H.P.; Lin, Y.K.; Huang, S.L. Thermal and mechanical properties of polyimide/nano-silica hybrid films. *Thin Solid Films* **2011**, *519*, 5238–5242. [CrossRef]
34. Chou, W.Y.; Ho, T.Y.; Cheng, H.L.; Tang, F.C.; Chen, J.H.; Wang, Y.W. Gate field induced ordered electric dipoles in a polymer dielectric for low-voltage operating organic thin-film transistors. *RSC Adv.* **2013**, *3*, 20267–20272. [CrossRef]
35. Huang, T.S.; Su, Y.K.; Wang, P.C. Study of organic thin film transistor with polymethylmethacrylate as a dielectric layer. *Appl. Phys. Lett.* **2007**, *91*. [CrossRef]
36. Miskiewicz, P.; Kotarba, S.; Jung, J.; Marszalek, T.; Mas-Torrent, M.; Gomar-Nadal, E.; Amabilino, D.B.; Rovira, C.; Veciana, J.; Maniukiewicz, W.; et al. Influence of SiO$_2$ surface energy on the performance of organic field effect transistors based on highly oriented, zone-cast layers of a tetrathiafulvalene derivative. *J. Appl. Phys.* **2008**, *104*. [CrossRef]
37. Savagatrup, S.; Chan, E.; Renteria-Garcia, S.M.; Printz, A.D.; Zaretski, A.V.; O'Connor, T.F.; Rodriquez, D.; Valle, E.; Lipomi, D.J. Plasticization of PEDOT:PSS by Common Additives for Mechanically Robust Organic Solar Cells and Wearable Sensors. *Adv. Funct. Mater.* **2015**, *25*, 427–436. [CrossRef]
38. Lu, C.; Lee, W.Y.; Shih, C.C.; Wen, M.Y.; Chen, W.C. Stretchable Polymer Dielectrics for Low-Voltage-Driven Field-Effect Transistors. *ACS Appl. Mater. Interfaces* **2017**, *9*, 25522–25532. [CrossRef] [PubMed]
39. Wang, C.; Lee, W.Y.; Nakajima, R.; Mei, J.G.; Kim, D.H.; Bao, Z.A. Thiol-ene Cross-Linked Polymer Gate Dielectrics for Low-Voltage Organic Thin-Film Transistors. *Chem. Mater.* **2013**, *25*, 4806–4812. [CrossRef]
40. Jiang, B.H.; Peng, Y.J.; Chen, C.P. Simple structured polyetheramines, Jeffamines, as efficient cathode interfacial layers for organic photovoltaics providing power conversion efficiencies up to 9.1%. *J. Mater. Chem. A* **2017**, *5*, 10424–10429. [CrossRef]
41. Huang, Z.Q.; Hu, X.T.; Liu, C.; Tan, L.C.; Chen, Y.W. Nucleation and Crystallization Control via Polyurethane to Enhance the Bendability of Perovskite Solar Cells with Excellent Device Performance. *Adv. Funct. Mater.* **2017**, *27*. [CrossRef]
42. Yu, Y.Y.; Huang, T.J.; Lee, W.Y.; Chen, Y.C.; Kuo, C.C. Highly transparent polyimide/nanocrystalline-zirconium dioxide hybrid materials for organic thin film transistor applications. *Org. Electron.* **2017**, *48*, 19–28. [CrossRef]
43. Yu, Y.Y.; Chiu, C.T.; Chueh, C.C. Solution-Processable, Transparent Polyimide for High-Performance High-k Nanocomposite: Synthesis, Characterization, and Dielectric Applications in Transistors. *Asian J. Org. Chem.* **2018**, *7*, 2263–2270. [CrossRef]
44. Yang, B.X.; Tseng, C.Y.; Chiang, A.S.T.; Liu, C.L. A sol-gel titanium-silicon oxide/organic hybrid dielectric for low-voltage organic thin film transistors. *J. Mater. Chem. C* **2015**, *3*, 968–972. [CrossRef]
45. Baeg, K.J.; Noh, Y.Y.; Sirringhaus, H.; Kim, D.Y. Controllable Shifts in Threshold Voltage of Top-Gate Polymer Field-Effect Transistors for Applications in Organic Nano Floating Gate Memory. *Adv. Funct. Mater.* **2010**, *20*, 224–230. [CrossRef]
46. Naber, R.C.G.; Tanase, C.; Blom, P.W.M.; Gelinck, G.H.; Marsman, A.W.; Touwslager, F.J.; Setayesh, S.; De Leeuw, D.M. High-performance solution-processed polymer ferroelectric field-effect transistors. *Nat. Mater.* **2005**, *4*, 243–248. [CrossRef]

 © 2020 by the authors. Licensee MDPI, Basel, Switzerland. This article is an open access article distributed under the terms and conditions of the Creative Commons Attribution (CC BY) license (http://creativecommons.org/licenses/by/4.0/).

Article

Electrically Conductive Nanocomposites Composed of Styrene–Acrylonitrile Copolymer and rGO via Free-Radical Polymerization

Eun Bin Ko [1], Dong-Eun Lee [2,*] and Keun-Byoung Yoon [1,*]

1. Department of Polymer Science and Engineering, Kyungpook National University, Daegu 41566, Korea; ebko@naver.com
2. School of Architecture & Civil Engineering, Kyungpook National University, Daegu 41566, Korea
* Correspondence: dolee@knu.ac.kr (D.-E.L.); kbyoon@knu.ac.kr (K.-B.Y.)

Received: 13 April 2020; Accepted: 26 May 2020; Published: 27 May 2020

Abstract: The polymerizable reduced graphene oxide (mRGO) grafted styrene–acrylonitrile copolymer composites were prepared via free radical polymerization. The graphene oxide (GO) and reduced graphene oxide (rGO) was reacted with 3-(tri-methoxysilyl)propylmethacrylate (MPS) and used as monomer to graft styrene and acrylonitrile on its surface. The successful modification and reduction of GO was confirmed using Fourier transform infrared spectroscopy (FT-IR), thermogravimetric analyzer (TGA), Raman and X-ray diffraction (XRD). The mRGO was prepared using chemical and solvothermal reduction methods. The effect of the reduction method on the composite properties and nanosheet distribution in the polymer matrix was studied. The thermal stability, electrical conductivity and morphology of nanocomposites were studied. The electrical conductivity of the obtained nanocomposite was very high at 0.7 S/m. This facile free radical polymerization provides a convenient route to achieve excellent dispersion and electrically conductive polymers.

Keywords: polymerizable reduced graphene oxide; in situ polymerization; electrical conductivity; dispersion of 2D nanosheets

---

1. Introduction

Graphene oxide (GO) is one of the most promising fillers for nanocomposites, due to its unique electrical, mechanical, optical and thermal properties [1–5]. GO sheets have been thoroughly investigated due to their good dispersity and possible post-functionalization. Because of they are easily reduced, they provide tunability of electrical conductivity [6,7].

The reduction of GO offers the potential to produce few-layered graphene sheets, i.e., reduced GO (rGO), in high amounts. It also offers the possibility to process advanced graphene-based materials, consisting of various oxygen containing functional groups and for any application concerning conductive polymer composites [8–10]. There are several reduction processes used to prepare rGO, including chemical reduction, solvothermal reduction, electrochemical reduction, thermal reduction and the use of surfactants or stabilizers. Among these, chemical reduction (especially with hydrazine) and solvothermal reduction are among the most effective and convenient approaches for preparing processable colloidal suspensions required for a broad field of applications [11–13].

Because rGO has high thermal and electrical conductivity and rapid transfer is possible between the effectively contacted rGO lamellae, the conductivity of nanocomposites can be significantly improved [14,15]. Therefore, the relative content of rGO within nanocomposites must be increased to form a continuous phase. Unfortunately, aggregation of rGO can easily occur with increasing rGO content. In earlier studies, solution blending, melt mixing and in situ polymerization have been used to produce these polymer/rGO composites [16–19]. In situ polymerization is commonly performed by mixing the filler in the presence of monomers, followed by subsequent polymerization. Upon this method, it is possible to obtain uniformly dispersed polymer/rGO nanocomposites.

In recent years, the grafting of GO sheets with polymer was studied for well-dispersed GO in polymer matrix. Yao et al. [20] grafted polyacrylonitrile using free radical polymerization on gamma irradiation. Yang et al. [21] prepared polystyrene/rGO nanocomposites with radical polymerization, the obtained nanocomposites were enhanced thermal stability on a uniform dispersion of rGO with polymer matrix. Voylov et al. [22] reported that poly(sodium 4-vinylbenzenesulfonate) obtained via RAFT had a controlled molecular weight with a very narrow polydispersity. Sánchez et al. [23] prepared rGO grafted with relatively short chains of poly(n-butyl methacrylate) for electrical field grading materials on high-voltage direct-current applications.

Such in situ polymerization and grafting polymerization methods provides the good dispersion and compatibility of rGO in polymer matrix, and thus excellent thermal, mechanical and electrical properties of the nanocomposites. Researches of polymeric nanocomposite using various monomers were reported, however few studies gave been done on GO-grafted radical copolymerization of styrene and acrylonitrile. Styrene–acrylonitrile copolymer (SAN) is especially tough and resistant to chemicals, it used a broad range of industries including the food, construction and electrical engineering.

In this work, we demonstrate an alternating approach for the growth of polymers from the surface of GO by covalently attaching polymerizable acrylate monomer and reducing them using reducing agents and solvothermal methods. GO was functionalized with 3-(trimethoxysilyl)propyl methacrylate (MPS), and subsequently, this polymerizable mGO was reduced using chemical and solvothermal reduction to prepare the modified reduced graphene oxides (mRGOs). The grafting of graphene with styrene–acrylonitrile copolymers was performed via conventional radical polymerization in the presence of mRGOs. The effects of each reduction method on the electrical conductivity, thermal stability and mechanical properties of the nanocomposites were examined.

2. Experimental

2.1. Materials

Expanded graphite (Timcal Graphite & Carbon, Paris, France, <100 μm, 99.9%), sulfuric acid (H_2SO_4, Duksan Chemical Co., Ansan, Korea, H_2SO_4, ≥95%), hydrochloric acid (HCl, Duksan Chemical Co., Ansan, Korea, HCl, 37%), hydrogen peroxide (H_2O_2, Duksan Chemical Co., Ansan, Korea, H_2O_2, 28%), sodium nitrate (Daejung Co., Seoul, Korea), potassium permanganate (Sigma-Aldrich, St. Louis, MO, USA, >99.0%), hydrazine monohydrate (Junsei, Tokyo, Japan, >98.0%), 3-(trimethoxysilyl)propyl methacrylate (MPS, TCI, Tokyo, Japan, >98.0%), dimethyl formamide (DMF, Duksan Chemical Co., Ansan, Korea, 99.9%) and 1-methyl-2-pyrrolidinone (NMP, Daejung Co., Seoul, Korea, 99.5%) were used without further purification. Styrene (Junsei, Tokyo, Japan, >99.5%) and acrylonitrile (Junsei, Tokyo, Japan, >99.0%) were purified by stirring with calcium hydride and distilling under reduced pressure. Azobisisobutyronitrile (AIBN, Sigma-Aldrich, St. Louis, MO, USA, 98%) was recrystallized from ethanol.

2.2. Synthesis of Graphene Oxide, Modified GO and Reduced mGO

GO was prepared from expanded graphite following Hummers oxidation method [24,25]. The polymerizable modified GO (mGO) was prepared as follows; GO (1 g) was dispersed in DMF (200 mL) via ultrasonication for 12 h and subsequently, MPS (20 mmol) in DMF was added to this dispersion, in a dropwise manner for 0.5 h. The mixture was stirred for 24 h at 60 °C to allow the formation of covalent linkage between the hydroxyl groups on GO and the silyl group. Subsequently, methanol (100 mL) was added to remove any unreacted MPS. Therefore, it was washed sequentially with methanol and DMF.

The mRGOs are produced via two different methods of reduction; chemical (mRGO1) and solvothermal (mRGO2) reduction methods, which are detailed as follows [26,27]. The chemical reduction was carried out by adding hydrazine monohydrate (1 mL) to the mGO suspension, which was stirred at 70 °C for 24 h. The color of suspension was observed to change from brown to black, indicating that the reduction of mGO had occurred. The reaction mixture was filtered and washed repeatedly with DMF to remove any excess, unbound hydrazine. A suspension of mRGO1 was finally obtained via ultrasonication in DMF (200 mL).

For the solvothermal reduction method, a suspension of GO (1 g) in DMF (200 mL) was heated to 180 °C and refluxed for 24 h. It was subsequently filtered and washed with DMF several times. The rGO (1 g) was re-dispersed in DMF (200 mL) via ultrasonication for 12 h and MPS (20 mmol) was subsequently added dropwise 0.5 h. The resulting mixture was stirred for 24 h at 60 °C to facilitate the occurrence of silylation. Subsequently, 100 mL of methanol was added to remove any residual silane molecules. The product was washed sequentially with methanol and DMF. The suspension of mRGO2 was finally obtained via ultrasonication in DMF (200 mL). The reaction procedure was illustrated in Scheme 1.

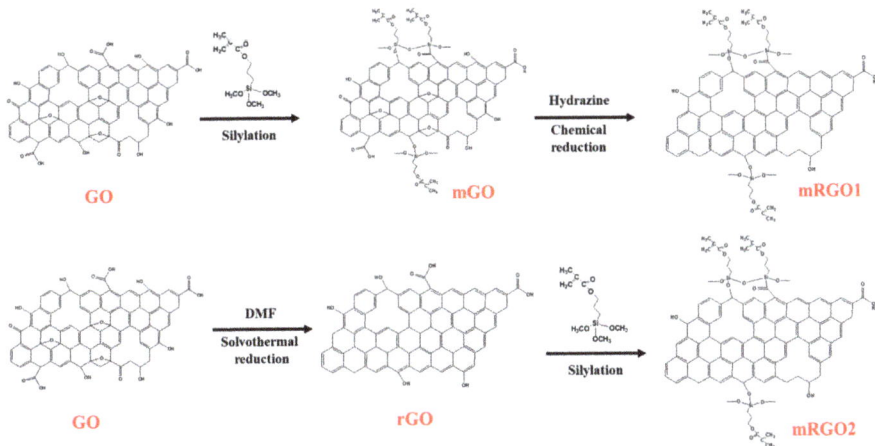

Scheme 1. Schematic of the preparation of mGO, rGO, mRGO1 and mRGO2.

2.3. In situ Copolymerization of Styrene and Acrylonitrile

Radical polymerization was conducted in a 500 mL glass reactor equipped with a magnetic stir bar. The reactor was back-filled three times with nitrogen and charged with styrene, acrylonitrile, mGO, mRGO1 and mRGO2. The mixture was heated to 80 °C and AIBN was added and subsequently, it was stirred under nitrogen for 24 h.

After the polymerization was complete, the mixture was poured into an excess of methanol to precipitate the polymer. The grayish powder was filtered, washed with methanol and dried overnight at 60 °C under vacuum.

The grafting ratio was determined by Soxhlet extraction. The molecular weight, thermal stability and electrical conductivity of the obtained copolymer were used without extraction.

2.4. Characterizations

The chemical structures of GO, mGO and mRGOs were examined using Fourier transform infrared (FT-IR) spectroscopy (Jasco 4100, Tokyo, Japan) and Raman spectroscopy (Nicolet Almega XR, Thermo Scientific, Waltham, MA, USA).

The X-ray diffraction (XRD) patterns were obtained using Philips X-Pert PRO MRD diffractometer (Malvern Panalytical Ltd., Malvern, UK) equipped with Cu-Kα radiation.

The thermal stability was tested at 20 °C/min from 30 to 800 °C under nitrogen atmosphere using a Setaram Labsys evo thermogravimetric analyzer (TGA, Setaram Instrumentation, Paris, France).

The electrical conductivity was measured by the four-point probe method using a Keithley 2400 semiconducting characterization system (Kiethiley, Cleveland, OH, U.S.) at 25 °C. The volume conductivities of the polymer and copolymers with resistance higher than 10^7 Ω were measured using CHI 660E electrochemistry workstation (CH Instruments, Inc., Beijing, China).

The molecular weight of the obtained polymer and copolymers were determined by using an Ubbelohde viscometer (Sigma-Aldrich Co., St. Louis, MO, USA) in tetrahydrofuran at 24 °C.

The morphologies of the GO, mRGOs and the fractured surfaces of nanocomposites were characterized by a field-emission scanning electron microscope (FE-SEM, JSM-6380LV, Joel, Tokyo, Japan).

3. Results and Discussion

3.1. Characterization of mGO and mRGOs

GO was synthesized from graphite using Hummers' method. MPS was used to modify GO to yield MPS-modified GO (mGO), which was subsequently reduced via chemical reduction with hydrazine (mRGO1). GO was also reduced via solvothermal reduction using DMF followed by silylation with MPS (mRGO2).

As shown in Figure 1, FT-IR spectroscopy was used to confirm the structures of GO, mGO and mRGOs and the formation of covalent bonds between the components. The GO exhibited peaks at 1040 cm^{-1}, 1610 cm^{-1}, 1720 cm^{-1} and 3300 cm^{-1} corresponding to the C–O, C=C, C=O and –OH groups, respectively, confirming that it contains a variety of oxygen-containing functional groups. For MPS modified materials, methylene C–H stretching peak appeared at 2920 cm^{-1} and 2850 cm^{-1}, methyl C–H deformation peak at 1465 cm^{-1} and Si–O–C stretching peak at 1130 cm^{-1}, also confirming that MPS was successfully introduced on the surface of GO. After the modification of GO with MPS and its reduction, the –OH stretching at 3300 cm^{-1} disappeared and the characteristic bands at 1300 cm^{-1} (C–Si stretching) and 1030 cm^{-1} (Si–O–C stretching) of the mGO appeared [28].

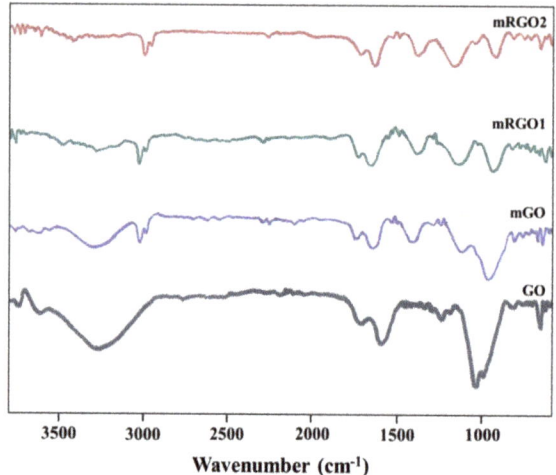

Figure 1. Fourier transform infrared (FT-IR) spectroscopy spectra of graphene oxide (GO), polymerizable modified GO (mGO), chemical (mRGO1) and solvothermal (mRGO2) reduction methods.

Raman spectroscopy and XRD were performed to determine the ordered and disordered carbonaceous materials and the results are shown in Figure S1 and S2, respectively. The D/G band intensity ratio (I_D/I_G) of mRGO1 and mRGO2 were 1.37 and 1.13, respectively, which were higher than that of GO (0.98). It indicated that the number of aromatic domains increased after the reduction step (in Figure S1) [29]. After the modification of GO with MPS, the diffraction angle in XRD patterns (in Figure S2) decreased from 11.3° to 10.4°, corresponding to an increase in the interlayer spacing from 0.78 to 0.85 nm. However, these peaks disappeared after the reduction of mGO. These phenomena indicate that MPS is bonded on the surface of GO successfully and is capable of enlarging the interlayer distance between GO sheets. For mRGO1 and mRGO2, a new broad peak can be observed at 22°–24°, which are close to that of pristine graphene nanosheet (25°) [30]. The results of Raman spectroscopy and XRD as shown in observation of FT-IR, MPS was successfully introduced on the surface of GO and the chemical and solvothermal method effectively reduced GO.

To observe the morphologies and layers of the obtained GO, rGO, mGO and mRGO, SEM was performed, and the images shown in Figure S3.

Optical observation is a direct way to see the dispersion of mRGOs. Figure 2 shows the dispersion states for mRGOs just sonication, after 18 h, 48 h and 144 h, respectively. All mRGOs showed good dispersion in DMF after just sonication. However, after 18 h, GO and mGO started to precipitate and after 144 h, they showed significant precipitation in DMF. The dispersion of mRGO1 and mRGO2 in the DMF did not precipitate even after 144 h from sonication, which attributed to the exfoliation of mRGO1 and mRGO2 layers.

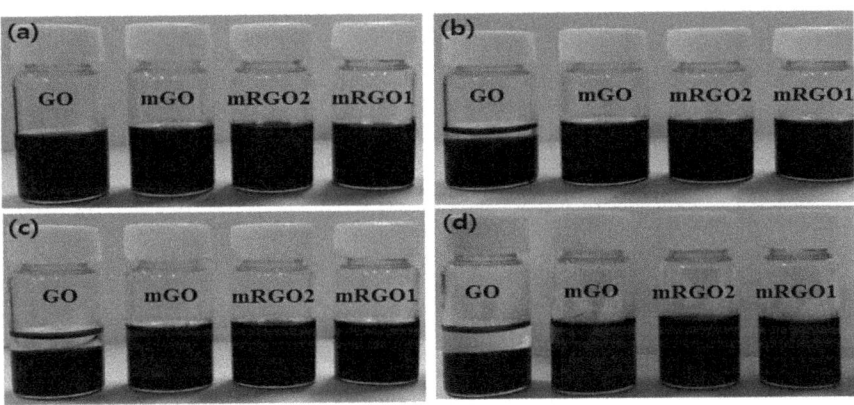

Figure 2. Colloidal suspensions of GO, mGO, mRGO2 and mRGO1 (**a**) just after sonication, (**b**) 18 h, (**c**) 48 h and (**d**) 144 h after sonication.

The thermal stability of GO, mGO and mRGOs was measured using TGA in a nitrogen atmosphere as shown in Figure 3. The GO had lower thermal stability than mGO and mRGOs and exhibited three distinct weight loss steps. The first step is the weight loss until 150 °C, and was caused by the residual evaporation of water (approximately 10%); the second step was from 150 to 300 °C (approximately 34%) and was attributed to the decomposition of labile oxygen-containing groups such as carboxylic, anhydride or lactone groups forming CO, CO_2 and steam; the third step above 300 °C (approximately 16%) was related to the pyrolysis of the carbon skeleton of GO [31]. Functionalizing the surface of GO with MPS and subsequent reduction processes caused the weight loss of approximately 12 wt % for mGO, 12 wt % for mRGO1 and 5 wt % for mRGO2 over 150–300 °C, mainly attributed to the removal of oxygen-containing functional groups. Further, mRGOs also exhibited higher char yields, 70 wt % for mRGO1 and 80 wt % for mRGO2. The solvothermally reduced GO (mRGO2) showed similar characteristics, but with lower amount of weight loss, compared to that of mRGO1. This could be explained by a smaller amount of oxygen functional groups in the surface.

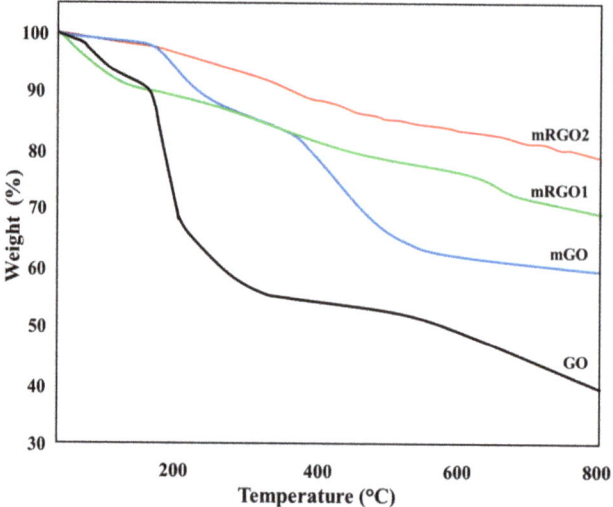

Figure 3. Thermogravimetric analyzer (TGA) thermograms of GO, mGO and mRGOs.

The electrical conductivity of mGO and mRGOs was measured using a four-point probe method; the results are shown in Figure 4. The reduction process increased the conductivity up to 9.8×10^3 S/m owing to the restored π-conjugated system in the mRGO film. Several factors are influenced by the electrical conductivity of graphene derivatives such as the amounts of defects, reduction atmosphere, film thickness and residual oxidation [32–34]. The electrical conductivities of mRGO1 and mGRO2 were 7.6×10^3 S/m and 9.8×10^3 S/m, respectively. The reduction methods did not significantly affect the conductivity, however, mRGO2 has a slightly higher electrical conductivity than mRGO1 because mRGO1 contains more oxygen functional groups in surface as observed in the TGA analysis.

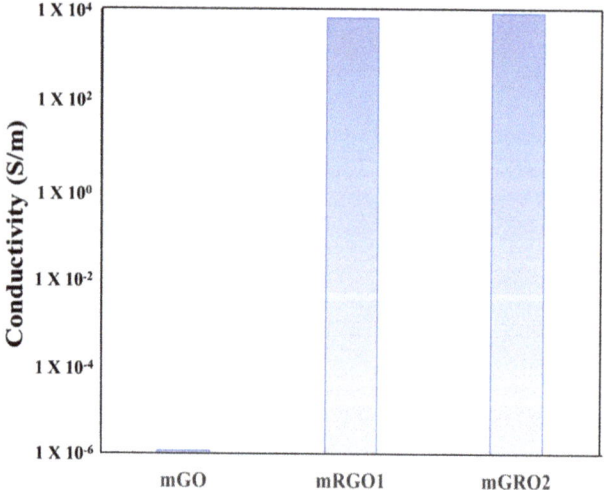

Figure 4. Electrical conductivity of mGO, mRGO1 and mRGO2.

3.2. Preparation of SAN/rGO Nanocomposites via in situ Radical Polymerization

Table 1 presents the results of the copolymerization of styrene and acrylonitrile in the absence or presence of mRGO initiated by AIBN. The grafted copolymer was named SAN–mGO, SAN–mRGO1 and SAN–mRGO2 for the obtained copolymer using mGO, mRGO1 and mRGO2, respectively.

Table 1. Results of the in situ copolymerization of styrene and arylonitrile.

Sample	Monomer feed Ratio (mol%) (Styrene/Acrylonitrile)	mrGO feed (wt %)	Composition [a] (mol%) Styrene	Acrylonitrile	mRGO Contents (wt %)	M_v [b]
SAN	70/30	–	69	31	–	58,000
SAN–mGO	70/30	1.5	68	32	3.7	80,000
		3.0	71	29	5.3	70,000
SAN–mRGO1	70/30	1.5	72	28	3.3	66,000
		3.0	74	26	7.0	54,000
SAN–mRGO3	70/30	0.5	72	28	1.8	52,000
		1.0	73	27	3.0	60,000
		2.5	74	26	4.7	50,000
		5.0	72	28	16.0	77,000

(a) Copolymers composition measured using FT-IR spectroscopy; (b) calculated from the experimental intrinsic viscosities ([η], THF at 24 °C, according to the [η] = k$(M_v)^a$ with k = 21.5 × 10^3 mL g^{-1} and a = 0.68. Polymerization conditions: [Monomer] = 8.7 × 10^{-1} mol/L, [AIBN]/[Monomer] = 1.5 × 10^{-3}, DMF= 100 mL, 80 °C, 24 h.

With the introduction of mRGOs, the conversion of the monomer decreased slightly and the mRGOs contents was varied between 3.3 wt % to 16.0 wt %. The concentration of mRGO and molecular weight of the copolymers did not affect the monomer conversion. The molecular weight was in the range of 5.0 × 10^4 to 8.0 × 10^4 g/mol.

The compositions of styrene and acrylonitrile were calculated using a calibration curve, obtained via FT-IR spectroscopy. The absorbance of the C=N stretching vibrational peak (2250 cm^{-1}) and the aromatic C=C stretching vibrational peak (1450 cm^{-1}) are characteristic of acrylonitrile and styrene, respectively. The calibration curve was obtained from the intensity ratio (I_{2250}/I_{1450}) of the mixture of polyacrylonitrile and polystyrene. The comonomer composition of the copolymers was calculated using the obtained calibration curve. (see Figure S4) The mole fraction of styrene in the copolymer was in the range of 68–74 mol%.

In order to measure the grafting efficiency, the nanocomposites obtained by free-radical polymerization was extracted as methyl ethyl ketone (MEK) using a Soxhlet apparatus [35]. Most of the copolymer that was grafted on mGO and mRGO was extracted only 5–6 wt %, while more than 98 wt % of copolymers without mGO and mRGO was extracted. Although there is a limit to calculating grafting efficiency with simple extraction, the obtained grafted copolymer was used to measure thermal stability and electrical conductivity because the extraction amount was only about 5 wt %.

The effect of the mRGOs on the thermal stability of the obtained styrene–acrylonitrile copolymers was examined using TGA in a nitrogen atmosphere.

TGA thermograms of SAN, SAN–mGO, SAN–mRGO1 and SAN–mRGO2 with different m(R)GOs contents are shown in Figure 5. The degradation temperature at 5% and 10% weight loss ($T_{d5\%}$ and $T_{d10\%}$) and the char yields are summarized in Table 2.

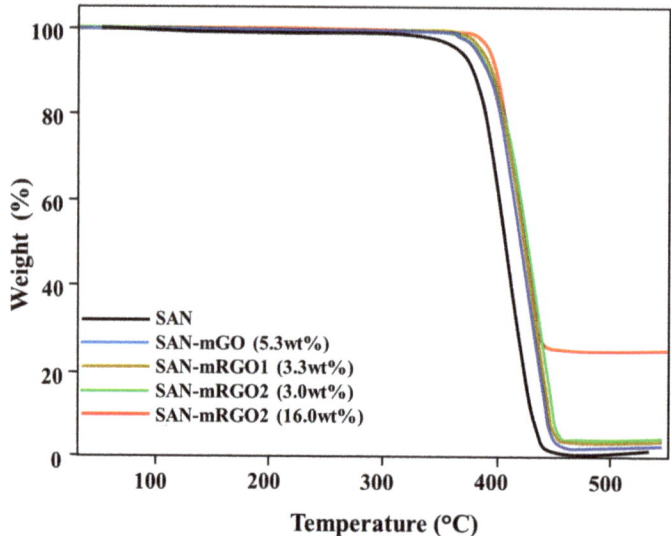

Figure 5. TGA curves of styrene–acrylonitrile copolymer (SAN), and grafted mGO, mRGO1 and mRGO2 copolymers (SAN–mGO, SAN–mRGO1 and SAN–mRGO2) at various contents of mRGOs.

Table 2. Thermal stabilities of SAN, SAN–mGO, SAN–mRGO1 and SAN–mRGO2 at various contents of mRGOs.

	m(R)GOs Contents (wt %)	Weight Loss (5 wt %, $T_{d5\%}$ °C)	Weight Loss (10 wt %, $T_{d10\%}$ °C)	Char Yield (wt %)
SAN	-	357.5	379.2	0.0
SAN–mGO	5.3	377.2	390.0	3.8
SAN–mRGO1	1.5	379.2	390.8	1.9
SAN–mRGO1	3.3	379.0	391.6	4.5
SAN–mRGO1	12.1	384.2	396.1	20.4
SAN–mRGO2	1.8	381.0	393.7	2.2
SAN–mRGO2	3.0	382.0	394.3	3.4
SAN–mRGO2	4.7	386.3	397.2	5.0
SAN–mRGO2	16.0	383.0	397.4	25.0

The weight loss curves exhibited a one-step degradation process and shifted slightly to higher temperatures with the introduction of mGO and mRGOs. The degradation temperatures of grafted nanocomposites ($T_{d5\%}$) are higher than that of neat copolymer. For 1.8, 3.0 wt % and 4.7 wt % mRGO2 contained SAN–mRGO2 nanocomposites, the $T_{d5\%}$ of composites was 23.5, 24.5 and 28.8 °C higher than that of SAN copolymer. In case of SAN–mRGO1, the $T_{d5\%}$ of composites increased by more than 20 °C with introduction of mRGO1. A distinct increase in thermal stability was due to mRGO being well-dispersed in the copolymer matrix. The improvement in thermal stability can be attributed to the delays the escape of volatile degradation products and thus slows down the initial degradation step [36]. The char yield of grafted nanocomposites was slightly higher than the amount of mRGO. The case of SAN–mGRO2 containing 16 wt % of mRGO2 was 25 wt %. The thermal stability of the obtained nanocomposites was almost same values even with mRGO1, which the loading amount of the MPS is twice as much as mGRO2. These results are not due to MPS modification, but to the high heat capacity and thermal conductivity of mRGO nanosheets, which can form the heat-resistant layers that act as mass transfer barriers [37].

The SAN–mRGO2 nanocomposites have excellent thermal stability, which can be attributed to good interfacial interactions between mRGO2 and SAN. From the SEM image of SAN–mRGO2, the well-dispersed status of mRGO2 in copolymer matrix with little aggregation, suggesting good adhesion between the fillers and matrix (Figure 7). Due to the interfacial interaction between mRGO2 and copolymer, the char was easy to form on the surface of mRGO sheets, which resulted in the high char yield of nanocomposites [38].

Superior electrical conductivity is the most important feature of graphene. It is well known that good dispersion of fillers is important for polymer nanocomposites to form a conducting network within the polymer matrix [39]. Polymer/graphene nanocomposites generally exhibit a nonlinear increase in electrical conductivity as a function of the filler concentration. At a certain filler loading fraction, which is known as the percolation threshold, the fillers form a network leading to a sudden rise in the electrical conductivity of the composites [40,41]. Patole et al. [42] prepared polystyrene/graphene films, containing 3.0 wt % graphene with a conductivity of 4.5×10^{-5} S/m. Tripathi et al. [43] also observed that the conductivity of in situ polymerized polymethylmethacrylate/rGO nanocomposites at 2.0 wt % loading of rGO was 9.9×10^{-5} S/cm.

Figure 6 shows the electrical conductivity of SAN–mGO, SAN–mRGO1 and SAN–mRGO2 as a function of mRGO concentrations. The electrical conductivity of SAN–mRGO1 and SAN–mRGO2 composites increased significantly, demonstrating a sharp transition from an electrical insulator to conductor with low percolation thresholds of 1.9 wt % for SAN–mRGO1 and 1.8 wt % for SAN–mRGO2. The electrical conductivity of SAN–mRGO1 significantly increased by five orders of magnitude with the increases in the concentration of mRGO1 from 1.9 wt % to 3.3 wt %, with a value of 7.5×10^{-3} S/m. In the case of SAN–mRGO2, the electrical conductivity significantly increased from 9.3×10^{-9} to 1.2×10^{-2} S/m by increasing the mRGO2 concentration from 1.8 wt % to 4.0 wt %. The electrical conductivity was 0.7 S/m at 7.0 wt % of mRGO1. When mGRO1 and mRGO2 were used, the maximum electrical conductivity and a percolation threshold of the obtained nanocomposites were similar values. These results suggested that the degree of reduction of mGO was more important than the reduction methods. To compare with mGRO2, SAN/rGO nanocomposite was prepared through melting mixing of SAN and rGO, the rGO was reduced by solvothermal method. The electrical conductivity of SAN/rGO nanocomposite was 4.5×10^{-3} S/m, which the concentration of 3.0 wt % rGO. The electrical conductivity of SAN/rGO was much low than that of SAN–mRGO2 at the same amount of nanosheets. The restacking of layers occurred when the unmodified GO was reduced, which did not effectively disperse in the polymer matrix. (in Figure S3 and Figure 7)

The conductivity of SAN–mRGO1 and SAN–mRGO2 starts to become saturated at approximately 7.0 wt % with the value of 0.7 S/m, which indicates that the electrical conductivity of the composites increased significantly once the conductive networks of rGO were formed above a certain critical concentration of rGO in the matrix [44]. Although the electrical properties of mRGOs are not comparable to those of pristine graphene owing to structural defects, there are still many advantages of mRGO for various important applications.

Figure 7 shows SEM images of the fracture surfaces of SAN/rGO and SAN–mRGO2 composites. When the rGO content was 3.1 wt %, the corresponding fracture was uneven however, the rGO sheets were obvious aggregation. In SAN–mRGO2, the mRGO2 sheets were randomly distributed within the SAN matrix even if high content mRGO2 was incorporated. SAN-mGRO2 was observed with only a few layers, but SAN/rGO composite was shown the aggregation of rGO sheets. In case of rGO, restacking occurred during chemical reduction process, however, mRGO2 is thought to have been effectively dispersed without agglomeration occurring due to surface modification of MPS.

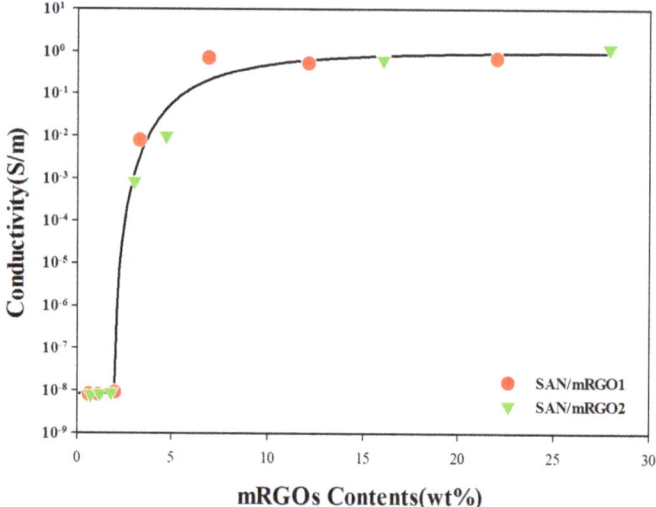

Figure 6. Electrical conductivity of SAN–mGO, SAN–mRGO1 and SAN–mRGO2 at various contents of mRGOs.

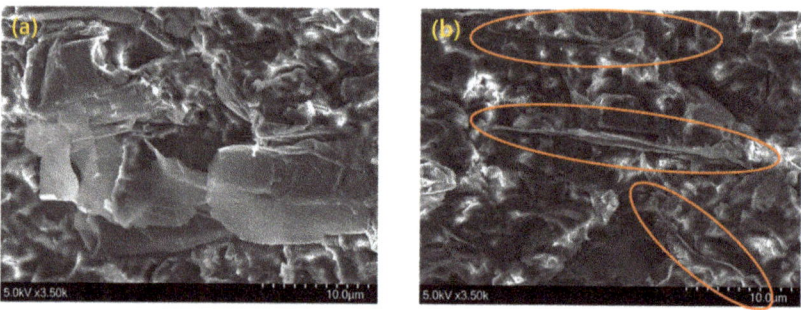

Figure 7. SEM images of (**a**) SAN/rGO (3.0 wt % rGO) and (**b**) SAN–mRGO2 (3.0 wt % mRGO2).

4. Conclusions

The polymerizable reduced graphene oxide, MPS modified reduced graphene oxides (mRGO) were obtained via the chemical and solvothermal reduction of GO. The structure of mRGOs was confirmed using FT-IR, Raman spectroscopy and XRD. The electrical conductivity of the mRGOs was approximately 1.0×10^4 S/m, which corresponds to half of the theoretical electrical conductivity of graphene (1.0×10^7 S/m).

The grafting of rGO with styrene–acrylonitrile copolymers was prepared via in situ radical polymerization in the presence of mRGOs. The thermal stability and electrical properties of the obtained rGO-grafted copolymers (SAN–mRGOs) were investigated. The degradation temperatures of grafted nanocomposites ($T_{d5\%}$) are higher than that of neat copolymer. The $T_{d5\%}$ of nanocomposites increased by more than 20 °C with introduction of mRGOs. A distinct increase in thermal stability was due to mRGO being well-dispersed in the copolymer matrix. The electrical conductivity of SAN–mRGO significantly increased from 9.3×10^{-9} to 0.7 S/m by increasing the mRGO2 concentration from 1.8 wt % to 7.0 wt %. The improved electrical and thermal properties are due to good dispersion and interfacial adhesion between graphene and the polymer matrix. When mGRO1 and mRGO2 were used, the thermal degradation temperature and the maximum electrical conductivity of the obtained nanocomposites were similar values. These results suggested that the degree of reduction of mGO was more important than the reduction methods. The radical polymerization process using mRGOs is a promising tool for the preparation of well-dispersed mRGOs within the polymer matrix. It is anticipated that this work will achieve a convenient route to achieving enhanced interfacial interaction. A distinctive and innovative strategy for the covalent functionalization of graphene derivatives has therefore been explored, which may widen the application of 2D nanosheet-based nanocomposites.

Supplementary Materials: The following are available online at http://www.mdpi.com/2073-4360/12/6/1221/s1, Figure S1: Raman spectra of GO, mGO, mRGO1 and mRGO2, Figure S2: XRD patterns of GO, mGO, mRGO1 and mRGO2, Figure S3: SEM images of (a) GO, (b) rGO, (c) mGO, and (d) mRGO2, Figure S4: Relationship between absorbance ratio and styrene contents of styrene-acrylonitrile copolymers.

Author Contributions: Conceptualization, D.-E.L.; K.-B.Y.; Investigation, E.B.K.; K.-B.Y.; Validation, D.-E.L.; K.-B.Y.; K.-B.Y.; Writing-original draft preparation, D.-E.L.; K.-B.Y.; Writing-review and editing, E.B.K.; D.-E.L.; K.-B.Y.; All authors have read and agreed to the published version of the manuscript.

Funding: This work was supported by the National Research Foundation of Korea (NRF) grant funded by the Korea government (MSIT) (No. NRF-2018R1A5A1025137 and No. NRF-2019R1A2 C3003890).

Conflicts of Interest: The authors declare no conflict of interest.

References

1. Liu, S.; Yao, J.; Liu, Q.; Hung, Y.; Kong, M.; Yang, Q.; Li, G. Tuning the physicochemical structure of graphene oxide by thermal reduction temperature for improved stabilization ability toward polymer degradation. *J. Phys. Chem. C* **2020**, *124*, 8999–9008. [CrossRef]
2. Kang, S.H.; Fang, T.H.; Hong, Z.H. Electrical and mechanical properties of graphene oxide on flexible substrate. *J. Phys. Chem. Solid* **2013**, *74*, 1783–1793. [CrossRef]
3. Li, J.C.; Zheng, L.F.; Sha, X.H.; Chen, P. Microstructural and mechanical characteristics of graphene oxide-fly ash cenosphere hybrid reinforced epoxy resin composites. *J. Appl. Polym. Sci.* **2020**, *137*, 47173. [CrossRef]
4. Wei, J.; Zang, Z.; Zhang, Y.; Wang, M.; Du, J.; Tang, X. Enhanced performance of light-controlled conductive switching in hybrid cuprous oxide/reduced graphene oxide (Cu2/rGO) nanocomposites. *Opt. Lett.* **2017**, *42*, 911–914. [CrossRef]
5. Yao, Y.; Zeng, X.; Wang, F.; Sun, R.; Xu, J.B.; Wong, C.P. Significant enhancement of thermal conductivity in bioinspired freestanding boron nitride paper filled with graphene oxide. *Chem. Mater.* **2016**, *28*, 1049–1057. [CrossRef]
6. Osicka, J.; Mrlik, M.; Ilcikova, M.; Hanulikova, B.; Urbanek, P.; Sedlacik, M.; Mosnacek, J. Reverible actuation ability upon light stimulation of the smart systems with controllably grafted graphene oxide with poly(glycidyl methacrylate) and PDMS elastomer: Effect of compatibility and graphene oxide reduction on the photo-actuation performance. *Polymers* **2018**, *10*, 832. [CrossRef] [PubMed]
7. Marcano, D.C.; Kosynkin, D.V.; Berlin, J.M.; Sinitskii, A.; Sun, Z.; Slesarev, A.; Alemany, L.B.; Lu, W.; Your, J.M. Improved synthesis of graphene oxide. *ACS Nano* **2010**, *4*, 4806–4814. [CrossRef]
8. Stankovich, S.; Dikin, D.A.; Dommett, G.H.; Kohlhaas, K.M.; Zimney, E.J.; Stach, E.A.; Piner, R.D.; Nguyen, S.T.; Ruoff, R.S. Graphene-based composite materials. *Nature* **2006**, *442*, 282–286. [CrossRef]
9. Li, D.; Müller, M.B.; Gilje, S.; Kaner, R.B.; Wallace, G.G. Processable aqueous dispersions of graphene nanosheets. *Nat. Nanotechol.* **2008**, *3*, 101–105. [CrossRef]
10. Wang, Z.; Nelson, J.K.; Hillborg, H.; Zhao, S.; Schadler, L.S. graphene oxide filled nanocomposite with novel electrical and dielectric properties. *Adv. Mater.* **2012**, *24*, 3134–3137. [CrossRef]
11. Tung, V.C.; Allen, M.J.; Yang, Y.; Kaner, R.B. High-throughput solution processing of large-scale graphene. *Nat. Nanotechnol.* **2009**, *4*, 25–29. [CrossRef] [PubMed]
12. Park, S.; An, J.; Jung, I.; Piner, R.D.; An, S.J.; Li, X.; Velamakanni, A.; Ruoff, R.S. Colloidal suspensions of highly reduced graphene oxide in a wide variety of organic solvents. *Nano Lett.* **2009**, *9*, 1593–1597. [CrossRef]
13. Kim, H.; Abdala, A.A.; Macosko, C.W. Graphene/polymer nanocomposites. *Macromolecules* **2010**, *43*, 6515–6530. [CrossRef]
14. Li, C.; Wang, X.; Xu, L.; Gao, A.; Xu, D. Preparation of graphene oxide/polyacrylonitrile fiber from graphene oxide solution with high dispersivity. *Compos. Interfaces* **2020**, *27*, 177–190. [CrossRef]

15. Chien, A.T.; Cho, S.; Joshi, Y.; Kumar, S. Electrical conductivity and Joule heating of polyacrylonitrile/carbon nanotube composite fibers. *Polymer* **2014**, *55*, 6896–6905. [CrossRef]
16. Phiri, J.; Johansson, L.S.; Gane, P.; Maloney, T. A comparative study of mechanical, thermal and electrical properties of graphene-, graphene oxide- and reduced graphene oxide-doped microfibrillated cellulose nanocomposites. *Compos. Part B Eng.* **2018**, *147*, 104–113. [CrossRef]
17. Botlhoko, O.J.; Ramontja, J.; Ray, S.S. Morphological development and enhancement of thermal, mechanical, and electronic properties of thermally exfoliated graphene oxide-filler biodegradable polylactide/poly(ε-caprolactone) blend composites. *Polymer* **2018**, *139*, 188–200. [CrossRef]
18. Hu, R.R.; Zhang, R.J.; He, Y.J.; Zhao, G.K.; Zhu, H.W. Graphene oxide-in-polymer nanofiltration membranes with enhanced permeability by interfacial polymerization. *J. Membr. Sci.* **2018**, *564*, 813–819. [CrossRef]
19. Peng, C.; Hu, W.B.; Zhou, Y.T.; Fan, C.H.; Huang, Q. Intracellular imaging with a graphene-based fluorescent probe. *Small* **2010**, *15*, 1686–1692. [CrossRef] [PubMed]
20. Nasir, A.; Raza, A.; Tahir, M.; Yasin, T. Free-radical graft polymerization of acrylonitrile on gamma irradiated graphene oxide: Synthesis and characterization. *Mater. Chem. Phys.* **2020**, *246*, 122807. [CrossRef]
21. Jin, M.; He, W.J.; Wang, C.M.; Yu, F.P.; Yang, W.M. Covalent modification of graphene oxide and applications in polystyrene composites. *React. Funct. Polym.* **2020**, *146*, 104437. [CrossRef]
22. Voylov, D.; Saito, T.; Lokitz, B.; Uhrig, D.; Wang, Y.; Agpov, A.; Holt, A.; Bocharova, V.; Kisliuk, A.; Sokolov, A.P. Graphene oxide as radical initiator: Free radical and controlled radical polymerization of sodium 4-vinylbenzenesulfonate with graphene oxide. *ACS Macro Lett.* **2016**, *5*, 199–202. [CrossRef]
23. Sánchez, C.C.; Wåhlander, M.; Karlsson, M.; Quintero, D.C.M.; Hillborg, H.; Malmstrom, E.; Nilsson, F. Characterization of reduced and surface-modified graphene oxide in poly(ethylene-co-butyl acrylate) composites for electrical applications. *Polymers* **2020**, *11*, 740. [CrossRef] [PubMed]
24. Hummers, W.S.; Offeman, R.E. Preparation of graphite oxide. *J. Am. Chem. Soc.* **1958**, *80*, 1339–1345. [CrossRef]
25. Perera, S.D.; Mariano, R.G.; Vu, K.; Nour, N.; Seitz, O.; Chabal, Y.; Balkus, K.J., Jr. Hydrothermal synthesis of graphene-TiO$_2$ nanotube composites with enhanced photocatalytic activity. *ACS Catal.* **2012**, *2*, 949–956. [CrossRef]
26. Sharma, N.; Arif, M.; Monga, S.; Shkir, M.; Mishra, Y.K.; Singh, A. Investigation of bandgap alteration in graphene oxide with different reduction routes. *Appl. Surf. Sci.* **2020**, *513*, 145396. [CrossRef]
27. Li, N.; Ming, J.; Ling, M.; Wu, K.L.; Ye, Y.; Wei, X.W. solvothermal synthesis of bi nanoparticles/reduced graphene oxide composites and their catalytic applications for dye degradation and fast aromatic nitro compounds hydrogenation. *Chem. Lett.* **2020**, *49*, 318–322. [CrossRef]
28. Cha, C.; Shin, S.R.; Gao, X.; Annabi, N.; Dokmeci, M.R.; Tang, X.; Khademhosseini, A. Controlling mechanical properties of Cell-Laden hydrogels by covalent incorporation of graphene oxide. *Small* **2014**, *10*, 514–523. [CrossRef]
29. Paredes, J.I.; Villar, R.S.; Solis, F.P.; Martinez, A.A.; Tascon, J. Atomic force and scanning tunneling micsroscopy imaging of graphene nanosheets derived from graphite oxide. *Langmuir* **2009**, *25*, 5957–5968. [CrossRef]
30. Lerf, A.; Buchsteiner, A.; Pieper, J.; Schöttl, S.; Dekany, I.; Szabo, T.; Boehm, H.P. Hydration behavior and dynamics of water molecules in graphite oxide. *J. Phys. Chem. Solids* **2006**, *67*, 1106–1110. [CrossRef]
31. Yang, Z.; Sun, Y.J.; Ma, F.; Lu, Y.; Zhao, T. Pyrolysis mechanism of graphene oxide reveled by ReaxFF molecular dynamic simulation. *Appl. Surf. Sci.* **2020**, *509*, 145247. [CrossRef]
32. Li, W.; Tang, X.Z.; Zhang, H.B.; Jiang, Z.G.; Yu, Z.Z.; Du, X.S.; Mai, Y.W. Simultaneous surface functionalization and reduction of graphene oxide with octadcylamine for electrically conductive polystyrene composites. *Carbon* **2011**, *49*, 4724–4730. [CrossRef]
33. Galindo, B.; Alcolea, S.G.; Gómez, J.; Navas, A.; Murguialday, A.O.; Fernandez, M.P.; Puelles, R.C. Effect of the number of layers of graphene on the electrical properties of TPU polymers. *ICP Conf. Ser. Mater. Sci. Eng.* **2014**, *64*, 012008. [CrossRef]
34. Gao, J.; Liu, C.; Miao, L.; Wang, X.; Chen, Y. Free-standing reduced graphene oxide paper with high electrical conductivity. *J. Electron. Mater.* **2016**, *45*, 1290–1295. [CrossRef]
35. Lin, Y.; Jin, J.; Song, M. Preparation and characterization of covalent polymer functionalized graphene oxide. *J. Mater. Chem.* **2011**, *21*, 3455–3461. [CrossRef]
36. Yang, N.; Xu, C.; Hou, J.; Yao, Y.; Zhang, Q.; Grami, M.E.; Qu, X. Preparation and properties of thermally conductive polyimide/boron nitride composites. *RSC Adv.* **2016**, *6*, 18279–18287. [CrossRef]

37. Ramezani, H.; Behzad, T.; Bagheri, R. Synergistic effect of graphene oxide nanoplatelets and cellulose nanofibers on mechanical, thermal, and barrier properties of thdrmoplastic starch. *Polym. Adv. Technol.* **2020**, *31*, 553–565. [CrossRef]
38. Si, J.J.; Li, J.; Wang, S.J.; Li, Y.; Jing, X.L. Enhanced thermal resistance of phenolic resin composites al low loading of graphene oxide. *Compos. Part A Appl. Sci. Manuf.* **2013**, *54*, 166–172. [CrossRef]
39. Divakaran, N.; Zhang, X.; Kale, M.B.; Senthil, T.; Mubarak, S.; Dhamodharan, D.; Wu, L.; Wang, J. Fabrication of surface modified graphene oxide/unsaturated polyester nanocomposites via in-situ polymerization: Comprehensive property enhancement. *Appl. Surf. Sci.* **2020**, *502*, 144164. [CrossRef]
40. Kuilla, T.; Bhadra, S.; Yao, D.; Kim, N.H.; Bose, S.; Lee, J.H. Recent advances in graphene based polymer nanocomposites. *Prog. Polym. Sci.* **2010**, *35*, 1350–1375. [CrossRef]
41. Galpaya, D.; Wang, M.; George, G.; Motta, N.; Waclawik, E.; Yan, C. Preparation of graphene oxide/epoxy nanocomposites with significantly improved mechanical properties. *J. Appl. Phys.* **2014**, *116*, 053518. [CrossRef]
42. Patrole, A.S.; Patole, S.P.; Kang, H.; Yoo, J.B.; Kim, T.H.; Ahn, J.H. A facile approach to the fabrication of graphene/polystyrene nanocomposite by in situ microemulsion polymerization. *J. Colloid Interface Sci.* **2010**, *350*, 530–537. [CrossRef] [PubMed]
43. Tripathi, S.N.; Saini, P.; Gupta, D.; Choudhary, V. Electrical and mechanical properties of PMMA/reduced graphene oxide nanocomposites prepared via in situ polymerization. *J. Mater. Sci.* **2013**, *48*, 6223–6232. [CrossRef]
44. Park, W.; Hu, J.; Jauregui, L.A.; Ruan, X.; Chen, Y.P. Electrical and thermal conductivities of reduced graphene oxide/polystyrene composites. *Appl. Phys. Lett.* **2014**, *104*, 113101. [CrossRef]

© 2020 by the authors. Licensee MDPI, Basel, Switzerland. This article is an open access article distributed under the terms and conditions of the Creative Commons Attribution (CC BY) license (http://creativecommons.org/licenses/by/4.0/).

Article

Performance Enhancement of Vanadium Redox Flow Battery by Treated Carbon Felt Electrodes of Polyacrylonitrile Using Atmospheric Pressure Plasma

Chien-Hong Lin [1,*], Yu-De Zhuang [1], Ding-Guey Tsai [1], Hwa-Jou Wei [1] and Ting-Yu Liu [2,*]

1. Institute of Nuclear Energy Research, Longtan 30546, Taiwan; zhuang@iner.gov.tw (Y.-D.Z.); dgtsai@iner.gov.tw (D.-G.T.); hwajou@iner.gov.tw (H.-J.W.)
2. Department of Materials Engineering, Ming Chi University of Technology, New Taipei City 24301, Taiwan
* Correspondence: chlin0805@iner.gov.tw (C.-H.L.); tyliu0322@gmail.com (T.-Y.L.)

Received: 31 May 2020; Accepted: 16 June 2020; Published: 18 June 2020

Abstract: A high-performance carbon felt electrode for all-vanadium redox flow battery (VRFB) systems is prepared via low-temperature atmospheric pressure plasma treatment in air to improve the hydrophilicity and surface area of bare carbon felt of polyacrylonitrile and increase the contact potential between vanadium ions, so as to reduce the overpotential generated by the electrochemical reaction gap. Brunauer-Emmett-Teller (BET) surface area of the modified carbon felt is, significantly, five times higher than that of the pristine felt. The modified carbon felt exhibits higher energy efficiency (EE) and voltage efficiency (VE) in a single cell VRFB test at the constant current density of 160 mA cm^{-2}, and also maintains good performance at low temperatures. Moreover, the cyclic voltammetry (CV) and electrochemical impedance spectroscopy (EIS) analysis results show that the resistance between electrolyte and carbon felt electrode decreased. As a result, owing to the increased reactivity of the vanadium ion on the treated carbon felt, the efficiency of the VRFB with the plasma-modified carbon felt is much higher and demonstrates better capacity under a 100-cycle constant current charge-discharge test.

Keywords: vanadium redox flow battery; carbon felt; atmospheric plasma; polyacrylonitrile

1. Introduction

The vanadium redox flow battery (VRFB) is a proven technology that has a number of key and promising advantages, which give it much promise as the future of energy storage systems with a good charge–discharge property, its long lifecycle, as well as being nonflammable and easily scalable with grid-scale potential [1–3]. The VRFB system consists of an energy management system (EMS) to control the power in and out, a battery management system (BMS) of two electrolyte tanks with V^{2+}/V^{3+} and VO^{2+}/VO_2^+ redox species in sulfuric or other acidic solutions with both negative and positive electrodes, and at least two pumps, as well as a battery stack where the key battery reaction takes place. The electrolyte is pumped into the stack and separated by the ion exchange membrane and fills the reaction area [4–6]. The electrode in the battery is used to conduct the electrons, provide the charge transfer platform, and make good contact with the electrolyte. For the VRFB system, the ideal electrode should have some essential properties, such as having good chemical stability in strong acid and redox reactions, good hydrophilicity, and lower electrochemical resistance [7–12], in order to obtain a reliable product that has higher voltage efficiency, charge capacity, and a longer lifecycle.

The electrode of VRFB often uses carbon materials such as carbon or graphite felt, paper, and cloth, which have excellent electronic conductivity and strong acid resistance because of their material composition [13,14]. The physical flexibility of the carbon material electrode can be compressed in the narrow electrode flow space and the good electronic properties mentioned above contribute to the low IR-drop (the voltage drop due to energy losses in a resistor) of the battery and the successful running of the battery during long operation cycles. Despite the graphite-based carbon electrodes having a number of benefits, there are still some drawbacks, including a highly hydrophobic surface and poor reaction surface area. The hydrophobic surface leads to poor contact with the vanadium ions in the aqua phase electrolyte, and the low reaction surface area limits the electron transfer efficiency between the electrode and the reaction species in the electrolyte [9], which can lead to an obvious decrease in voltage efficiency (VE), energy efficiency (EE), and capacity of the VRFB in operational conditions. It is possible to improve the performance and efficiency of the VRFB by increasing the surface water affinity or the surface area. Surface modifications to make the carbon material surface hydrophilic can be achieved by wet (acid, alkali), dry (plasma), and radiation treatments (laser, radiations), without affecting the supporting structural properties. Various carbon electrode modification procedures have been documented in past literature, including oxidative methods to increase the surface oxygen functional groups, such as acidic treatment [8], heat treatment [7,11,12], and electrochemical active treatment [15], or surface decoration methods to improve the reaction surface area or spots, such as pasting Bi nanoparticles [16] or carbon nanotube immobilization [17]. There are also other special methods, such as carving out laser pinholes [18], water–gas reaction methods, and plasma treatment methods [19,20].

Atmosphere plasma treatment is an effective method for surface modification because it is solvent-free, dry, controllable, and easy to operate, with low or no waste [20]. For carbon materials, the main purpose of plasma treatment is the physical bombardment effect of the accelerated molecules, which effectively brings up the etching fragment and forms the carbon radicals on the carbon material surface [21–23], while still keeping the graphite backbone stable, as shown in Figure 1A. The etched carbon fiber surface significantly increases the surface area to improve the contact rate of the reaction species in the electrolyte of the batteries. The radicals formed by the plasma treatment on the carbon surface will change into oxygen-containing functional groups to increase the affinity of the aqua solution [20,24]. All of these benefits serve as a solution for improving the performance of the VRFB system, but an overetched electrode can lead to decreased conductivity and decreased performance. Thus, the modified conditions or methods are still being further investigated and developed.

In this work, the raw carbon felt was first treated by a nitrogen plasma jet under air and then the treated felt was exposed to air for a few minutes post-treatment. The radicals formed by the nitrogen plasma treatment on the surface of carbon felt will transfer into oxygen-containing functional groups after exposed in the air. This process is focused on solving the low electrochemical reactivity and the poor aqua affinity of the carbon felt electrode. Brunauer-Emmett-Teller (BET) examination showed that the treated felt had approximately 20 times higher BET surface area than the pristine felt, had become more hydrophilic, and had better reactivity within the vanadium electrolyte, shown using a water dropping test and electrochemical analysis methods such as cyclic voltammetry (CV), electro impedance spectrum (EIS), and single-cell VRFB test. Therefore, the atmosphere plasma jet treatment for preparing the modified carbon electrode is a very simple, well-established, and inexpensive technique, which can directly improve the performance of the VRFB cell without the need for other hardware changes.

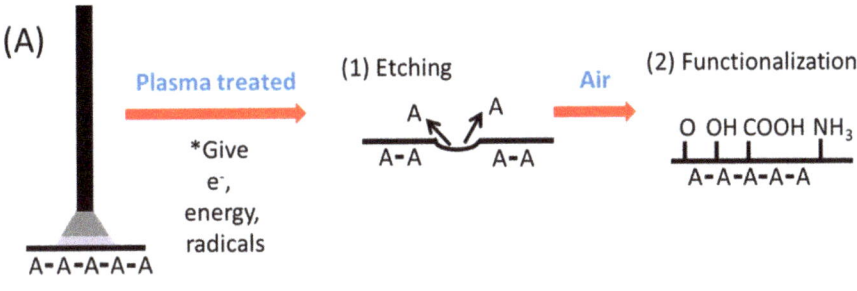

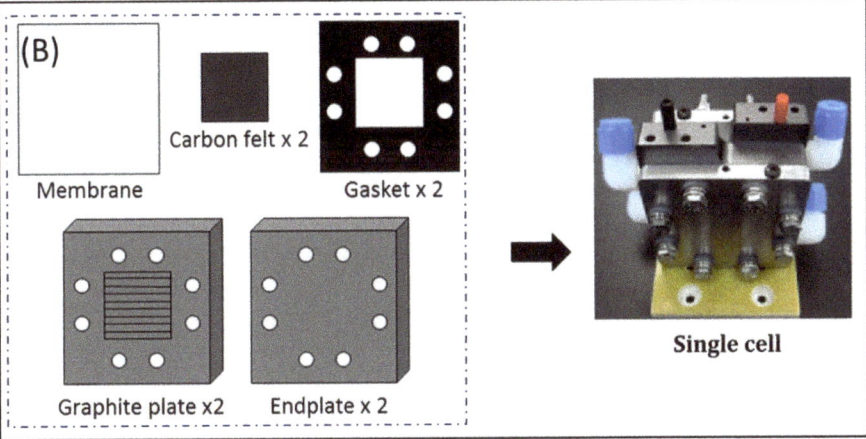

Figure 1. (**A**) Schematic diagram of how the plasma jet affects the surface of carbon felt. (**B**) Schematic illustration of single cell construction.

2. Materials and Methods

2.1. Materials

Bare carbon felt material was purchased from CeTech co. Ltd., Taichung, Taiwan. Vanadium electrolyte for a single-cell test was purchased from Hong Jing environment, Pingtung, Taiwan. Vanadyl sulfate (VOSO4) for CV tests was purchased from Echo Chemical co. Ltd., Miaoli, Taiwan. The 99.999% pure nitrogen gas for the plasma treatment process was purified by the pressure swing adsorption (PSA) system (United Air System Co. Ltd., New Taipei City, Taiwan). All other chemicals, except for gases, were used as-received without further purification.

2.2. Preparation of Plasma-Treated Carbon Felt

Figure 1A depicts how the plasma jet affects the surface of the materials. The plasma-treated carbon felt electrode was prepared using the atmosphere plasma jet system with a rotating nozzle. Prior to modification, the plasma was generated using an atmosphere plasma generator (Plasmatreat GmbH, Steinhagen, Germany) at room temperature and the atmospheric environment. The purified nitrogen gas with an output at a pressure of 7 bars and a volume concentration of 99.999% as a further plasma gas source was produced from the PSA machine (United Air System Co. Ltd., New Taipei City, Taiwan) to prevent an unpredictable oxygen side effect. Bare carbon felt was placed under the fixed plasma jet nozzle at a distance of 10 mm and moved by a moving plate system at a constant speed of at least 2 mm sec^{-1}.

2.3. Hydrophilicity Characterization

The surface hydrophilicity of the felt electrode was first tested by water drop. The data of contact angles were observed using a FTA-1000B contact angle goniometer (Ten Angstroms, Folio Instruments, Kitchener, ON, USA) at 25 °C.

2.4. BET Surface Area Analysis'

For the BET surface area test, a total of 10 g felt was cut into pieces to form the sample. ASAP2020 micromeritics® (Micromeritics Instrument Corp., Norcross, GA, USA) was used as the measuring tool. The process of isothermal absorption line condition started from the degas process, followed by a measuring process set from relative pressure 0.1 to 1 under 77 K. The desorption process operated under reverse, at room temperature. The results were transferred to surface area data using the Brunauer-Emmett-Teller (BET) calculation model.

2.5. CV (Syclic Voltammetry) Analysis

For the cyclic voltammetry test, the felt was cut to 1 mm² size and 6.5 mm thick as the sample, and Autolab® (swiss) was used as the current provider and data collector. The three-electrode system was built for a positive test and negative test.

In the positive electrode test, 0.2M $VOSO_4$ (purchased from Alfa Aesar, Echo Chemical Co. Ltd., Miaoli, Taiwan) solution was used as electrolyte, saturated calomel electrode (SCE) as the reference electrode (RE), and 1 mm thick platinum wire as the working electrode (WE) and counter electrode (CE). The starting voltage was autodetected by the tool, the voltage range was set from −0.5 V to 1.5 V, and a scan rate of 5 mV s^{-1} was chosen for the data collection.

In the negative electrode test, the electrolyte had a concentration of 0.17 M vanadium and a valance of 2.8–3, and a 2 M sulfuric acid diluted solution was made from the electrolysis reaction of the VRFB single-cell test. For the electrodes, the SCE was used as the RE, and 2 mm of thick glassy carbon electrode was used as the WE and CE to prevent the fast evolution of hydrogen within the analysis process. The starting voltage was also autodetected by the tool, the voltage range was set from −1.5 V to 0.5 V, and a scan rate of 5 mV s^{-1} was selected.

2.6. EIS Analysis

For electrical impedance spectrometer (EIS) test, the felt and the electrode features were the same as the previous CV tests, but used a different process. In the test, 0.2 M $VOSO_4$ (purchased from Alfa) solution was used as the positive test electrolyte, with 0.17 M and 2.8–3 valance, and the vanadium and sulfuric acid mix solution made from the electrolysis reaction of the VRFB single cell test was used as the negative electrolyte. The SCE was used as the RE and a 2 mm thick glassy carbon rod used for the WE and CE. The starting voltage was automatically detected by the tool. EIS was performed, wherein an alternating current (AC) voltage of 10 mV in the frequency range of 10^5–10^{-2} Hz was applied at the open circuit potential.

2.7. The Construction of the Single Cell of VRFB

Figure 1B depicts the construction of the single cell of VRFB. The single cell comprises two plastic plates with a 3 mm depth of flow field, two copper plates with 3 mm thickness, two embedded graphite plates of 6 mm thickness, and two stainless steel plates, which served as the endplates. There are also two gaskets that are 1 mm in thickness, two 25 cm² treated or pristine carbon felt electrodes with a thickness of 6.5 mm, and an N212 Nafion® membrane (purchased form Chemours, Taipei, Taiwan) for electrolyte separation.

2.8. VRFB Single-Cell Test

The VRFB single cell, as described above, was used for this test. In the charge-discharge tests, the solutions of 1.7 M V^{3+}/VO^{2+} (with valance 3.5) and 5 M H_2SO_4 were used as the starting electrolyte in both the negative and positive electrodes. The carbon felt served as the electrode, and the graphite plates and copper plates served as the current collector. The active area of the electrode in the cell was 25 cm^2. The volume of electrolyte in each half cell was 80 mL. The VRFB single cell was charged and discharged within the current density range of 80–200 mA cm^{-2} depending on the need. To protect the carbon felt and graphite plates from breaking under the high power, the VRFB cell was charged and discharged within the voltage limit of 1.6–0.7 V. The lifecycle test was conducted under a current density set to 120 mA cm^{-2} and the other described conditions, for at least 50 cycles.

3. Results and Discussion

3.1. The Plasma-Treated Process and Condition Decision

Carbon felt is an inert electrode that is difficult to modify. To break down the smooth carbon fiber surface or to introduce a functional group on it requires relatively high-energy reactions, such as the widely used plasma treatment methods, water-gas reactions, or electrochemical reactions between the carbon and chemicals. In this work, we used atmosphere pressure plasma as the treatment method because of its advantages of low temperature working conditions, being a fast treatment process, post-treatment free, and inexpensive. After treatment by the moving plasma jet at a velocity of 5 mm/s, keeping a 10 mm distance between the surface of the felt and the nozzle of the plasma jet, the surface hydrophilicity of the treated felt was determined by the water dropping method. Figure 2 indicates 118° ± 2° (Figure 2B) and ~0° (Figure 2C) of the contact angle on the pristine and treated felt surface, respectively, which may be attributed to the functional groups and defects formed by the free radical species reaction between the plasma species and carbon surface in the plasma jet. The result demonstrates how hydrophilic the treated felt had become. Moreover, it would be a great help to improve the pump loss of the VRFB stacks. In addition, the weight loss of the treated felt is less than 1%, which shows that the treated felt had broken down in some structures. The ash stacked in the plasma treatment process inside the chamber also proved that some destruction of the felt occurred. The thicknesses of the felt remained unchanged after plasma treatment, and it is thus directly ready to use.

Furthermore, while the treatment may increase the hydrophilicity, it decreases the electronic conductivity of the felt. In other words, there is a tradeoff between electronic conductivity and electrochemical reactivity that should be carefully managed. In Figure 3, the single-cell was measured by carrying out 100 cycles of charge-discharge at a current density of 120 and 140 mA cm^{-2}. The results of the average of EE suggest the modification process with the plasma jet at the relative velocity of 5 mm/s ($EE_{Avg.}$ = 84.2 ± 0.08%@120 mA cm^{-2} and $EE_{Avg.}$ = 82.8 ± 0.08%@140 mA cm^{-2}) to be the best. Double speed plasma treatment ($EE_{Avg.}$ = 80.0 ± 0.05%@120 mA cm^{-2}) and running the plasma treatment two times ($EE_{Avg.}$ = 81.7 ± 0.07%@140 mA cm^{-2}) or three times ($EE_{Avg.}$ = 81.9 ± 0.08%@140 mA cm^{-2}) did not deliver a better result.

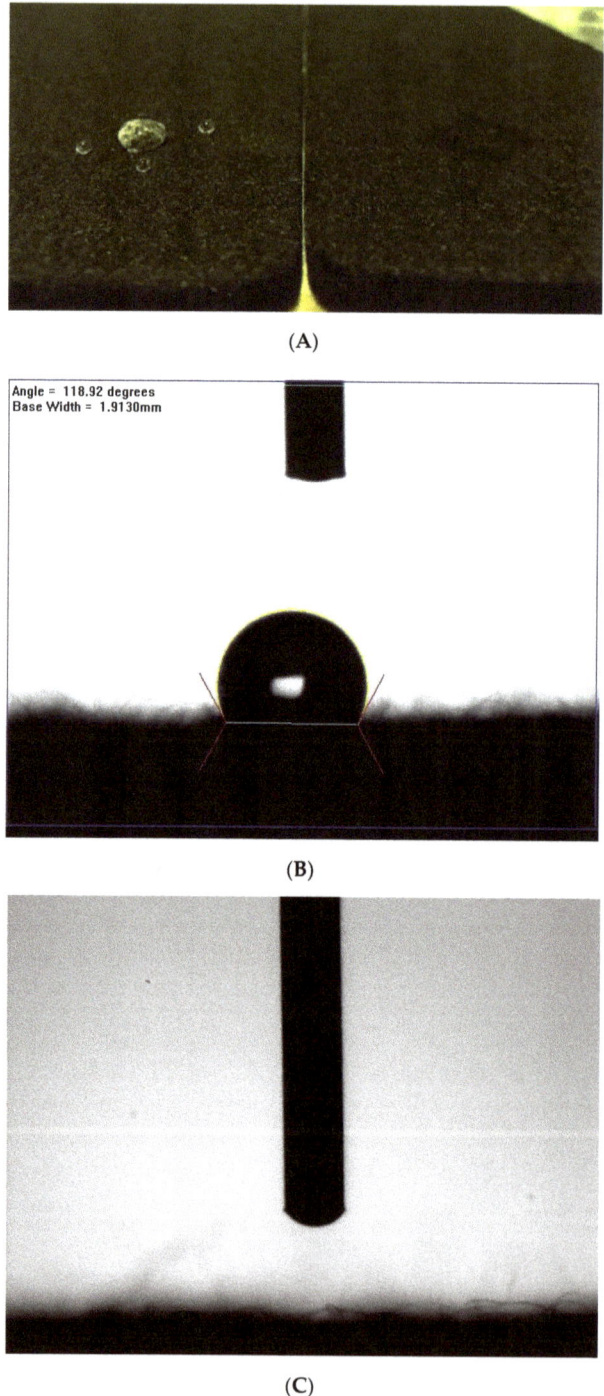

Figure 2. (**A**) The water dropping image of (left) the pristine felt and (right) the atmospheric plasma treated felt, the contact angle of (**B**) the pristine felt, and (**C**) the atmospheric plasma treated felt.

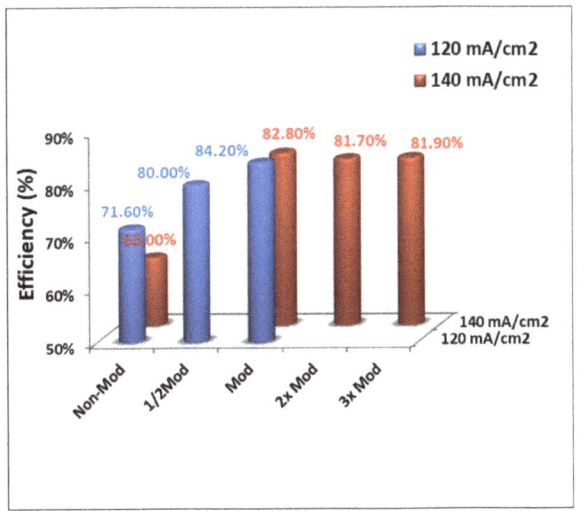

Figure 3. Efficiency (%) performance of the single cell equipped with the different parameters of atmospheric plasma treated carbon felt.

3.2. The Surface Morphology Analysis

To check the morphology changes in the pristine and treated felt surface, scanning electron microscope (SEM) and transmission electron microscopy (TEM) tools were used as the observation methods. Figure 4A–D depict the SEM images and Figure 4E,F depict the TEM images of the pristine felt and plasma-treated felt. The image (10,000 times zoom) of the plasma-treated felt (Figure 4D) shows that the defects on the carbon fiber surface were increased after the plasma treatment process. By contrast, the image of the pristine felt shows a smoother surface on the carbon fiber. Therefore, the roughness of the fiber surface increased after the plasma treatment, owing to the bombardment of accelerated heavy plasma species from the plasma jet.

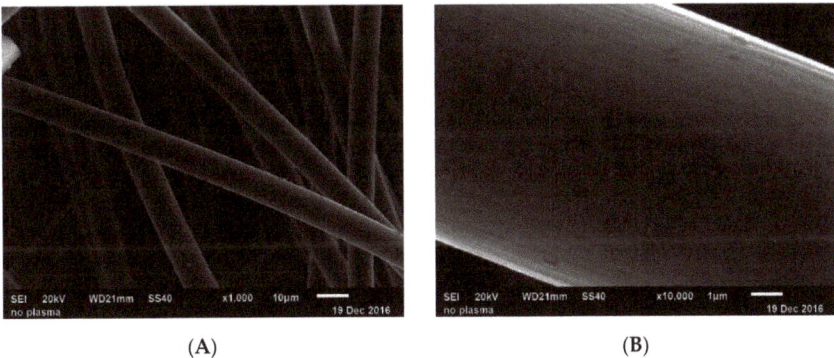

(A) (B)

Figure 4. *Cont.*

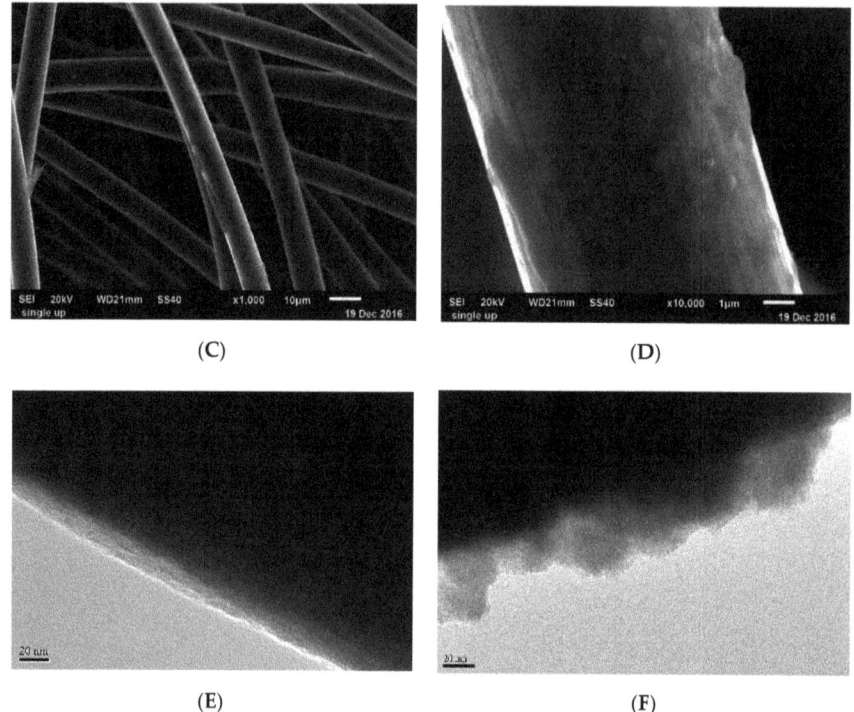

Figure 4. The SEM images of the pristine felt in (**A**) 1000 times zoom and (**B**) 10,000 times zoom; the atmospheric plasma treated felt in (**C**) 1000 times zoom and (**D**) 10,000 times zoom. The TEM images of (**E**) the pristine felt and (**F**) the atmospheric plasma treated felt.

3.3. BET Surface Area Analysis

The electrode reactive surface is an important issue as it affects the resistance of the electrochemical reaction, especially in a nonselective reaction system. In order to improve the electrochemical reaction efficiency between carbon felt electrode and vanadium ions in the electrolyte, we chose to increase the surface area of the felt. Figure 5 gives the comparisons of the pristine and the treated felt. The results of the tests, which were carried out under the same conditions, show that the BET surface area of the plasma-treated felt was approximately five times that of the pristine one. The measured surface area of the treated felt was 0.74 ± 0.06 m^2 g^{-1} and the pristine one was only 0.13 ± 0.01 m^2 g^{-1}, although the surface area was very low for the BET model. The increasing surface area may also be attributed to the bombardment of the heavy plasma species in the plasma jet.

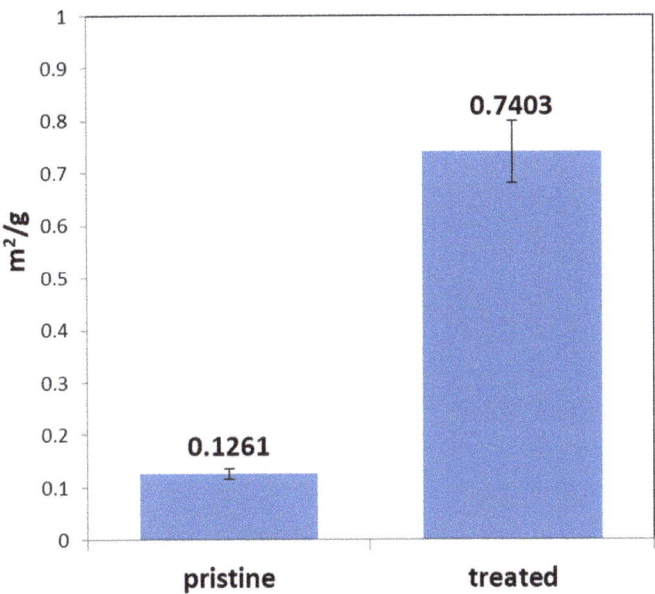

Figure 5. The diagram of the Brunauer–Emmett–Teller (BET) surface area results of the pristine and the atmospheric plasma treated felt.

3.4. CV and EIS Analysis

To observe the electrochemical property of the treated felt, both cyclic voltammetry (CV) and electro impedance spectrum (EIS) are good testing methods. The CV plot are the anodic peak current I_{pa}, cathodic peak current I_{pc}, anodic peak potential E_{pa}, and cathodic peak potential E_{pc}. Previous studies show that improved performance of the VRFB electrode is often indicated with an I_{pc}/I_{pa} ratio close to 1 and a decreased ΔE value in CV examination, meaning that the reversibility of the redox reaction is improved [19]. Moreover, the Nyquist Plot by EIS analysis would have a smaller curve radius because of the decreased impedance of the felt or electrode after modification. In this case, Figure 6 shows the (A) positive and (B) negative electrode CV curves of the treated and pristine felts, which indicate a similar result to previous studies [15–20]. The positive electrode test result shows the decrease of the I_{pc}/I_{pa} ratio from 1.93 to 1.34 and ΔE value from 0.532 V to 0.508 V, and the negative electrode shows the same trends, with the I_{pc}/I_{pa} ratio increased from 0.424 to 0.669 and the ΔE value decreased from 1.582 V to 1.311 V. All of the results provide the evidence that the reversibility of the redox reaction to the felt electrode was improved after the plasma treatment process.

The Nyquist plots contain one semicircle in the high frequency range arising from charge transfer reactions at the electrolyte-electrode interface. The radius of the semicircle reflects the charge transfer resistance, with a smaller radius indicating a lower charge transfer resistance, which in turn indicates a faster electron transfer reaction [19]. EIS results (Figure 7) show that a smaller curve radius was found in the treated felt from the Nyquist plot compared with that in the pristine felt. It provides the evidence that the resistance of the felt used in the electrolyte system was decreased.

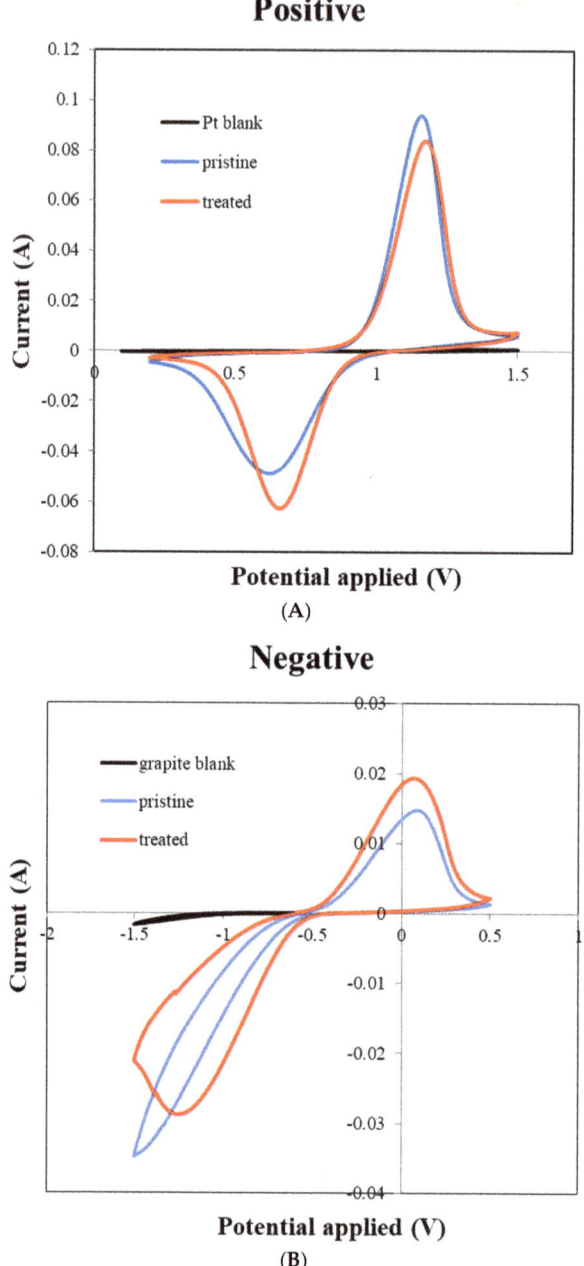

Figure 6. Cyclic voltammetry (CV) results of (**A**) positive electrodes (**B**) negative electrodes.

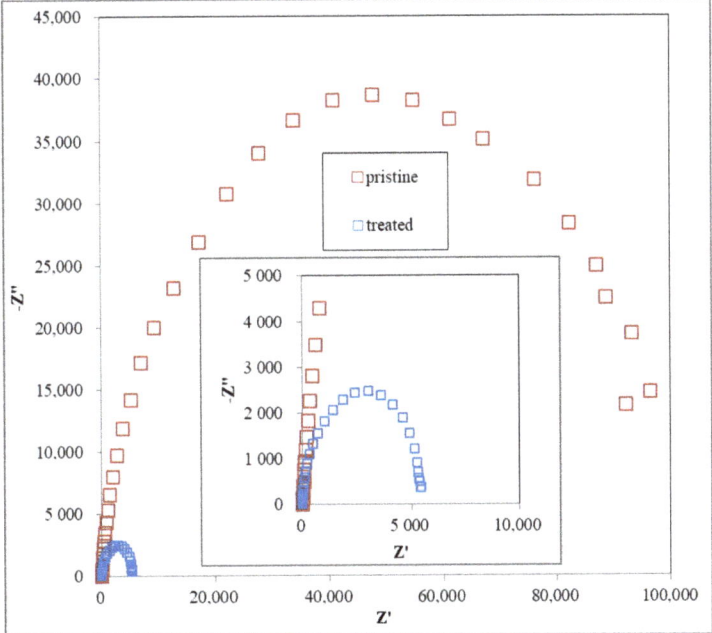

Figure 7. The Nyquist plots of the compared electrochemical impedance spectroscopy (EIS) results.

3.5. Charge-Discharge Curves

The charge-discharge curves of the second charge-discharge test cycle using the VRFB single cell often become the indication for cell performance comparison. Choosing the data of the second cycle of the test is owing to the unsteady electrolyte state in the first cycle, with a starting 3.5 valence vanadium electrolyte on both sides of the electrode. The test cell composed of Nafion 212 membrane was combined with the plasma-treated or the pristine carbon felt electrode to obtain comparable results.

Figure 8 shows the charge-discharge curves of the second cycle of VRFB single cell with plasma-treated or the pristine carbon felt at 160 mA cm^{-2}. It is obvious that the charge voltage of VRFB with plasma-treated felt is lower than that of the VRFB with pristine felt, while the discharge voltage of VRFB with plasma-treated felt is higher than that of the VRFB with pristine felt. While the discharge voltage trend is reversed, both results are attributed to the smaller IR drop of the treated felt. This result is likely caused by the lower area of resistance of the treated felt. This is because plasma treatment produces numerous oxygen-containing functional groups (such as –OH groups) on the surface of the carbon felts fibers, which are known to be electrochemically active sites for vanadium redox reaction. Furthermore, an increase of hydroxyl and carboxyl groups on the carbon felts fiber surface enhances its hydrophilicity, which makes it favorable for electrochemical reaction.

In addition, the data of the treated felt in a higher current density test provided a decreased capacity and EE owing to the stronger polarization effect, but it was still better than the pristine felt. Figure 8 shows the increased CE and VE results of the VRFB with the treated felt, which were 97.0% and 79.9% at the current density of 160 mA cm^{-2}, respectively.

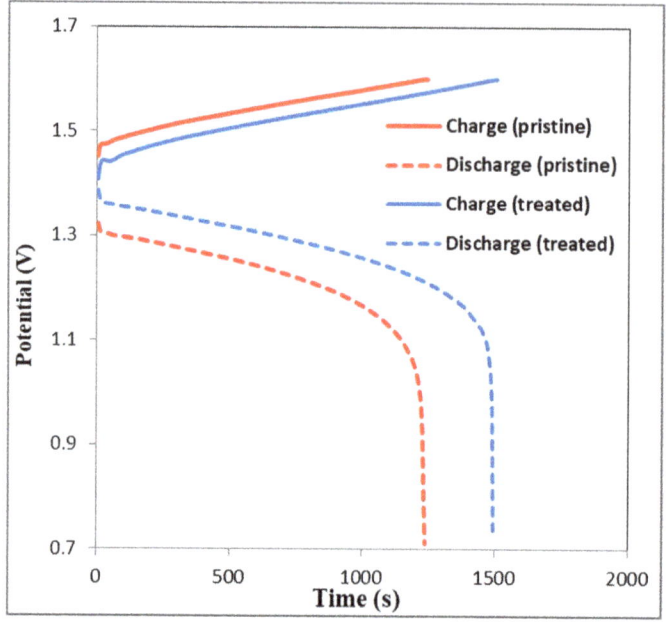

Figure 8. The comparison of second cycle charge-discharge curves of the carbon felt with and without atmospheric plasma treatments.

3.6. VRFB Single-Cell Performance

A charge-discharge test was performed using a VRFB single cell to further demonstrate the effect of carbon felt on the electrochemical performance of the cell before and after plasma treatment. The in situ stability and performance test of the plasma-treated felt was carried out by a 100 cycle charge-discharge test using the VRFB single cell at the current density of 160 mA cm^{-2}. The results shown in Figure 9A give the key performance values for the battery, which are the EE, CE, and VE. The curves of the above performance results remained smooth and stable for 100 cycles. The lack of decline in the performance indicated the high stability of the treated felt and also proved that the treated felt can remain stable in the strongly acidic and relatively high-oxidative vanadium electrolyte.

The performance of the EE value is the product of the VE and CE values. The increased VE value indicates the lower IR drop and thus the overpotential of the cell, and the increased CE value indicates the lower self-discharge that occurred in the test. The VE of the VRFB with the treated felt was higher than that of the VRFB with the pristine felt, at all current densities, which could be attributed to the reduced electrochemical resistance. The improved resistance of the felt electrode depends on two factors from the previous work the increased active surface area and the reduced area resistance [10]. Both of these aspects in the treated felt were improved, as shown by the surface area test, CV, and EIS analysis, thus demonstrating that the treated felt exhibited higher VE. The VRFB cell equipped with the treated felt has a greater VE than the pristine felt at all tested current densities, especially higher current densities, owing to the plasma treatment producing large amounts of oxygen-containing functional groups on the felt surface and promoting faster charge transfer, leading to improved electrode performance. In addition, the VE of the VRFB decreased with increasing charge-discharge current densities owing to the increase of ohmic resistance and the overpotential caused by the increase of current densities. The VE and EE are considerably higher for the VRFBs containing the plasma treated electrodes than the containing the pristine electrodes. Notably, these high efficiencies are maintained even at higher current densities.

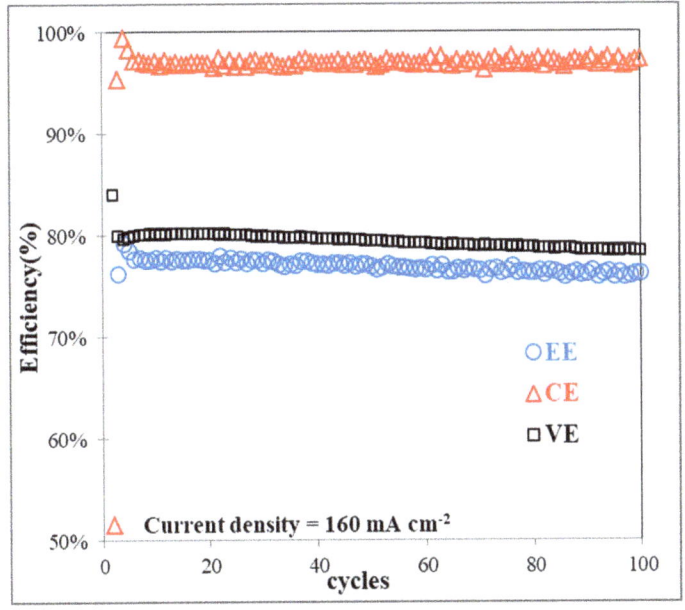

Figure 9. Cont.

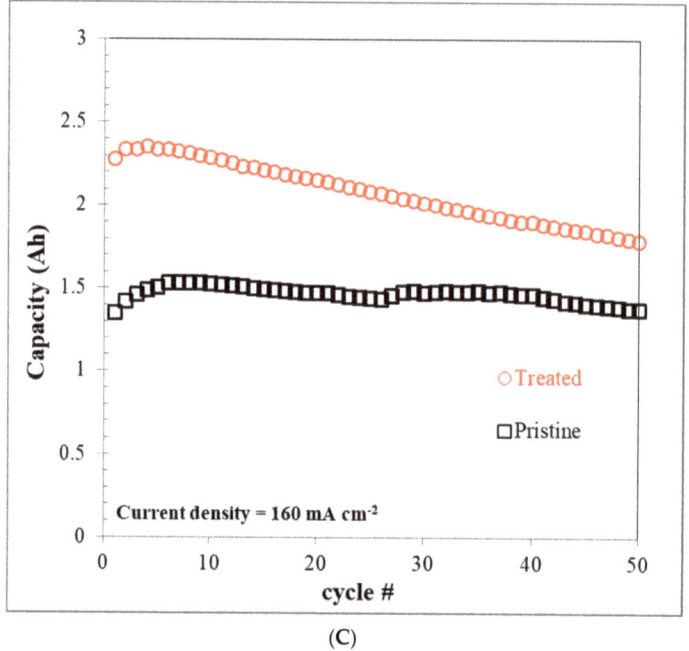

(C)

Figure 9. Diagrams of the performance of vanadium redox flow battery (VRFB). (**A**) One-hundred cycles of efficiency (%) performance of VRFB with the atmospheric plasma treated carbon felt. Fifty cycles of (**B**) efficiency (%) and (**C**) capacity (Ah) performance with and without atmospheric plasma treated carbon felts. EE, energy efficiency; VE, voltage efficiency; CE, coulombic efficiency.

In Figure 9B, the cell with the treated felt exhibited a higher performance than the pristine one in the same test conditions. It had the greater EE, which increased from 67.9% to 77.6%, and the capacity increased from 1.47 Ah to 2.08 Ah under the same constant current density and the other test conditions, which is more than 10% improvement. The higher average capacity of the 50 cycles test can be attributed to the improved hydrophilicity leading to the higher utilization rate of electrolyte and leading to the higher capacity of the VRFB under the same charge-discharge conditions.

The capacity curve (Figure 9C) of the treated felt showed a larger decreasing trend in the results, owing to the increasing migration of vanadium ions, hydrogen ions, and water in the electrolytes [9]. The imbalance of electrolytes increased faster than in the pristine felt by cycle number, because of the increased number of cycles completed on the treated felt. Therefore, the comprehensive performance increase in the VRFB single-cell test with the treated felt can be seen as important for future use in the scale-up stacks, as it will reduce costs because of requiring less electrolyte maintenance and having a higher electrolyte usage rate. Simple surface treatment of carbon felts using plasma treatment is thus promising for the assemblage of high-performance VRFBs, and we consider that this method is suitable for large-scale production of economical carbon felts electrodes.

4. Conclusions

In this study, the carbon felt electrode used for the VRFB cell was treated by an atmosphere plasma jet via a specific process and exhibited higher comprehensive cell performance than the pristine felt, thanks to its five times larger surface area and lower electrochemical resistance. The plasma treatment can also improve the hydrophilicity owing to the additional temperate water affinity functional group on the felt surface, which can reduce the contact angle to 0° and reduce the pumping loss when the VRFB system is operating. The single-cell test results with the treated felt from the charge-discharge cycling test shows that, even though the CE only had a small decrease, owing to the more than 20% improved capacity in the same test condition, the VE and the EE still increased significantly-up to 10% higher than the pristine felt under 160 mA cm^{-2} test conditions. The chemical stability of the treated felt tested by the 100 in situ charge-discharge cycle tests show the treated felt has high chemical stability in the vanadium electrolyte working environment. The results indicated that the hydrophilicity and electrochemical reaction of plasma-treated carbon felt electrodes can be greatly increased, which can improve the energy efficiency and capacity of carbon felt electrodes for VRFB. The facile and rapid surface treatment of carbon felt electrodes using atmospheric plasma would have potential to be applied in constructing the high-performance VRFB. Furthermore, we believe that the novel method is suitable for large-scale production of carbon felt electrodes, because the atmospheric plasma treatment industry is already well established.

Author Contributions: Conceptualization, C.-H.L. and T.-Y.L.; Data curation, C.-H.L., Y.-D.Z., D.-G.T. and H.-J.W.; Funding acquisition, C.-H.L. and Y.-D.Z.; Investigation, C.-H.L., Y.-D.Z., D.-G.T., H.-J.W. and T.-Y.L.; Methodology, C.-H.L. and T.-Y.L.; Validation, C.-H.L. and Y.-D.Z.; Formal analysis, Y.-D.Z., D.-G.T. and H.-J.W.; Visualization, C.-H.L. and T.-Y.L.; Project administration, C.-H.L. and Y.-D.Z.; Resources, C.-H.L., Y.-D.Z. and T.-Y.L.; Writing—original draft, C.-H.L., Y.-D.Z. and T.-Y.L. Writing-revised manuscript, C.-H.L., Y.-D.Z. and T.-Y.L. All authors have read and agreed to the published version of the manuscript.

Funding: This study was financially supported by Bureau of Energy, Ministry of Economic Affairs (109 D0114) and Ministry of Science and Technology of Taiwan (MOST 108-2622-E-131-002-CC3).

Conflicts of Interest: The authors declare no conflict of interest.

References

1. Skyllas-Kazacos, M.; Chakrabarti, M.H.; Hajimolana, S.A.; Mjalli, F.S.; Saleem, M. Progress in flow battery research and development. *J. Electrochem. Soc.* **2011**, *158*, R55. [CrossRef]
2. Poullikkas, A. A comparative overview of large-scale battery systems for electricity storage. *Renew. Sustain. Energy Rev.* **2013**, *27*, 778–788. [CrossRef]
3. Yang, Z.; Zhang, J.; Kintner-Meyer, M.C.; Lu, X.; Choi, D.; Lemmon, J.P.; Liu, J. Electrochemical energy storage for green grid. *Chem. Rev.* **2011**, *111*, 3577–3613. [CrossRef] [PubMed]
4. Cho, H.; Atanasov, V.; Krieg, H.M.; Kerres, J.A. Novel Anion Exchange Membrane Based on Poly (Pentafluorostyrene) Substituted with Mercaptotetrazole Pendant Groups and Its Blend with Polybenzimidazole for Vanadium Redox Flow Battery Applications. *Polymers* **2020**, *12*, 915. [CrossRef] [PubMed]
5. Cui, Y.; Chen, X.; Wang, Y.; Peng, J.; Zhao, L.; Du, J.; Zhai, M. Amphoteric ion exchange membranes prepared by preirradiation-induced emulsion graft copolymerization for vanadium redox flow battery. *Polymers* **2019**, *11*, 1482. [CrossRef] [PubMed]
6. Weber, A.Z.; Mench, M.M.; Meyers, J.P.; Ross, P.N.; Gostick, J.T.; Liu, Q. Redox flow batteries: A review. *J. Appl. Electrochem.* **2011**, *41*, 1137. [CrossRef]
7. Sun, B.; Skyllas-Kazacos, M. Modification of graphite electrode materials for vanadium redox flow battery application-I. Thermal treatment. *Electrochim. Acta* **1992**, *37*, 1253–1260. [CrossRef]
8. Sun, B.; Skyllas-Kazacos, M. Chemical modification of graphite electrode materials for vanadium redox flow battery application-part II. Acid treatments. *Electrochim. Acta* **1992**, *37*, 2459–2465. [CrossRef]
9. Yue, L.; Li, W.; Sun, F.; Zhao, L.; Xing, L. Highly hydroxylated carbon fibres as electrode materials of all-vanadium redox flow battery. *Carbon* **2010**, *48*, 3079–3090. [CrossRef]
10. Kim, K.J.; Kim, Y.J.; Kim, J.H.; Park, M.S. The effects of surface modification on carbon felt electrodes for use in vanadium redox flow batteries. *Mater. Chem. Phys.* **2011**, *131*, 547–553. [CrossRef]
11. Flox, C.; Skoumal, M.; Rubio-Garcia, J.; Andreu, T.; Morante, J.R. Strategies for enhancing electrochemical activity of carbon-based electrodes for all-vanadium redox flow batteries. *Appl. Energy* **2013**, *109*, 344–351. [CrossRef]
12. Pezeshki, A.M.; Clement, J.T.; Veith, G.M.; Zawodzinski, T.A.; Mench, M.M. High performance electrodes in vanadium redox flow batteries through oxygen-enriched thermal activation. *J. Power Sources* **2015**, *294*, 333–338. [CrossRef]

13. Kaneko, H.; Nozaki, K.; Wada, Y.; Aoki, T.; Negishi, A.; Kamimoto, M. Vanadium redox reactions and carbon electrodes for vanadium redox flow battery. *Electrochim. Acta* **1991**, *36*, 1191–1196. [CrossRef]
14. Joerissen, L.; Garche, J.; Fabjan, C.; Tomazic, G. Possible use of vanadium redox-flow batteries for energy storage in small grids and stand-alone photovoltaic systems. *J. Power Sources* **2004**, *127*, 98–104. [CrossRef]
15. Zhang, W.; Xi, J.; Li, Z.; Zhou, H.; Liu, L.; Wu, Z.; Qiu, X. Electrochemical activation of graphite felt electrode for VO^{2+}/VO_2^+ redox couple application. *Electrochim. Acta* **2013**, *89*, 429–435. [CrossRef]
16. Li, B.; Gu, M.; Nie, Z.; Shao, Y.; Luo, Q.; Wei, X.; Wang, W. Bismuth nanoparticle decorating graphite felt as a high-performance electrode for an all-vanadium redox flow battery. *Nano Lett.* **2013**, *13*, 1330–1335. [CrossRef]
17. Li, W.; Liu, J.; Yan, C. Multi-walled carbon nanotubes used as an electrode reaction catalyst for VO^{2+}/VO_2^+ for a vanadium redox flow battery. *Carbon* **2011**, *49*, 3463–3470. [CrossRef]
18. Mayrhuber, I.; Dennison, C.R.; Kalra, V.; Kumbur, E.C. Laser-perforated carbon paper electrodes for improved mass-transport in high power density vanadium redox flow batteries. *J. Power Sources* **2014**, *260*, 251–258. [CrossRef]
19. Kabtamu, D.M.; Chen, J.Y.; Chang, Y.C.; Wang, C.H. Water-activated graphite felt as a high-performance electrode for vanadium redox flow batteries. *J. Power Sources* **2017**, *341*, 270–279. [CrossRef]
20. Chen, J.Z.; Liao, W.Y.; Hsieh, W.Y.; Hsu, C.C.; Chen, Y.S. All-vanadium redox flow batteries with graphite felt electrodes treated by atmospheric pressure plasma jets. *J. Power Sources* **2015**, *274*, 894–898. [CrossRef]
21. Boudou, J.P.; Paredes, J.I.; Cuesta, A.; Martınez-Alonso, A.; Tascon, J.M.D. Oxygen plasma modification of pitch-based isotropic carbon fibres. *Carbon* **2003**, *41*, 41–56. [CrossRef]
22. Okajima, K.; Ohta, K.; Sudoh, M. Capacitance behavior of activated carbon fibers with oxygen-plasma treatment. *Electrochim. Acta* **2005**, *50*, 2227–2231. [CrossRef]
23. Chen, C.; Ogino, A.; Wang, X.; Nagatsu, M. Oxygen functionalization of multiwall carbon nanotubes by Ar/H_2O plasma treatment. *Diam. Relat. Mater.* **2011**, *20*, 153–156. [CrossRef]
24. Hong, S.M.; Kim, S.H.; Kim, J.H.; Hwang, H.I. Hydrophilic surface modification of PDMS using atmospheric RF plasma. *J. Phys. Conf. Ser.* **2006**, *34*, 656–661. [CrossRef]

© 2020 by the authors. Licensee MDPI, Basel, Switzerland. This article is an open access article distributed under the terms and conditions of the Creative Commons Attribution (CC BY) license (http://creativecommons.org/licenses/by/4.0/).

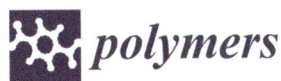

Article

Evaluation of Carbon Dioxide-Based Urethane Acrylate Composites for Sealers of Root Canal Obturation

Hao-Hueng Chang [1,2], Yi-Ting Tseng [3], Sheng-Wun Huang [1], Yi-Fang Kuo [1], Chun-Liang Yeh [1], Chien-Hsin Wu [3], Ying-Chi Huang [3], Ru-Jong Jeng [3,*], Jiang-Jen Lin [4,*] and Chun-Pin Lin [1,2,*]

[1] Graduate Institute of Clinical Dentistry, School of Dentistry, National Taiwan University, Taipei 100, Taiwan; changhh@ntu.edu.tw (H.-H.C.); yelsaint@gmail.com (S.-W.H.); franca_kay@hotmail.com (Y.-F.K.); staryeh0524@gmail.com (C.-L.Y.)
[2] Department of Dentistry, National Taiwan University Hospital, Taipei 100, Taiwan
[3] Institute of Polymer Science and Engineering and Advanced Research Center for Green Materials Science and Technology, National Taiwan University, Taipei 100, Taiwan; kop11239@gmail.com (Y.-T.T.); chwuoliver@gmail.com (C.-H.W.); inzmehuang@gmail.com (Y.-C.H.)
[4] Department of Materials Science and Engineering, National Chung Hsing University, Taichung 400, Taiwan
* Correspondence: rujong@ntu.edu.tw (R.-J.J.); jianglin@ntu.edu.tw (J.-J.L.); pinlin@ntu.edu.tw (C.-P.L.); Tel.: +886-2-3366-5884 (R.-J.J.); +886-4228-57261 (J.-J.L.); +886-2-2312-3456 (ext. 67400) (C.-P.L.)

Received: 24 January 2020; Accepted: 18 February 2020; Published: 21 February 2020

Abstract: A new root canal sealer was developed based on urethane acrylates using polycarbonate polyol (PCPO), a macrodiol prepared in the consumption of carbon dioxide as feedstock. The superior mechanical properties and biostability nature of PCPO-based urethane acrylates were then co-crosslinked with a difunctional monomer of tripropylene glycol diarylate (TPGDA) as sealers for resin matrix. Moreover, nanoscale silicate platelets (NSPs) immobilized with silver nanoparticles (AgNPs) and/or zinc oxide nanoparticles (ZnONPs) were introduced to enhance the antibacterial effect for the sealers. The biocompatibility and the antibacterial effect were investigated by Alamar blue assay and LDH assay. In addition, the antibacterial efficiency was performed by using *Enterococcus faecalis* (*E. faecalis*) as microbial response evaluation. These results demonstrate that the PCPO-based urethane acrylates with 50 ppm of both AgNP and ZnONP immobilized on silicate platelets, i.e., Ag/ZnO@NSP, exhibited great potential as an antibacterial composite for the sealer of root canal obturation.

Keywords: root canal obturation; composites; urethane acrylates; nanoscale silicate platelets; carbon dioxide-based resins

1. Introduction

A tooth with damaged or injured dental pulp requires root canal therapy to cure or prevent apical periodontitis. However, the root canal system has complex internal anatomy with high prevalence [1]. The successful root canal therapy includes correct diagnosis, adequate debridement, and dense filling of three-dimensional space. The root canal obturation is to fill up the root canal and to prevent microorganisms from re-entering into the canal system [2]. Obturation is the method used to fill and seal a cleaned and shaped root canal using core filling materials and root canal sealer. The root canal filling materials are usually composed of cone and sealer (Figure 1). Gutta-percha (GP) cone is the most commonly used material for the obturation of the root canal therapy. Due to the insufficient dentinal adhesion to GP cone, the uses of endodontic sealers are required to provide cohesive strength between core material interface and root canal dentin wall to hold the obturation material together [3]. The root canal sealers are the filling materials that cover all areas of the canal, while acting as lubricants to reduce the friction resistance between the cone material and the canal wall [4,5]. As a result, the key factor to the success of root canal therapy relies on the choice of the appropriate filling material that can completely cover the root canal to prevent the infection.

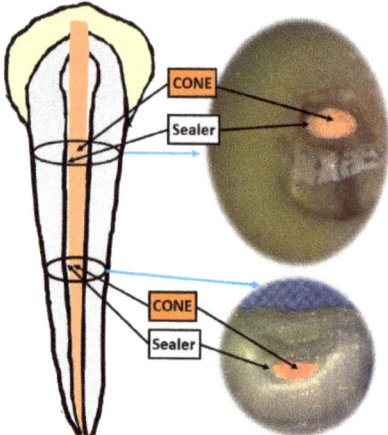

Figure 1. The illustration of filling materials for root canal therapy.

Typical root canal sealers are composed of zinc-oxide eugenol, calcium hydroxide, or glass ionomer. Many studies have suggested that the conventional root canal sealers are easily dissolved [6] and have no dentinal adhesion [7]. Recently, bioceramics such as tricalcium silicate-, dicalcium silicate-, calcium phosphates-, colloidal silica-, or calcium hydroxide-based root canal sealers were developed [8,9]. The moisture within the dentinal tubules was utilized to solidify the calcium silicate hydrate phase [10]. Subsequently, hydroxyapatite was precipitated within the hydrate phase to produce a composite-like structure with reinforcing effects and good sealing ability. Moreover, the pH of the bioceramic sealer during the setting process was usually higher than 12, which increased its bactericidal properties under such circumstances. As compared to the typical root canal sealers, the presence of water is usually a requirement in the canal space for the hydration reaction in order to enable the materials' solidification in the bioceramic sealers such as calcium silicate-based endodontic sealers [11]. However, the difficulty on controlling the precise water content in the bioceramic sealers would result in uncertain setting times along with microhardness. Furthermore, Pawar et al. [12] also suggested that the use of the new bioceramic sealer usually exhibits a lower bond strength than the commonly used root filling materials such as AH-Plus® during the push-out bond strength test. The development of composites for dental-clinical performance still a challenging topic nowadays.

Apart from that, a series of novel endodontic sealers has been derived from synthetic polymers, such as epoxy-, methacrylate-, and silicon-based resins. [13–16] Moreover, urethane-dimethacrylate-based resins were also developed in order to achieve adhesion with dentin and resin properties [13,14]. As one of the key components in the success of advanced technologies, the urethane-based resins can be easily obtained from a rapid formation between the isocyanate chemistry and raw materials with active hydrogens. A large variety of properties can be tailored in the biomedical applications such as wound dressings, artificial organs, and tissue scaffolds [17]. Although the improved bonding strength between the sealers and dentin has been realized, the adhesive bonding with cone materials is still a challenging issue. The gaps along with the sealer-dentin interfaces would attribute to the polymerization shrinkage of the sealer, leading to the re-infection result after the root canal therapy.

The incorporation of nanoscale antibacterial materials is one of the solutions to produce an antibacterial root canal sealer material. The nanoscale silicate platelets (NSP) previously developed by our research group were used as supports for nanoparticle dispersion, such as silver nanoparticles

(AgNPs) [18,19]. The nanometer-thin silicate platelets were intended for surface interactions in order to achieve NSPs immobilized with AgNPs (Ag@NSP) [20]. In fact, Ag@NSP demonstrated excellent antimicrobial activities against common bacteria. Their antibacterial activities depends heavily on the particle size of AgNPs along with the weight ratios of the mobilized AgNPs on NSPs [21]. Moreover, Ag@NSP with low cytotoxicity would further promote wound healing [22]. Apart from that, ZnONPs are employed extensively in a variety of areas such as health products, cosmetics, and catalyst industry owing to their optoelectronic properties, high catalytic efficiency, and antibacterial activities [23,24]. A considerable enhancement on the antibacterial bioactivity toward microorganisms along with biocompatibility to human cells can be achieved by ZnONPs [25]. Therefore, ZnONPs were also immobilized on the nanometer-thin silicate platelets for further enhancing antibacterial properties. Based on the above, nanohybrids such as Ag@NSP, ZnO@NSP or Ag/ZnO@NSP are have great potential as antibacterial agents for the root canal applications.

The purpose of this study was to develop a novel root canal sealer, using a biocompatible resin matrix with NSPs as antibacterial components. The urethane resins were first prepared from the aliphatic polycarbonate polyol (PCPO), which is a macrodiol prepared in the consumption of carbon dioxide [26,27]. The urethane acrylates prepared from PCPO exhibited not only excellent resistance to hydrolysis and good mechanical properties but significant improvement in biostability and nontoxicity in many cases [28,29]. It was also reported that the NSPs immobilized with nanoparticles such as AgNPs could effectively prevent the nanoparticles from agglomeration into Ag clusters, while exhibiting antimicrobial activity through dramatically decreasing glucose uptake and hindering adenosine triphosphate (ATP) synthesis for microbial growth [21]. As a result, the root canal sealers were then prepared by urethane acrylate (UA) composites comprising the immobilized nanoparticles on nanoscale platelets for the root canal obturation evaluations.

2. Materials and Methods

2.1. Synthesis of Urethane-acrylate (UA)

As shown in Figure 2, UAs were prepared from the reaction between the isocyanate groups of isophorone diisocyanate (IPDI) in the presence of 1,8-diazabicyclo[5.4.0]undec-7-ene (DBU) as catalyst, and hydroxyl groups of polyols to achieve urethane prepolymer end-capped with isocyanates, followed by the addition of hydroxyethylmethacrylate (HEMA). In this study, UA prepared from polyols by using polytetramethylene ether glycol with a Mw of 650 (PTMEG 650) is denoted as UAT65, while UA prepared from polycarbonate polyol with a Mw of 500 (PCPO 500) is denoted as UAC50. The introduction of acrylate group was prepared from using HEMA as reagent in the molar ratio (2/1/1) of IPDI/polyol/HEMA. The synthesis of UAs was monitored by using IR spectrum (Jasco 4600 FT-IR Spectrophotometer with a Jasco ATR Pro 450-S accessory, Jasco, Tokyo, Japan). The molecular weights were measured by using gel permeation chromatography (GPC) with samples dissolved in 1 wt % THF. (Waters chromatography system, two Waters Styragel linear columns, and polystyrene as the standard with Mw 104, 2560, 7600, 18,000, and 37,900 by using THF as eluent at the flow rate of 1 mL/min.)

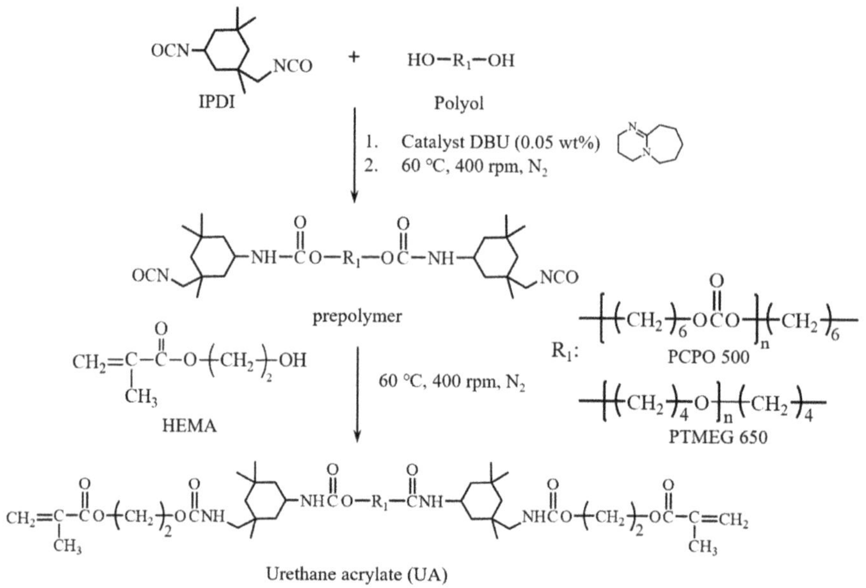

Figure 2. The preparation of urethane acrylate (UA).

2.2. Preparation of Antibacterial Sealer Based on UA

The UA samples (UAT65 and UAC50) were dissolved in acetone at 80 °C and mixed with nanoscale silicate materials (Ag@NSP, ZnO@NSP, or Ag/ZnO@NSP) to prepare UA-based composites. The preparation and characterization of Ag@NSP, ZnO@NSP, and Ag/ZnO@NSP were performed according to the literature reported by our groups [21]. The sealers were then prepared by the introduction of a diluent, tripropylene glycol diacrylate (TPGDA), and an initiator containing camphorquinone (CQ) as a photosensitizer with ethyl-4-dimethylaminobenzoate (EDMA) for binary initiation processes with a ratio of CQ/EDMAB = 1/2 (w/w), together with 0.5 wt % azobisisobutyronitrile (AIBN) for ternary initiation processes. The characterizations of composites were carried out by using the thermogravimetric analysis (TGA; TGA Q50 TA instrument, TA Instrument, New Castle, DE, USA) to investigate the inorganic contents. For the curing process, the sealer samples were first cast into Teflon molds in the dimension of 20-mm long, 2-mm wide, and subsequently cured with a light emitting diode curing unit at an intensity of 800 mW/cm^2 for 40 s from the coronal aspect (SmartLite, Dentsply, PA). The degrees of curing conversions were investigated by using FT-IR to calculate the peak area transition at 1638 cm^{-1} of aliphatic C=C double bond of acrylate group according to the following equation (1) [30]:

$$\text{Degree of conversion \% (DC\%)} = \left[1 - \frac{[A_{1638}]_{\text{after curing}}}{[A_{1638}]_{\text{before curing}}}\right] \times 100\% \tag{1}$$

2.3. Curing Condition and Depth Test

To evaluate the feasibility of the sealers for root canal, the flow and viscosity tests were carried out for UA resins according to ISO 6876:2001. Based on the specifications, the mixture (0.05 mL) of UA resin and TPGDA (UAT65/TPGDA or UAC50/TPGDA) was placed on the center of glass plate using a graduated syringe. After the initial mixing for 180 s (±5 s), another glass plate was placed on top of the sealer, following a weight providing a total mass of 120 g (±2 g). After 10 min, the weight was removed in order to measure the maximum and minimum diameter of the compressed disks of the sealers [31]. Apart from that, the viscosities of the sealers were measured by using a syringe-based viscometer based on Instron 3360 series universal testing system [32]. In Figure 3, the measurement of maximal curing increment thickness for resin composites was conducted by using a ISO 4049:2009 method [33]. The curing process was performed with a 2 × 2 × 20 mm re-usable stainless-steel mold with a top cover to prevent materials' leakage. After curing process under a light source of halogen lamp for 40 s, all samples were kept in saturated humidity at 37 °C for 24 h to measure the curing depths. The mechanical properties were also measured by using the Universal Testing Instrumentals according to ASTM D412-98a with a stretching rate at 100 mm/min. The tensile tests were conducted for five times on an average.

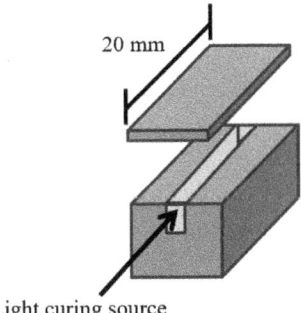

Figure 3. The test method for curing depths according to ISO 4049:2009 [33].

2.4. Biocompatibility Test

Biocompatibility tests were performed by using the ISO 10993-5 regulation through Alamar Blue assay and lactate dehydrogenase (LDH) assay [34]. Alamar Blue assay was carried out according to the following steps: first, 3T3 cells were cultured under 10% fetal bovine serum (FBS) Dulbecco's modified eagle's medium (DMEM) at 37 °C in the presence of 5% CO_2. All Cells were subcultured twice before the following experiment. In the next step, the specimens were polymerized under photo-polymerization machine on top of a round Teflon mold, which is 8 mm in diameter and 2 mm in height, in a distance of 0.5 cm for 40 s. After the polymerization, the UA composites were then removed from the mold, sterilized by UV-light for 24 h, and soaked in fresh medium at 37 °C for 24 h. In the final step, the medium containing 3T3 cells were added to 96-well plates with a cell number of 104/well. After the cells fully adhered on the plate, the original cell culture medium was replaced by the extract material. The cell incubation for Alamar blue assay cell activity analysis was conducted at 37 °C for 1, 3, and 7 days in the presence of 5% CO_2. Moreover, the evaluation of LDH assay for cytotoxicity were also conducted for 1, 3, and 7 days by using 3T3 cells under similar culture condition in 10% FBS DMEM at 37 °C in the presence of 5% CO_2.

2.5. Antibacterial Test

The antibacterial activity of UAT65/TPGDA or UAC50/TPGDA with various Ag@NSP, ZnO@NSP, or Ag/ZnO@NSP contents was performed according to the Japanese Industrial Standard as shown in Figure 4 (JIS Z 2801-2000) [32]. Gram-positive *E. faecalis* (ATCC 29212, Super Laboratory Co., Taiwan) was the bacterial strain cultivated in brain heart infusion (BHI) broth. The number of bacterial suspensions was adjusted to 104 colony-forming units per milliliter (CFU/mL). The antibacterial test was carried out by the following steps. First, colonies were added into 5 mL HBI broth, cultivated at 37 °C for 16–18 h. The medium of bacteria fluid of lysogeny broth (LB) was replaced by phosphate-buffered saline (PBS) by using the centrifugation for three times (8,000 xg, 3 min), and diluted to the concentration of 10^{-5}–10^{-7} (CFU/mL). Subsequently, the as-prepared suspension was spread onto the BHI agar plates

(10 µL) and incubated cells at 37 °C for 16–18 h. The 0.5 McFarland (10^8 CFU/mL) bacterial fluid was produced by using *E. faecalis* (ATCC 29212). Specimens were shaped to square (5 cm × 5 cm), wiped with alcohol, and sterilized by the exposure of UV-light for 24 h. Then, 400 µL bacterial fluid (10^4 CFU/mL) was added to the specimens, covered with a sterile square PE film (4 cm × 4 cm) and incubated at 37 °C under relative humidity 90% for 24 h. After bacterial adhesion or proliferation, the specimens were rinsed for three times by using PBS and then transformed to a new 24-well plate. Each specimen was soaked in 2.5% glutaraldehyde and reacted at 4 °C for 1 h, followed by the removal of glutaraldehyde by rinsing with PBS for three times. Before the investigation of morphology for the evaluation of antibacterial activity, the samples were prepared by using ethanol-wet bonding technique [35]. The freeze-drying processes were conducted for 24 h by the increased concentration of (w/w) ethanol from 30%, 50%, 70%, 90%, 95%, 99% to 100%. In addition, the statistical analysis was performed by Statistical Analysis Software (SAS). One-way analysis of variance (ANOVA) was used to analyze the difference between bacterial groups, and Duncan's multiple tests were used to distinguish various bacterial groups. The *p*-value <0.05 is regarded as statistically significant.

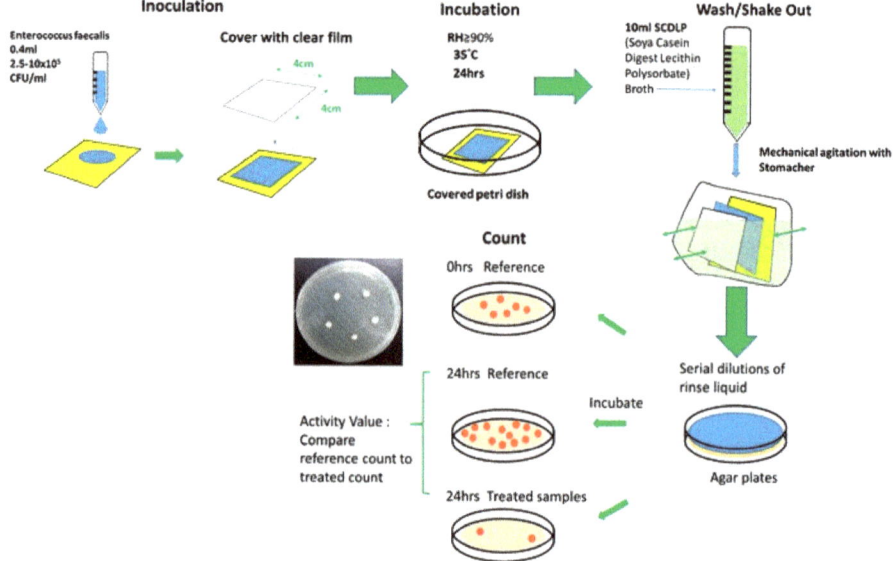

Figure 4. The process of antibacterial evaluation.

3. Results and Discussion

3.1. Synthesis of Urethan-acrylate (UA)

The preparation of urethane acrylate was monitored by using the IR spectra as shown in Figure 5. For the spectrum of the initial mixture of IPDI and PCPO 500, two distinct peaks at 2260 and 1742 cm^{-1} were present for isocyanate, and carbonyl group of carbonate ester, respectively. After 2 h into the reaction, a newly emerged shoulder at 1718 cm^{-1} was observed, indicating the formation of urethane carbonyl group [36,37]. Moreover, the urethane prepolymer end-capped with isocyanates was further reacted with HEMA to provide the product of UAC50 as evident by the near disappearance of isocyanate group and the formation of a new adsorption peak at 1638 cm^{-1} of vinyl group. The additional TPGDA in the mixture of UAC50/TPGDA = 70/30 (w/w) resulted in a stronger absorption intensity of vinyl group. The weight average molecular weights were measured by using GPC, leading to the results of 3320 g/mol for UAC50 and 3620 g/mol for UAT65.

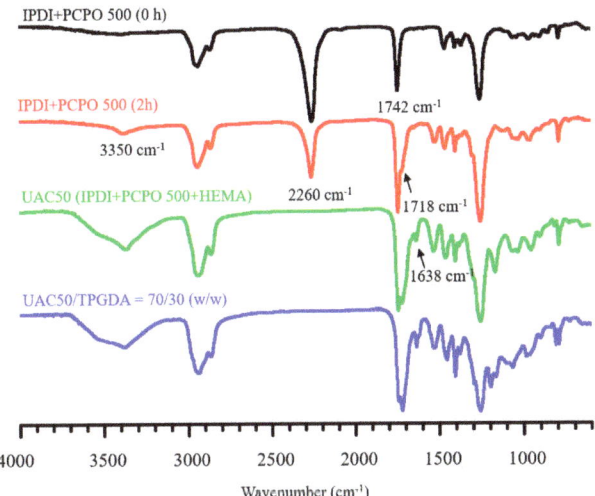

Figure 5. FTIR spectra of UAC50 and UAC50/TPGDA.

3.2. Preparation and Thermal Properties of UA Composites

The determination of the processing effectiveness and the quantitative fillers within the matrix are usually investigated by using thermogravimetric analysis (TGA) for the polymer composites [38]. The UA composites were prepared by mixing 5 wt % or 10 wt % in total composites by using the nanometer-thin silicate platelets immobilized with nanoparticles such as AgNPs (Ag@NSP), ZnONPs (ZnO@NSP), or AgNPs together with ZnONPs (Ag/ZnO@NSP) [20,22,25,39,40]. In TGA thermograms (Figure 6), the UA composites exhibited 5% weight loss at about 300 °C, indicating that good thermal stability was available for the following tests. Moreover, the addition of nanoparticles immobilized on the nanometer-thin silicate platelets provided higher char yield (%) dependent on the addition of inorganic contents.

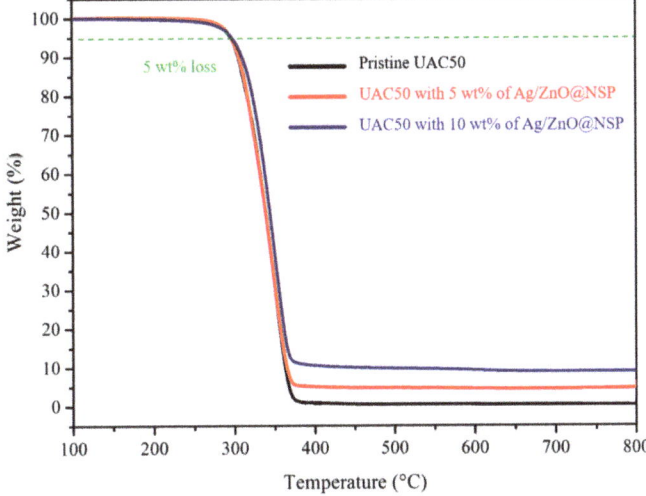

Figure 6. Thermogravimetric analysis (TGA) thermograms of UAC50- and UAC50-based composites.

3.3. Curing Conditions and Depths for UA Composites

The endodontic sealers are required to exhibit several properties to meet the desired performance. The flow behavior is one of the most important factors for sealers to penetrate into small irregularities of the root canal system and dentinal tubules as shown in Figure 7. According to the regulation of ISO 4049:2009, the flow diameter should be over 20 mm after compression under a weight disc for 10 min. The tests were carried out by the introduction of difunctional TPGDA in order to increase flow diameter and crosslinking density. As a result, the samples with higher ratio of TPGDA exhibited larger flow diameter, such as UAT65/TPGDA = 80/20, 70/30, or 60/40; or UAC50/TPGDA = 70/30 or 60/40, since the difunctional TPGDA monomer is a diluent with a lower molecular weight when compared with the UA resins.

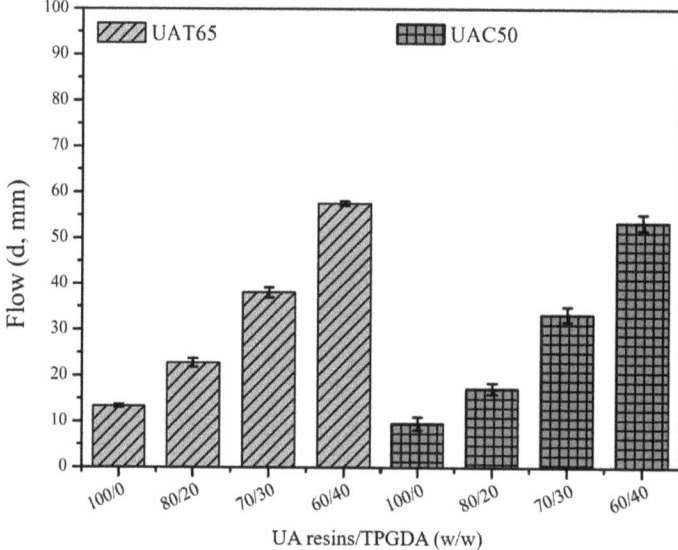

Figure 7. Flow analysis of UA resins with various weight ratios of UAT65/tripropylene glycol diacrylate (TPGDA) or UAC50/TPGDA according to ISO 4049:2009.

Moreover, viscosity is another important factor for the root canal sealer to establish connection between the root canal, periodontal ligament, and the apical foramen. According to ISO 4049:2009, the viscosity should range from 1000 to 2000 cp to meet the desired operation [32]. In Figure 8, UAT65/TPGDA (70/30) and UAC50/TPGDA (70/30) were the samples of choice for further investigations since the addition of the optimized TPGDA content would achieve the viscosity range mentioned above. It is important to note that flow and viscosity properties depends on not only the ratios between UA resins and TPGDA but the use of polyols such as PTMEG or PCPO. The PCPO-based UA resins exhibited somewhat lower flow diameter and higher viscosity under the same ratio of UA/TPGDA composition. This is because the incorporation of the PCPO-based polyol with carbonate groups brought about more hydrogen bonding interactions with the urethane linkages. This would restrain the molecular mobility of the UA segments [41,42].

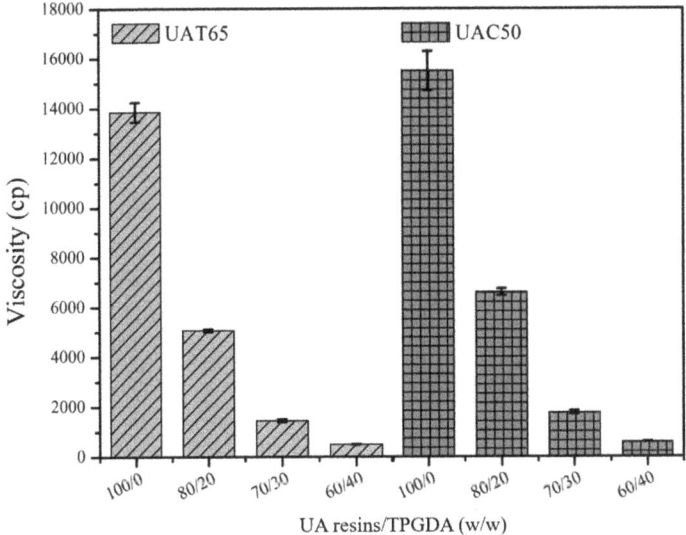

Figure 8. Viscosity analysis of UA resins with various weight ratios of UAT65/TPGDA or UAC50/TPGDA according to ISO 4049:2009.

The curing conditions of UAT65/TPGDA (70/30) or UAC50/TPGDA (70/30) were carried out by using ternary initiation processes for photo crosslinking. The degree of acrylate conversion (DC%) would be based on the peak area at 1638 cm^{-1} of aliphatic C=C double bond of acrylate content as a function of various initiator concentrations (1, 3, 6, and 9 phr) before and after curing (Table 1). Optimum mechanical properties with a tensile strength of 56.15 ± 3.26 MPa and a Young's modulus of 361.88 ± 32.38 MPa were achieved for the sample UAT65/TPGDA (70/30) cured with 3 phr initiator concentration. It is important to note that the DC% values were close to 70% for the above-mentioned samples. This is because the low conversion degree of acrylate functional groups with a low concentration initiator resulted in insufficient crosslinking density for obtaining good mechanical performance, whereas the excessive photoinitiators would provide the localized absorption and crosslinking density over the percolation threshold, leading to the adverse effect on the mechanical properties [43].

Since the acrylate-based photopolymerization depends deeply on the sealer transparency, the curing depths of UAT65/TPGDA resin (w/w = 70/30) and UAC50/TPGDA resin (w/w = 70/30) were assessed by using various contents of antibacterial NSPs immobilized with nanoparticles including Ag@NSP, ZnO@NSP, and Ag/ZnO@NSP (Figures 9–11). It is reported that the photo crosslinking activity relies on the use of different types of metal nanoparticles and the rate of photo crosslinking, and the DC% also varies with the additives under similar processing condition [44]. In this study, the curing depths decreased with increasing content of antibacterial NSP agents.

These UA composite composed of various antibacterial nanomaterials are denoted as Ag@NSP-n, ZnO@NSP-n, or Ag/ZnO@NSP-n, where the "n" is denoted as the parts per million (ppm) to the total weight of UAT65/TPGDA or UAC50/TPGDA. According to the regulation of ISO 4049:2009, the curing depth should be larger than 10 mm. As a result, the limitation for the addition of a maximum amount of Ag@NSP is 500 ppm (Ag@NSP-500) as shown in Figure 9. Similar results were also obtained for the composites incorporated with ZnO@NSP or Ag/ZnO@NSP (Figures 10 and 11). This indicates that the presence of different nanomaterials such as Ag@NSP, ZnO@NSP, or Ag/ZnO@NSP in the UA composites did not influence curing depths much.

Table 1. Mechanical properties and DC% for the UAT65/TPGDA (w/w = 70/30) and TPGDAC50/TPGDA (w/w = 70/30) samples after curing with different dosages of photoinitiator.

Entry	UA Formulation	Photoinitiator (phr [1])	Tensile Strength (MPa)	Young's Modulus (MPa)	DC [2] (%)
1	UAT65/TPGDA (w/w = 70/30)	1	51.71 ± 3.03	362.81 ± 37.10	46.27 ± 3.07
2	UAT65/TPGDA (w/w = 70/30)	3	56.15 ± 3.26	361.88 ± 32.38	64.91 ± 1.06
3	UAT65/TPGDA (w/w = 70/30)	6	45.96 ± 2.22	33.49 ± 1.65	67.58 ± 2.27
4	UAT65/TPGDA (w/w = 70/30)	9	36.78 ± 1.65	1.92 ± 0.42	70.35 ± 1.76
5	UAC50/TPGDA (w/w = 70/30)	1	46.30 ± 3.18	1072.25 ± 46.54	57.89 ± 0.41
6	UAC50/TPGDA (w/w = 70/30)	3	60.54 ± 4.72	1042.02 ± 39.62	72.44 ± 1.57
7	UAC50/TPGDA (w/w = 70/30)	6	61.09 ± 1.15	14.01 ± 2.65	79.03 ± 2.57
8	UAC50/TPGDA (w/w = 70/30)	9	59.04 ± 2.75	8.61 ± 0.41	80.08 ± 2.79

[1]: parts per hundred; [2]: after the exposure under UV light for 40 s.

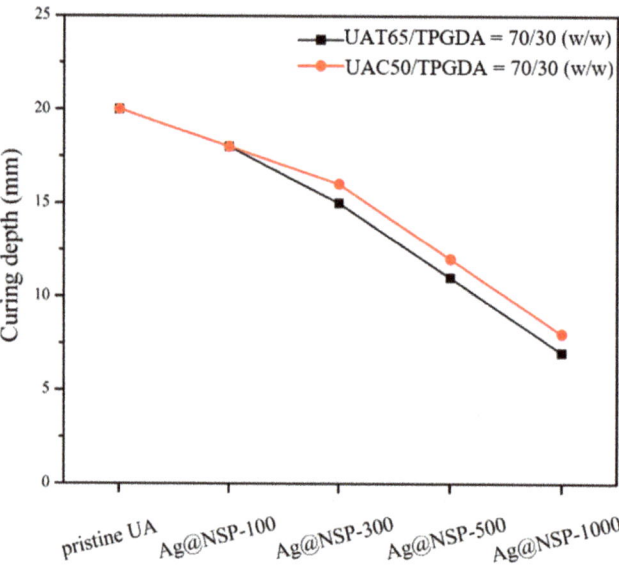

Figure 9. Curing depths of UAT65/TPGDA resins (w/w = 70/30) and UAC50/TPGDA resins (w/w = 70/30) with various Ag@NSP contents (ppm).

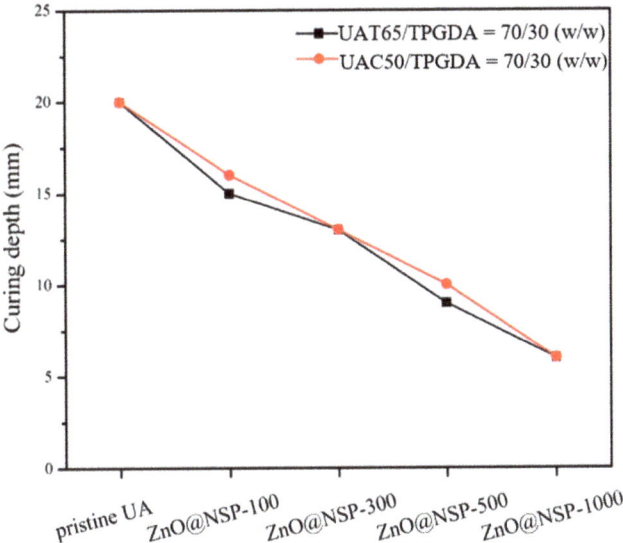

Figure 10. Curing depths of UAT65/TPGDA resins (w/w = 70/30) and UAC50/TPGDA resins (w/w = 70/30) with various ZnO@NSP contents (ppm).

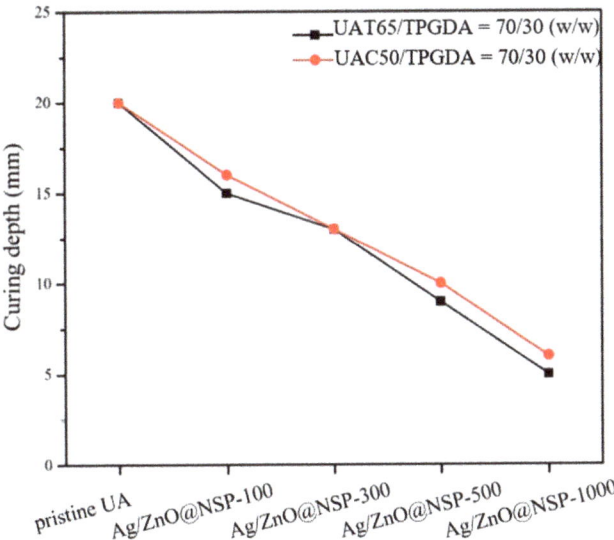

Figure 11. Curing depths of UAT65/TPGDA resins (w/w = 70/30) and UAC50/TPGDA resins (w/w = 70/30) with various Ag/ZnO@NSP contents (ppm).

3.4. Biocompatibility Analysis

Grossman [45] advocated that an ideal root canal filling material should not irritate periradicular tissues. In addition, Faccioni et al. found that root canal materials with metal ions might influence cell metabolism and differentiation [46]. Other studies found that the incomplete photopolymerization reaction resulted in the release of uncured monomers and initiators, which would affect the mitochondrial enzyme activity [47,48]. Therefore, the concentration of antibacterial agent in sealers depends deeply on the biocompatibility of composites. In this study, the tests of Alamar Blue assay and LDH assay were conducted for the biocompatibility tests for UAT65/TPGDA (*w/w* = 70/30) and UAC50/TPGDA resins (*w/w* = 70/30) with various concentrations of antimicrobial agents such as Ag@NSP, ZnO@NSP, and Ag/ZnO@NSP as shown in Figures 12 and 13, respectively.

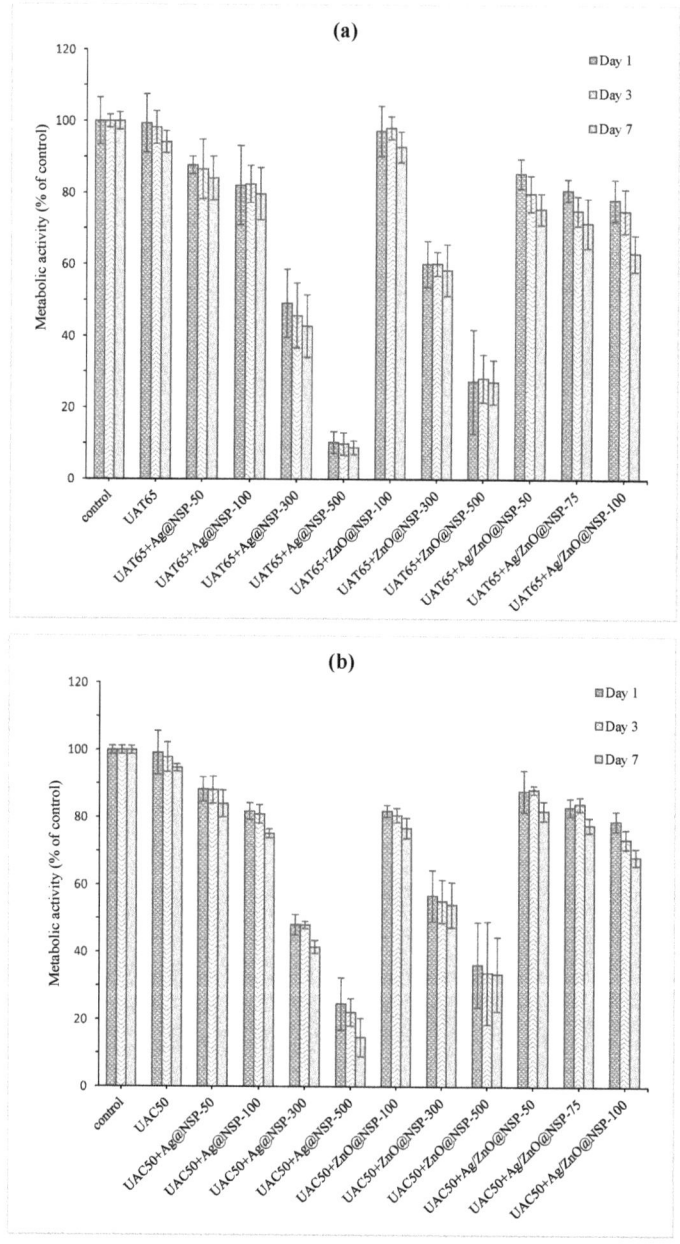

Figure 12. Alamar blue assay of (**a**) UAT65/TPGDA (*w/w* = 70/30) and (**b**) UAC50/TPGDA (*w/w* = 70/30) resins with various concentrations of Ag@NSP, ZnO@NSP, or Ag/ZnO@NSP ($p < 0.05$).

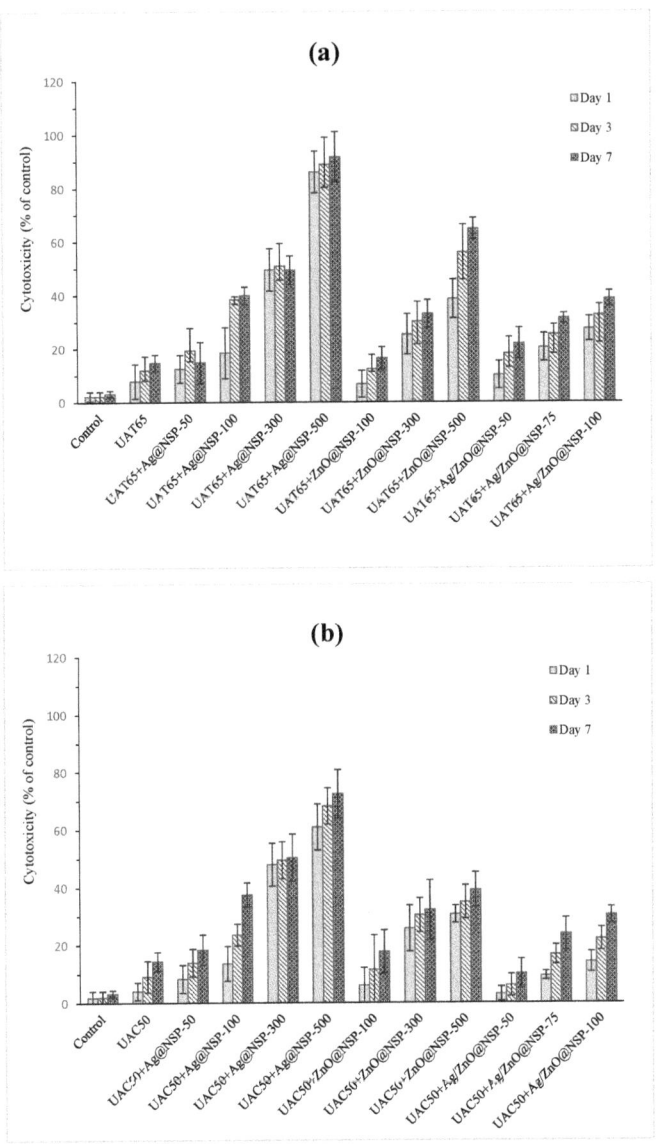

Figure 13. Lactate dehydrogenase (LDH) assay of (**a**) UAT65/TPGDA (*w/w* = 70/30) and (**b**) UAC50/TPGDA (*w/w* = 70/30) with various concentrations of Ag@NSP, ZnO@NSP, or Ag/ZnO@NSP ($p < 0.05$).

In the Alamar blue assay test, both pristine resins of UAT65/TPGDA (*w/w* = 70/30) (Figure 12a) and UAC50/TPGDA (*w/w* = 70/30) (Figure 12b) were substantially free of cytotoxicity. As the antimicrobial agents were incorporated into the resins to form UA composites, reduced metabolic activities were observed, as shown in Figure 12. Furthermore, the composite with Ag@NSP exhibited the poorest biocompatibility when compared to other samples, especially in the example for both UAT65/TPGDA and UAC50/TPGDA incorporated with 500 ppm additives. The metabolic activities were higher than 70% for all the UA composites with 100 ppm or less than 100 ppm antibacterial NSP agents. This indicates that good biocompatibility could be achieved with the addition of a certain content of the antimicrobial agents such as Ag@NSP, ZnO@NSP, or Ag/ZnO@NSP to the composites.

In the LDH assay test, the cytotoxicity is also dependent on the addition of various concentrations of Ag@NSP, ZnO@NSP, and Ag/ZnO@NSP to the composites (Figure 13). Given the fact that UA composites with 100 ppm would exhibit good biocompatibility, the cytotoxicity of the UA composites with 100 ppm antimicrobial agents on 3T3 cells was investigated and observed in the following order: Ag@NSP-100 > Ag/ZnO@NSP-100 > ZnO@NSP-100. This implies that the composites comprising AgNPs would exhibit poor cytotoxicity performance. In addition, the cytotoxicity of the composites based on the UAC50/TPGDA (w/w = 70/30) are lower than that of the composites based on UAT65/TPGDA (w/w = 70/30), especially in the example for both UAT65/TPGDA (~90% of control) and UAC50/TPGDA (~80% of control) incorporated with 500 ppm Ag@NSP. As a matter of fact, polycarbonate-based polyurethanes (PUs) with better biocompatibility correspond to the weaker immune response when compared with polyether-based PUs according to the literature [49]. This is because the α-carbon atoms of the polyether-based PUs (such as UAT65) are highly susceptible to oxidation by oxygen radicals to form esters, which result in unstable chemical structures for polymers [50]. As a result, the carbonate containing UAC50/TPGDA system would be the material of choice for dental root canal sealers instead of the ether-containing UAT65/TPGDA resins.

3.5. Antibacterial Analysis

The antibacterial activities of the composites based on UAC50/TPGDA (w/w = 70/30) with nanoparticles-on-platelet nanohybrids (Ag@NSP, ZnO@NSP, Ag/ZnO@NSP) were analyzed by treating the composites with Gram-positive *E. faecalis* (ATCC 29212) according to Japanese Industrial Standard (JIS Z 2801-2000) as shown in Table 2. According to the biocompatibility tests in previous section, the use of antibacterial NSP agents should not be higher than 100 ppm for UAC50/TPGDA (w/w = 70/30) resins. Indeed, the AgNPs-based composites with 75 ppm (Ag@NSP-75) or 100 ppm (Ag@NSP-100) inorganic additives exhibited antibacterial activity. However, ZnONPs-based composites even with 1000 ppm (ZnO@NSP-1000) were not able to exhibit antibacterial activity. In fact, the antibacterial activity was feasible only when the use of high concentration of ZnONPs over 2000 ppm (ZnO@NSP-2000 and ZnO@NSP-3000), indicating that the Ag@NSP-based composites would be better candidates for the root canal sealer applications. In addition, the simultaneous immobilization of AgNPs and ZnONPs on silicate platelets, i.e., Ag/ZnO@NSP, led to much better antimicrobial results even for the composite with an antibacterial agent concentration as low as 50 ppm (Ag/ZnO@NSP-50).

The direct contact of UAC50/TPGDA composites with the bacterial population could be visualized by using scanning electron microscopy (SEM) as shown in Figure 14. According to the investigation above, this study was conducted by using *E. faecalis* on the surfaces of UA/TPGDA composites with 50 ppm Ag@NSP, 75 ppm Ag@NSP, and 50 ppm Ag/ZnO@NSP after 6h and 24h. For the UAC50/TPGDA resin without the use of antibacterial agents, a rapid growth of bacterial in number was observed as shown in Figure 14a,e. For the composite (Ag@NSP-50) with 50 ppm of Ag@NSP, a certain amount of bacteria appeared first after 6h (Figure 14b). Subsequently, these bacteria increased in number and aggregated after 24 h (Figure 14f). For the composite (Ag@NSP-75) with 75 ppm of Ag@NSP, a certain amount of bacteria appeared to be aggregated and deformed after 6h, and subsequently these bacteria remained aggregated without the sign of number increase as shown in Figure 14c,g, respectively. It is likely that the nanoparticles can re-charge from NSPs to inactivate and rupture bacterial aggregates [51].

It was reported that the combined use of different types of nanoparticles could adsorb onto the cytoderm of the bacteria and even penetrate the cytomembrane to disturb the normal function of cells, leading to cell apoptosis [50]. This is evidenced by the presence of only 50 ppm Ag/ZnO@NSP in the UAC50/TPGDA composite capable of exhibiting a satisfactory antibacterial effect against *E. faecalis* (Table 2). For the composites (Ag/ZnO@NSP-50) with 50 ppm of Ag/ZnO@NSP, once again a certain amount of bacteria appeared to be aggregated and deformed after 6h, and subsequently these bacteria remained aggregated without the sign of number increase as shown in Figure 14d,f. The simultaneous immobilization of AgNPs and ZnONPs on silicate platelets could not only enhance the antibacterial activities and reduce the dose of AgNPs, but act as a promoter in the antibacterial effect for the Ag/ZnO@NSP-based composites as well.

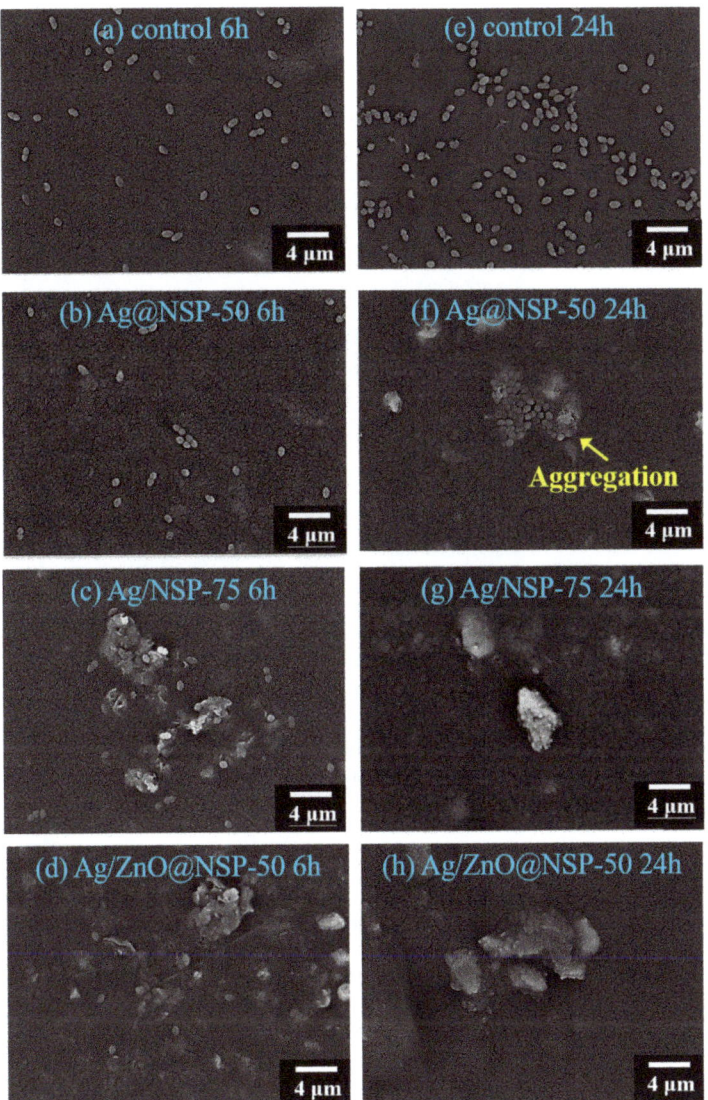

Figure 14. SEM images (2500x) of *E. faecalis* on the surfaces of UAC50/TPGDA = (w/w = 70/30) with various antibacterial agents: control (**a,e**); 50 ppm Ag@NSP (**b,f**); 75 ppm Ag@NSP (**c,g**); 50 ppm Ag/ZnO@NSP (**d,h**) for 6 h and 24 h, respectively.

Table 2. Antibacterial activity of the composites based on UAC50/TPGDA (w/w = 70/30) resins with various concentrations of Ag@NSP, ZnO@NSP, or Ag/ZnO@NSP.

Entry	Type of Composites [1]	Antibacterial Activity [2]
1	Ag@NSP-50	×
2	Ag@NSP-75	○
3	Ag@NSP-100	○
4	ZnO@NSP-100	×
5	ZnO@NSP-200	×
6	ZnO@NSP-300	×
7	ZnO@NSP-500	×
8	ZnO@NSP-1000	×
9	ZnO@NSP-2000	○
10	ZnO@NSP-3000	○
11	Ag/ZnO@NSP-10	×
12	Ag/ZnO@NSP-25	×
13	Ag/ZnO@NSP-50	○
14	Ag/ZnO@NSP-75	○
15	Ag/ZnO@NSP-100	○

(1): composites using UAC50 as polymer matrix; (2): "○" indicating the inhibition of bacteria; "×" indicating the growth of bacteria.

4. Conclusions

The purpose of this study is to develop a new root canal sealer with a biocompatible resin based on a macrodiol of polycarbonate (PCPO) that could be prepared through the utilization of carbon dioxide as feedstock. With good biocompatibility and biostability, these carbonate-containing urethane acrylate resins would be able to exhibit superior performance to the reference, polyether (PTMEG)-based resins in the preparation of root canal obturation sealers. The successful incorporation of nanoparticles immobilized on nanoscale platelets in the resin matrix resulted in the root canal obturation sealers with satisfactory biocompatibility and antibacterial effect. As a result, the PCPO-based urethane acrylate was selected to be the resin sealer matrix. Moreover, the incorporation of ZnONPs and AgNPs simultaneously immobilized on silicate platelets into the PCPO-based urethane acrylates would not only enhance the antibacterial activities, but also serve as a promoter in the antibacterial effect. Based on the above, the UAC50/TPGDA (70/30 = w/w) resin with 50 ppm Ag/ZnO@NSP has a great potential as an antibacterial root canal sealer.

Author Contributions: Conceptualization, H.-H.C., R.-J.J., and C.-P.L.; methodology, Y.-T.T.; validation, H.-H.C. and C.-P.L.; data curation, S.-W.H. and Y.-F.K.; formal analysis, Y.-C.H., H.-H.C. and Y.-T.T.; writing—original draft preparation, Y.-T.T.; writing—review and editing, C.-L.Y. and C.-H.W.; supervision, J.-J.L. and C.-P.L. All authors have read and agreed to the published version of the manuscript.

Funding: This work was financially supported by the "Advanced Research Center for Green Materials Science and Technology" from The Featured Area Research Center Program within the framework of the Higher Education Sprout Project by the Ministry of Education (108L9006) and the Ministry of Science and Technology in Taiwan (MOST 108-3017-E-002-002, MOST 106-2221-E-002-189-MY3).

Conflicts of Interest: The authors declare no conflict of interest. The funders had no role in the design of the study; in the collection, analyses, or interpretation of data; in the writing of the manuscript, or in the decision to publish the results.

References

1. Yang, H.; Tian, C.; Li, G.; Yang, L.; Han, X.; Wang, Y. A cone-beam computed tomography study of the root canal morphology of mandibular first premolars and the location of root canal orifices and apical foramina in a chinese subpopulation. *J. Endod.* **2013**, *39*, 435–438. [CrossRef] [PubMed]
2. Carrotte, P. Endodontics: Part 3 Treatment of endodontic emergencies. *Br. Dent. J.* **2004**, *197*, 299–305. [CrossRef] [PubMed]

3. Hsieh, K.-H.; Liao, K.-H.; Lai, E.H.-H.; Lee, B.-S.; Lee, C.-Y.; Lin, C.-P. A novel polyurethane-based root canal-obturation material and urethane acrylate-based root canal sealer—Part I: Synthesis and evaluation of mechanical and thermal properties. *J. Endod.* **2008**, *34*, 303–305. [CrossRef]
4. Carrotte, P. Endodontics: Part 5 Basic instruments and materials for root canal treatment. *Br. Dent. J.* **2004**, *197*, 455–464. [CrossRef] [PubMed]
5. Al-Hiyasat, A.S.; Alfirjani, S.A. The effect of obturation techniques on the push-out bond strength of a premixed bioceramic root canal sealer. *J. Dent.* **2019**, *89*, 103169. [CrossRef] [PubMed]
6. Schäfer, E.; Zandbiglari, T. Solubility of root-canal sealers in water and artificial saliva. *Int. Endod. J.* **2003**, *36*, 660–669. [CrossRef] [PubMed]
7. Wennber, A.; Niom, D.Ø. Adhesion of root canal sealers to bovine dentine and gutta-percha. *Int. Endod. J.* **1990**, *23*, 13–19. [CrossRef]
8. Washio, A.; Morotomi, T.; Yoshii, S.; Kitamura, C. Bioactive glass-based endodontic sealer as a promising root canal filling material without semisolid core materials. *Materials* **2019**, *12*, 3967. [CrossRef]
9. Teoh, Y.-Y.; Athanassiadis, B.; Walsh, L.J. Sealing ability of alkaline endodontic cements versus resin cements. *Materials* **2017**, *10*, 1228. [CrossRef]
10. Tyagi, S.; Mishra, P.; Tyagi, P. Evolution of root canal sealers: An insight story. *Eur. J. Gen. Dent.* **2013**, *2*, 199–218. [CrossRef]
11. De-Deus, G.; Canabarro, A.; Alves, G.G.; Marins, J.R.; Linhares, A.B.R.; Granjeiro, J.M. Cytocompatibility of the ready-to-use bioceramic putty repair cement iRoot BP Plus with primary human osteoblasts. *Int. Endod. J.* **2012**, *45*, 508–513. [CrossRef] [PubMed]
12. Pawar, A.M.; Pawar, S.; Kfir, A.; Pawar, M.; Kokate, S. Push-out bond strength of root fillings made with C-Point and BC sealer versus gutta-percha and AH Plus after the instrumentation of oval canals with the self-adjusting file versus waveone. *Int. Endod. J.* **2016**, *49*, 374–381. [CrossRef] [PubMed]
13. Lee, B.-S.; Lai, E.H.-H.; Liao, K.-H.; Lee, C.-Y.; Hsieh, K.-H.; Lin, C.-P. A novel polyurethane-based root canal-obturation material and urethane-acrylate-based root canal sealer-Part 2: Evaluation of Push-out Bond Strengths. *J. Endod.* **2008**, *34*, 594–598. [CrossRef] [PubMed]
14. Lee, B.-S.; Wang, C.-Y.; Fang, Y.-Y.; Hsieh, K.-H.; Lin, C.S. A novel urethane acrylate-based root canal sealer with improved degree of conversion, cytotoxicity, bond strengths, solubility, and dimensional stability. *J. Endod.* **2011**, *37*, 246–249. [CrossRef] [PubMed]
15. Chen, C.-Y.; Huang, C.-K.; Lin, S.-P.; Han, J.-L.; Hsieh, K.-H.; Lin, C.-P. Low-shrinkage visible-light-curable urethane-modified epoxy acrylate/SiO2 composites as dental restorative materials. *Compos. Sci. Technol.* **2008**, *68*, 2811–2817. [CrossRef]
16. Tay, F.R.; Loushine, R.J.; Weller, R.N.; Kimbrough, W.F.; Pashley, D.H.; Mak, Y.-F.; Shirley Lai, C.-N.; Raina, R.; Williams, M.C. Ultrastructural evaluation of the apical seal in roots filled with a polycaprolactone-based root canal filling material. *J. Endod.* **2005**, *31*, 514–519. [CrossRef]
17. Delebecq, E.; Pascault, J.-P.; Boutevin, B.; Ganachaud, F. On the versatility of urethane/urea bonds: Reversibility, blocked isocyanate, and non-isocyanate polyurethane. *Chem. Rev.* **2013**, *113*, 80–118. [CrossRef]
18. Lin, J.-J.; Chu, C.-C.; Chiang, M.-L.; Tsai, W.C. First isolation of individual silicate platelets from clay exfoliation and their unique self-assembly into fibrous arrays. *J. Phys. Chem. B* **2006**, *110*, 18115–18120. [CrossRef]
19. Chu, C.-C.; Chiang, M.-L.; Tsai, C.-M.; Lin, J.-J. Exfoliation of montmorillonite clay by mannich polyamines with multiple quaternary salts. *Macromolecules* **2005**, *38*, 6240–6243. [CrossRef]
20. Dong, R.-X.; Chou, C.-C.; Lin, J.-J. Synthesis of immobilized silver nanoparticles on ionic silicate clay and observed low-temperature melting. *J. Mater. Chem.* **2009**, *19*, 2184–2188. [CrossRef]
21. Su, H.-L.; Lin, S.-H.; Wei, J.-C.; Pao, I.C.; Chiao, S.-H.; Huang, C.-C.; Lin, S.-Z.; Lin, J.-J. Novel nanohybrids of silver particles on clay platelets for inhibiting silver-resistant bacteria. *PLoS ONE* **2011**, *6*, e21125. [CrossRef] [PubMed]
22. Chu, C.-Y.; Peng, F.-C.; Chiu, Y.-F.; Lee, H.-C.; Chen, C.-W.; Wei, J.-C.; Lin, J.-J. Nanohybrids of silver particles immobilized on silicate platelet for infected wound healing. *PLoS ONE* **2012**, *7*, e38360. [CrossRef]
23. Janaki, S.; Sailatha, E.; Gunasekaran, S. Synthesis, characteristics and antimicrobial activity of ZnO nanoparticles. *Spectrochim. Acta Part A Mol. Biomol. Spectrosc.* **2015**, *144*, 17–22. [CrossRef] [PubMed]

24. Wahid, F.; Yin, J.-J.; Xue, D.-D.; Xue, H.; Lu, Y.-S.; Zhong, C.; Chu, L.-Q. Synthesis and characterization of antibacterial carboxymethyl Chitosan/ZnO nanocomposite hydrogels. *Int. J. Biol. Macromol.* **2016**, *88*, 273–279. [CrossRef] [PubMed]
25. Sirelkhatim, A.; Mahmud, S.; Seeni, A.; Kaus, N.H.M.; Ann, L.C.; Bakhori, S.K.M.; Hasan, H.; Mohamad, D. Review on zinc oxide nanoparticles: Antibacterial activity and toxicity mechanism. *Nano-Micro Lett.* **2015**, *7*, 219–242. [CrossRef]
26. Cao, N.; Pegoraro, M.; Severini, F.; Di Landro, L.; Zoia, G.; Greco, A. Crosslinked polycarbonate polyurethanes: Preparation and physical properties. *Polymer* **1992**, *33*, 1384–1390. [CrossRef]
27. Naik, P.U.; Refes, K.; Sadaka, F.; Brachais, C.-H.; Boni, G.; Couvercelle, J.-P.; Picquet, M.; Plasseraud, L. Organo-catalyzed synthesis of aliphatic polycarbonates in solvent-free conditions. *Polym. Chem.* **2012**, *3*, 1475–1480. [CrossRef]
28. Khan, I.; Smith, N.; Jones, E.; Finch, D.; Cameron, R. Analysis and evaluation of a biomedical polycarbonate urethane tested in an in vitro study and an ovine arthroplasty model. Part II: In vivo investigation. *Biomaterials* **2005**, *26*, 633–643. [CrossRef]
29. Zhu, W.; Wang, Y.; Sun, S.; Zhang, Q.; Li, X.; Shen, Z. Facile synthesis and characterization of biodegradable antimicrobial poly(ester-carbonate). *J. Mater. Chem.* **2012**, *22*, 11785–11791. [CrossRef]
30. Rueggeberg, F.A. Determination of resin cure using infrared analysis without an internal standard. *Dent. Mater.* **1994**, *10*, 282–286. [CrossRef]
31. Almeida, J.F.A.; Gomes, B.P.F.A.; Ferraz, C.C.R.; Souza-Filho, F.J.; Zaia, A.A. Filling of artificial lateral canals and microleakage and flow of five endodontic sealers. *Int. Endod. J.* **2007**, *40*, 692–699. [CrossRef] [PubMed]
32. Zhou, H.-m.; Shen, Y.; Zheng, W.; Li, L.; Zheng, Y.-f.; Haapasalo, M. Physical properties of 5 root canal sealers. *J. Endod.* **2013**, *39*, 1281–1286. [CrossRef] [PubMed]
33. Goracci, C.; Cadenaro, M.; Fontanive, L.; Giangrosso, G.; Juloski, J.; Vichi, A.; Ferrari, M. Polymerization efficiency and flexural strength of low-stress restorative composites. *Dent. Mater.* **2014**, *30*, 688–694. [CrossRef] [PubMed]
34. Schmalz, G. Concepts in biocompatibility testing of dental restorative materials. *Clin. Oral Investig.* **1998**, *1*, 154–162. [CrossRef] [PubMed]
35. Shin, T.P.; Yao, X.; Huenergardt, R.; Walker, M.P.; Wang, Y. Morphological and chemical characterization of bonding hydrophobic adhesive to dentin using ethanol wet bonding technique. *Dent. Mater.* **2009**, *25*, 1050–1057. [CrossRef] [PubMed]
36. Wu, C.-H.; Lin, Y.-R.; Yeh, S.-C.; Huang, Y.-C.; Sun, K.-H.; Shih, Y.-F.; Su, W.-C.; Dai, C.-A.; Dai, S.A.; Jeng, R.-J. A facile synthetic route to ether diols derived from 1,1-cyclopentylenylbisphenol for robust cardo-type polyurethanes. *Macromolecules* **2019**, *52*, 959–967. [CrossRef]
37. Wu, C.-H.; Chen, L.-Y.; Jeng, R.-J.; Dai, S.A. 100% Atom-economy efficiency of recycling polycarbonate into versatile intermediates. *ACS Sustain. Chem. Eng.* **2018**, *6*, 8964–8975. [CrossRef]
38. Nabinejad, O.; Sujan, D.; Rahman, M.E.; Davies, I.J. Determination of filler content for natural filler polymer composite by thermogravimetric analysis. *J. Therm. Anal. Calorim.* **2015**, *122*, 227–233. [CrossRef]
39. Hajipour, M.J.; Fromm, K.M.; Akbar Ashkarran, A.; Jimenez de Aberasturi, D.; Larramendi, I.R.d.; Rojo, T.; Serpooshan, V.; Parak, W.J.; Mahmoudi, M. Antibacterial properties of nanoparticles. *Trends Biotechnol.* **2012**, *30*, 499–511. [CrossRef]
40. Mustapha, S.; Ndamitso, M.M.; Abdulkareem, A.S.; Tijani, J.O.; Shuaib, D.T.; Ajala, A.O.; Mohammed, A.K. Application of TiO_2 and ZnO nanoparticles immobilized on clay in wastewater treatment: A review. *Appl. Water Sci.* **2020**, *10*, 1–36. [CrossRef]
41. Chen, L.; Qin, Y.; Wang, X.; Zhao, X.; Wang, F. Plasticizing while toughening and reinforcing poly(propylene carbonate) using low molecular weight urethane: Role of hydrogen-bonding interaction. *Polymer* **2011**, *52*, 4873–4880. [CrossRef]
42. Ma, X.; Yu, J.; Wang, N. Compatibility characterization of poly(lactic acid)/poly(propylene carbonate) blends. *J. Polym. Sci. Part B Polym. Phys.* **2006**, *44*, 94–101. [CrossRef]
43. Xu, H.; Qiu, F.; Wang, Y.; Wu, W.; Yang, D.; Guo, Q. UV-curable waterborne polyurethane-acrylate: Preparation, characterization and properties. *Prog. Org. Coat.* **2012**, *73*, 47–53. [CrossRef]
44. Balan, L.; Schneider, R.; Lougnot, D.J. A new and convenient route to polyacrylate/silver nanocomposites by light-induced cross-linking polymerization. *Prog. Org. Coat.* **2008**, *62*, 351–357. [CrossRef]

45. Gopikrishna, V.; Suresh Chandra, B. *Grossman's Endodontic Practice*, 13th ed.; Wolters Kluwer Health: Philadelphia, PA, USA, 2014.
46. Faccioni, F.; Franceschetti, P.; Cerpelloni, M.; Fracasso, M. In vivo study on metal ion release fixed orthodontic appliances and DNA damage in oral mucosa cells. *Am. J. Orthod. Dentofac. Orthop.* **2003**, *124*, 687–693. [CrossRef]
47. Eldeniz, A.; Mustafa, K.; Ørstavik, D.; Dahl, J. Cytotoxicity of new resin-, calcium hydroxide- and silicone-based root canal sealers on fibroblasts derived from human gingiva and L929 cell lines. *Int. Endod. J.* **2007**, *40*, 329–337. [CrossRef]
48. Al-Hiyasat, A.; Tayyar, M.; Darmani, H. Cytotoxicity evaluation of various resin based root canal sealers. *Int. Endod. J.* **2010**, *43*, 148–153. [CrossRef]
49. Mathur, A.B.; Collier, T.O.; Kao, W.J.; Wiggins, M.; Schubert, M.A.; Hiltner, A.; Anderson, J.M. In vivo biocompatibility and biostability of modified polyurethanes. *J. Biomed. Mater. Res.* **1997**, *36*, 246–257. [CrossRef]
50. Schubert, M.A.; Wiggins, M.J.; Schaefer, M.P.; Hiltner, A.; Anderson, J.M. Oxidative biodegradation mechanisms of biaxially strained poly(etherurethane urea) elastomers. *J. Biomed. Mater. Res.* **1995**, *29*, 337–347. [CrossRef]
51. Wei, J.-C.; Yen, Y.-T.; Su, H.-L.; Lin, J.-J. Inhibition of bacterial growth by the exfoliated clays and observation of physical capturing mechanism. *J. Phys. Chem. C* **2011**, *115*, 18770–18775. [CrossRef]

© 2020 by the authors. Licensee MDPI, Basel, Switzerland. This article is an open access article distributed under the terms and conditions of the Creative Commons Attribution (CC BY) license (http://creativecommons.org/licenses/by/4.0/).

Article

Effect of Type and Concentration of Nanoclay on the Mechanical and Physicochemical Properties of Bis-GMA/TTEGDMA Dental Resins

J. J. Encalada-Alayola [1], Y. Veranes-Pantoja [2], J. A. Uribe-Calderón [1], J. V. Cauich-Rodríguez [1] and J. M. Cervantes-Uc [1,*]

1. Centro de Investigación Científica de Yucatán, A.C. Unidad de Materiales, Calle 43 No. 130 x 32 y 34, Col. Chuburná de Hidalgo C.P. Mérida 97205, Mexico; juan.encalada@cicy.mx (J.J.E.-A.); jorge.uribe@cicy.mx (J.A.U.-C.); jvcr@cicy.mx (J.V.C.-R.)
2. Centro de Biomateriales, Universidad de La Habana, Avenida Universidad, s/n, e/G y Ronda, C.P. La Habana 10600, Cuba; yayma@biomat.uh.cu
* Correspondence: manceruc@cicy.mx; Tel.: +52-999-981-3966

Received: 7 February 2020; Accepted: 3 March 2020; Published: 6 March 2020

Abstract: Bis-GMA/TTEGDMA-based resin composites were prepared with two different types of nanoclays: an organically modified laminar clay (Cloisite® 30B, montmorillonite, MMT) and a microfibrous clay (palygorskite, PLG). Their physicochemical and mechanical properties were then determined. Both MMT and PLG nanoclays were added into monomer mixture (1:1 ratio) at different loading levels (0, 2, 4, 6, 8 and 10 wt.%), and the resulting composites were characterized by scanning electron microscopy (SEM), thermogravimetric analysis (TGA), dynamic mechanical analysis (DMA) and mechanical testing (bending and compressive properties). Thermal properties, depth of cure and water absorption were not greatly affected by the type of nanoclay, while the mechanical properties of dental resin composites depended on both the variety and concentration of nanoclay. In this regard, composites containing MMT displayed higher mechanical strength (both flexural and compression) than those resins prepared with PLG due to a poor nanoclay dispersion as revealed by SEM. Solubility of the composites was dependent not only on nanoclay-type but also the mineral concentration. Dental composites fulfilled the minimum depth cure and solubility criteria set by the ISO 4049 standard. In contrast, the minimum bending strength (50 MPa) established by the international standard was only satisfied by the dental resins containing MMT. Based on these results, composites containing either MMT or PLG (at low filler contents) are potentially suitable for use in dental restorative resins, although those prepared with MMT displayed better results.

Keywords: dental resin composite; montmorillonite; palygorskite

1. Introduction

Dental resin-based composites (RBCs) for both direct and indirect dental restorations have been in use for the past 50 years [1]. These materials are composed of inorganic particles (conferring most of the mechanical properties to the final material), which are embedded into an organic matrix (consisting of a mixture of crosslinking monomers, a photoinitiator system and other additives, which form a dense cross-linked polymer upon a free-radical copolymerization) [1,2].

Bisphenol A glycidyl methacrylate (Bis-GMA) is the base-monomer most frequently used in the formulations of dental restorative materials since it reduces polymerization shrinkage and enhances both modulus and thermal stability of the resulting materials; however, it also exhibits a high viscosity, which yields usually a heterogeneous material and problems during handling and application of the product. Thus, in order to achieve high filler loading in dental resin composites, low-viscosity diluent monomers such as triethylene glycol dimethacrylate (TEGDMA) are commonly used [3–5]. Other monomers have been also used to improve specific properties [1].

Monomers used in the commercial formulations of dental composites have remained largely unchanged, whereas the type, shape, size and distribution of inorganic fillers have undergone significant changes [1]. Recently, the incorporation of filler particles with nanometric dimensions into dental resins has attracted great attention, as it is possible to obtain materials with improved properties with it. Wear resistance, gloss retention, elastic modulus, flexural strength, diametral tensile strength and reduced polymerization shrinkage have been improved by the addition of nanoparticles to dental composites [5,6].

It is worth mentioning that these improvements are generally achieved through addition of a small amount of nanoclay (contrary to what has been observed in conventional inorganic macro- and micro-fillers as they constitute up to 75–85 wt.%). This fact is due to the extremely large interface area provided by the nano-size particles of the suspended filler [7–9].

At present, there is an increasing interest in the incorporation of nanoclays into dental composites. Montmorillonite (MMT) is the most commonly used clay, both in its natural form and organically modified. Its use has been reported in Cloisite® Na+ [6,7,9]; Cloisite® 10A [10]; Cloisite® 20A [6] and Cloisite® 93A [7], although Cloisite® 30B is the organoclay most employed [5–7,11].

In contrast, there are few studies in the literature involving the use of alternative clay minerals. Tian et al. and Zhang et al. reported the used of palygorskite (attapulgite) in dental resins, while Weidenbach et al. studied materials containing functional halloysite–nanotube filler [12–14].

Therefore, the aim of this work is to investigate the influence of nanoclay type at different loading levels (0, 2, 4, 6, 8 and 10 wt.%) on properties of the dental resins composites. Tetraethylene glycol dimethacrylate (TTEGDMA) was used as co-monomer instead of triethylene glycol dimethacrylate (TEGDMA) monomer, as Rüttermann et al. suggested that viscosities of both monomers are similar but the molecular mass of TTEGDMA is higher than TEGDMA; in consequence, it was expected to be more advantageous regarding shrinkage [15]. The TTEGDMA is often used not only in commercial resins but also in experimental dental restorative resins [15,16].

2. Materials and Methods

2.1. Materials

The monomers used for the preparation of the dental composites were Bisphenol A glycidyl methacrylate (Bis-GMA) and tetraethylene glycol dimethacrylate (TTEGDMA). Camphorquinone (CQ) and N,N-dimethyl aminoethyl methacrylate (DMAEMA) were used as photo-initiator and co-initiator, respectively. All reagents were purchased from Sigma–Aldrich Co. (Milwaukee, WI, USA) and used as received without further purification. The clay minerals used in this study were a commercial montmorillonite (Cloisite® 30B, abbreviated as MMT) modified with a quaternary ammonium salt (90 meq/100 g of clay) from Southern Clay Products (Gonzáles, TX, USA) and an HCl purified palygorskite (PLG) extracted from a mineral deposit in Chapab, Yucatán, México.

2.2. Preparation of Dental Composite

Bis-GMA was mixed manually with TTEGDMA in a glass container at 50:50 wt.% ratio. The photo-initiator CQ and the tertiary amine DMAEMA were then added to the Bis-GMA/TTEGDMA mixture (both materials at 0.5 wt.% to the monomer mixture). The container was covered with aluminum foil (to avoid premature curing) and refrigerated until its use. Monomers and initiator system were manually mixed, and the nanoclay (MMT or PLG) was slowly added at different loading levels in small portions to avoid the formation of agglomerates; formulations were mixed until no filler agglomerations were visually observed in the monomer mixture. Resin composites were cured with a light emitting diode unit (LED H, Woodpecker, Guilin, China) with a wavelength range of 420–480 nm and a light intensity of 1000 mW/cm². Table 1 summarizes the dental resin composites prepared in this study.

Table 1. Composition of dental composites.

Nanoclay Content (wt.%)	Composites	
	MMT	PLG
0	Unfilled	
2	MMT-2	PLG-2
4	MMT-4	PLG-4
6	MMT-6	PLG-6
8	MMT-8	PLG-8
10	MMT-10	PLG-10

2.3. Characterization of Dental Composites

2.3.1. Fourier Transform Infrared Spectroscopy (FTIR)

Infrared spectra were obtained with a Thermo Scientific Nicolet 8700 spectrometer (Madison, WI, USA) by the attenuated total reflectance (ATR) technique using a germanium crystal. Samples were analyzed in the 4000 to 650 cm^{-1} wavenumber range with a resolution of 4 cm^{-1} and averaging 100 scans.

2.3.2. Thermogravimetric Analysis (TGA)

Thermogravimetric analysis was performed in a Perkin Elmer TGA-7 thermogravimetric analyzer (Norwalk, CT, USA), from 45 to 750 °C at a heating rate of 10 °C/min, under nitrogen atmosphere. From the first derivative curve, the decomposition temperature of composites was determined.

2.3.3. Dynamic Mechanical Analysis (DMA)

The glass transition temperature (Tg) of composites was determined by dynamic mechanical analysis using a Perkin Elmer DMA-7 (Norwalk, CT, USA) in bending mode. Bars of 30 × 10 × 0.5 mm were heated from 35 to 200 °C at 3 °C/min, under nitrogen atmosphere, using a frequency of 1 Hz.

2.3.4. Mechanical Properties

An overlapping irradiation regime was applied to photo-polymerize the specimens for mechanical properties dependent on size and shape; five overlapped irradiations were employed on both sizes (40 s each irradiation; 400 s in total) to cure bending specimens, while one irradiation was used on both sizes (40 s each irradiation; 80 s in total) in compression specimens. All photo-cured specimens were immersed in distilled water at 37 ± 1 °C until mechanical testing (7 days).

In order to establish the influence of both nanoclay-type and filler content on the mechanical properties of dental resins, flexural tests were carried out according to the ISO 4049 standard [17] while compressive tests were performed according to the ASTM D695 standard [18]. For 3-point bending tests, five rectangular samples (25 mm length, 2 mm width and 2 mm thickness) were tested at 1 mm/min in a Shimadzu Autograph AGS-X (Kyoto, Japan) Universal Testing Machine, using a load cell of 1 kN and a distance between supports of 20 mm. For compression tests, five cylindrical specimens (8mm height and 4mm diameter) were deformed at 1 mm/min in a Shimadzu Autograph AG-I (Kyoto, Japan) Universal Testing Machine employing a load-cell of 5 kN.

Flexural strength (σ, MPa) and modulus (E, MPa) were calculated according to Equations (1) and (2), respectively.

$$\sigma = \frac{3Fl}{2bh^2} \tag{1}$$

$$E = \frac{Fl^3}{4bh^3d} \tag{2}$$

where F is the maximum load recorded (N), l is the span between the supports (mm), b is the width of the specimen (mm), h is the height of the specimen (mm) and d is the deflection at load F (mm).

Compressive strength (σ, MPa) was determined according to Equation (3) and the elastic modulus (E, MPa) was calculated as the slope of the elastic part of the stress–strain curve.

$$\sigma = \frac{4F}{\pi D^2} \tag{3}$$

where F is the maximum load recorded (N) and D is the diameter of the specimen (mm).

Fracture surfaces from flexural test samples were analyzed by Scanning Electron Microscopy (SEM) using a JEOL JSM-6360 LV (Akishima, Tokyo, Japan) operated at 20 kV. Samples were adhered on aluminum cylinders using a double-sided tape of copper and coated with a thin layer of gold in a Denton Vacuum Desk-II (Moorestown, NJ, USA) sputter coater system for 1 min prior to examination.

2.3.5. Depth of Cure

Depth of cure (Dc, mm) tests were performed on cylindrical specimens (6 mm height; 4 mm diameter) according to Section 7.10 of the ISO 4049 standard [17]. Three samples of each composite were irradiated during 40 s on one side, and the Dc was calculated according to Equation (4)

$$Dc = \frac{l}{2} \tag{4}$$

where l is the height of the specimen (mm) after removing the uncured material.

2.3.6. Sorption and Solubility

For water sorption and solubility tests, disc-shaped specimens (1 mm thickness; 15 mm diameter) were used according to Section 7.12 of the ISO 4049 standard [17]. Nine overlapped irradiations of 40 s were applied to specimens on one side (360 s in total). Water sorption (Wsp, µg/mm^3) and solubility (Wsl, µg/mm^3) were calculated according to Equations (5) and (6), respectively.

$$Wsp = \frac{m_2 - m_1}{V} \tag{5}$$

$$Wsl = \frac{m_1 - m_3}{V} \tag{6}$$

where m_1 is the mass (µg) of the conditioned specimen, m_2 is the mass of the specimen (µg) after immersion in distilled water at 37 ± 1 °C for 7 days, m_3 is the mass of the reconditioned specimen (µg) and V is the volume of the specimen (mm3).

2.4. Statistical Analysis

One-way analysis of variance (ANOVA) and Tukey's test ($P < 0.05$) were used to determine significant differences between properties of dental resin composites prepared with either MMT or PLG.

3. Results and Discussion

3.1. Fourier Transform Infrared Spectroscopy (FTIR)

FTIR spectra of neat resin and its nanocomposites prepared with either MMT or PLG are shown in Figure 1a,b, respectively. As noted, spectra of nanocomposites were very similar to those obtained for pure resin, probably due to the low filler content. Thus, a broad band was observed at 3344 cm^{-1}, attributed to O-H stretching vibration of hydroxyls in Bis-GMA structure; bands at 2951 and 2877 cm^{-1} related to asymmetric and symmetric stretching vibrations of the methylene group, and an intense band at 1723 cm^{-1} owing to C=O stretching of ester groups from dimethacrylates (Bis-GMA and TTEGDMA). Spectra also showed bands at 1640 and 1608 cm^{-1} which correspond to the stretching vibration of the aliphatic (from vinyl group of monomers) and aromatic (from benzene ring) C=C bonds, respectively [19]; in fact, the height ratio of these bands is generally used to determine the degree of conversion of dental resins [20]. Finally, an intense band around 1130 cm^{-1} was also detected and associated with symmetric vibration of C-O-C linkage from TTEGDMA structure.

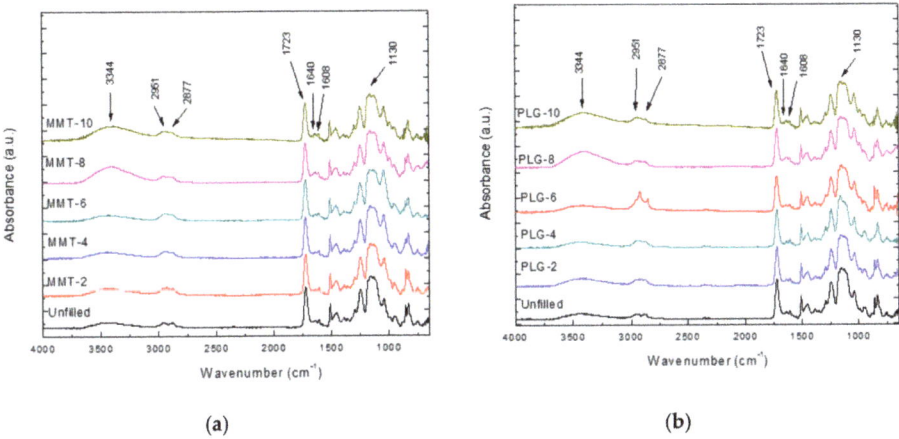

Figure 1. FTIR spectra of nanocomposites prepared with either (**a**) montmorillonite (MMT) or (**b**) palygorskite (PLG).

3.2. Thermogravimetric Analysis (TGA)

Figure 2 shows the DTGA (First derivative of the TGA curve) curves for pure resin, obtained from Bis-GMA and TTEGDMA, and its composites prepared with either MMT (Figure 2a) or PLG (Figure 2b). As can be seen, pure resin presented two well-defined degradation stages (Td) at 380 and 435 °C, although a broad transition at lower temperatures (320 °C) was also observed. Teshima et al. studied the thermal degradation of resins prepared from Bis-GMA and TEGDMA and detected three degradation steps during thermal decomposition of this material. They found that methacrylic acid and 2-hydroxyethyl methacrylate are released during the first and second stages, whereas propionic acid and phenol are produced in the final stage [21].

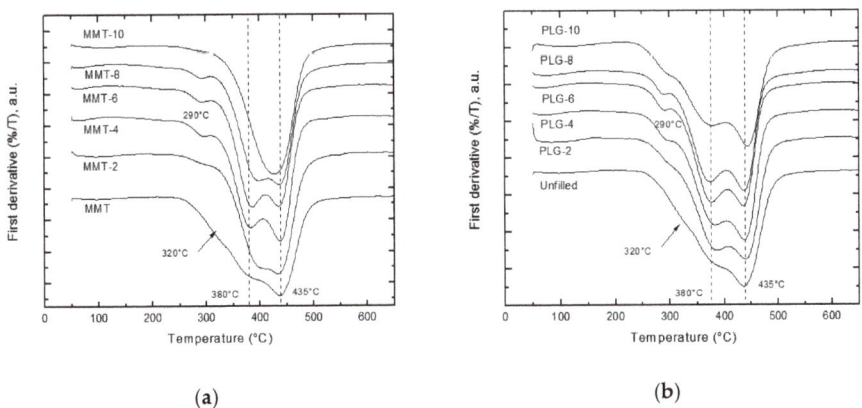

Figure 2. DTGA curves for nanocomposites prepared with either (**a**) MMT or (**b**) PLG.

On the other hand, it has been suggested that addition of nanoclays into polymeric matrices could improve the thermal stability of the resulting material [22]. This fact was not observed in nanocomposites prepared in this study; i.e., temperatures observed for dental resin composites were practically similar to those obtained for neat resin. The latter is in agreement with Munhoz et al., who pointed out that it was not possible to identify an impact of the presence of the modified clay on the thermal stability of the composites [23]. However, a close inspection of thermograms allowed detection of small changes in thermal degradation behavior in some composites. For instance, the degradation stage at 380 °C in some MMT nanocomposites was shifted to higher temperatures overlapping with that of 435 °C in composites containing 10 wt.%. This could be associated with interaction between surfactant and crosslinking polymer structures. Mahnoodian et al. studied the thermal degradation of Bis-GMA/TEGDMA/Cloisite 30B nanocomposites and found that the pure resin exhibited two decomposition stages at 292 and 392 °C. The first temperature did not change when MMT was added to resins, although temperature corresponding to the second decomposition step was reduced with increasing clay content [7].

Finally, it should be also mentioned that the broad transition observed at 320 °C was shifted to lower temperatures (290 °C) in nanocomposites, being more evident when nanoclay content was increased. This event could be related to the emission of volatiles such as water and fragments of organic modifier from PLG and MMT nanoclays respectively.

3.3. Dynamic Mechanical Analysis (DMA)

Glass transition temperatures of dental composites were obtained from the maximum of the tan d versus temperature curves and their maxima are summarized in Table 2. Thermograms exhibited a one broad peak at around 110 °C, which suggests a homogeneous polymeric network; this value is in agreement with that reported by Terrin et al. for dental resins composed mainly of Bis-GMA and TEGDMA [9]. Composites containing PLG showed slightly higher values than those obtained for the corresponding MMT composites. It is also noted that dental resins prepared with PLG exhibited an increase in this parameter as nanoclay content was increased. In contrast, when MMT was added to dental resin formulations, values decreased slightly from 110 to 108 °C returning to higher Tg values (112 °C) at nanoclay contents of 8 and 10 wt.%; a similar trend was reported by Terrin et al., who studied resin-based composites with organically modified MMT [9]. The shift of the relaxation temperature in composites containing nanoclays towards higher temperatures is attributed to immobilized polymeric chains at the polymer/filler interphase [23].

Table 2. Glass transition temperatures (Tg) of dental composites.

Nanoclay Content (%)	Tg (°C)	
	MMT	PLG
0		110
2	108	111
4	110	112
6	110	116
8	112	116
10	112	116

3.4. Mechanical Properties

Figures 3 and 4 show the flexural and compressive properties of the nanocomposites prepared in this study, respectively. As can be seen, the mechanical properties of dental resin composites varied depending on the type and concentration of the inorganic aggregate.

Regarding flexural properties (Figure 3), it was observed that strength decreased monotonically with the MMT loading; on the other hand, samples containing PLG exhibited a drastic decrease of strength when the additive was added (i.e., 2 wt.%) and remained almost constant at higher filler concentration. It is worth mentioning that the minimum flexural strength set by the ISO 4049 standard [17] (50 MPa) was fulfilled by all composites containing MMT and that prepared with PLG at 2 wt.%. Typical stress–strain curves for nanocomposites prepared with either MMT or PLG are presented in Figure 3c,d, respectively.

The flexural modulus tended to increase with filler concentration regardless of type of nanoclay employed; MMT composites seemed to exhibit slight increments, but the differences were not statistically significant). PGL composites showed higher modulus with PLG up to 4 wt.% clay content, and no statistical differences were found for composites with higher clay content. Several authors [4,24,25] have also reported that flexural strength of nanocomposites decreased and modulus increased [8,10] as filler content increased.

It is well known that the dispersion of nanoparticles and compatibility between phases (filler and matrix) play a key role in the enhancement of mechanical properties of nanocomposites. Thus, it is probable that MMT presented better results than PLG as it was organically modified to improve its interaction with organic polymers and increase the interlayer spacing favoring its dispersion, as observed in the SEM analysis (see Figure 5).

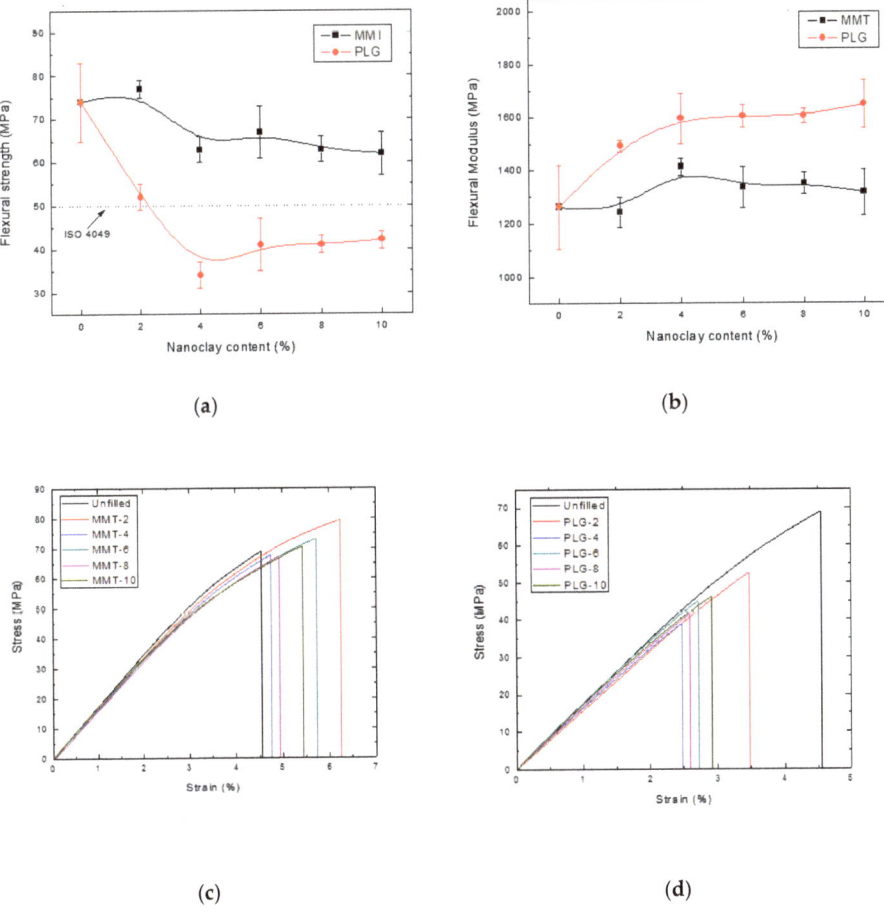

Figure 3. Flexural properties of dental resin composites: (**a**) Strength; (**b**) modulus; (**c**) stress–strain curves for nanocomposites prepared with MMT and (**d**) stress–strain curves for nanocomposites prepared with PLG.

Compressive properties (Figure 4) were not greatly affected with addition of nanoclay content, except compressive strength of materials containing PLG; these properties decreased with nanoclay concentration as reported by Mucci et al. [8]. Interestingly, the strength of composites containing MMT also showed higher values than corresponding PLG materials. Typical stress–strain curves for nanocomposites prepared with MMT or PLG are presented in Figure 4c,d, respectively.

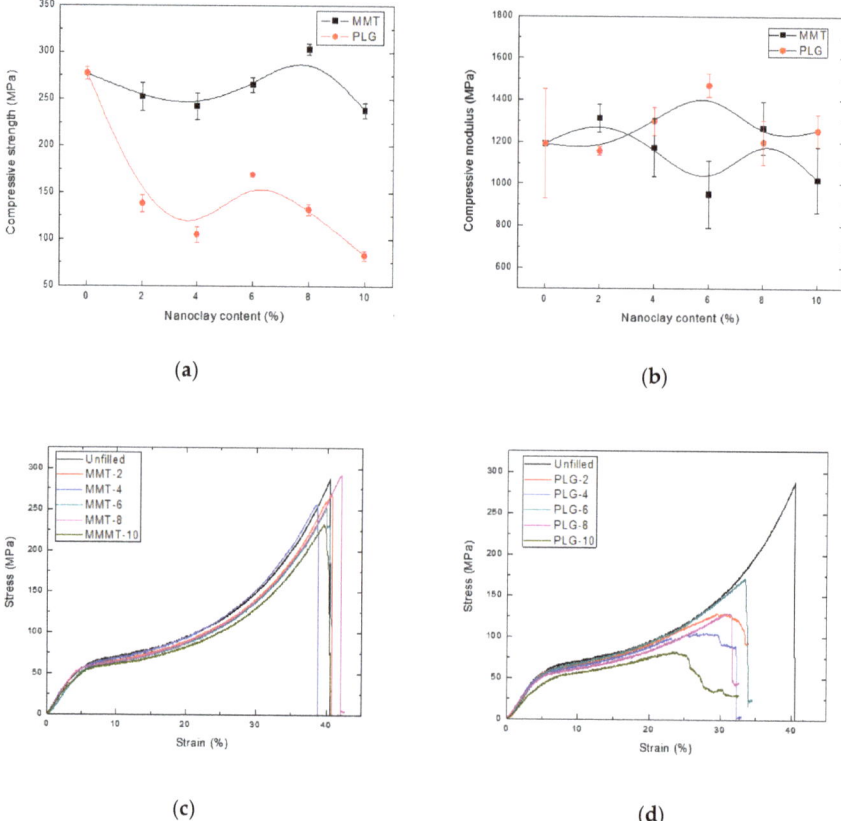

Figure 4. Compressive properties of dental resin composites: (**a**) Strength; (**b**) modulus; (**c**) stress–strain curves for nanocomposites prepared with MMT and (**d**) stress–strain curves for nanocomposites prepared with PLG.

Figure 5 shows the nanoclay dispersion within the dental composites. Unloaded dental resin exhibited a clear brittle fracture producing a smooth surface; the presence of nanoclays produced a different fracture mechanism generating a rough surface upon breaking. SEM observation at higher magnification revealed that nanoclays were dispersed differently within dental resin; MMT seems to be better distributed in the resin due to its surface modification than PLG, which was distributed as agglomerates.

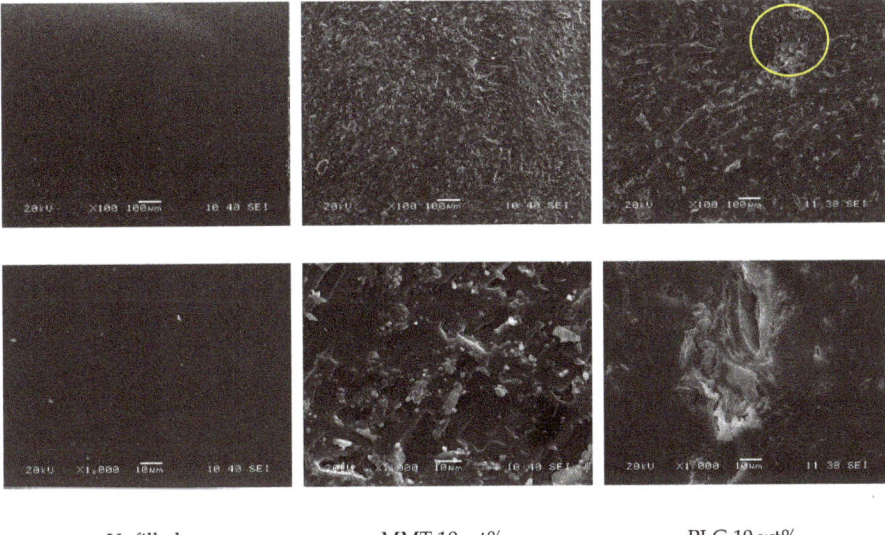

Figure 5. SEM micrographs of fracture surface from bending specimens.

3.5. Depth of Cure

Higher curing depth of restorative materials is one of the factors that determine the quality of the material [26]. Curing depth results obtained for dental composites prepared with either MMT or PLG are displayed in Table 3. It can be seen that the curing depth was not affected either by the nanoclay-type or filler content. Further, all composites exhibited higher values than that established in the ISO 4049 standard [17] (1.5 mm), and even better values than those reported by other authors [26].

Table 3. Depth of cure of dental composites.

Nanoclay Content (%)	Depth of cure (mm)	
	MMT	PLG
0	3.0 + 0.01	
2	2.99 + 0.01	2.99 + 0.02
4	2.99 + 0.01	2.99 + 0.01
6	2.99 + 0.02	2.99 + 0.02
8	2.99 + 0.03	2.99 + 0.02
10	2.99 + 0.04	2.99 + 0.05

3.6. Sorption and Solubility

The water sorption for neat resin and its composites prepared with either MMT or PLG at different contents are shown in Figure 6. As expected, results showed that clays induce water sorption in the composites (although this parameter seems not to be affected by the nanoclay type) as it is well known that natural clays have a hydrophilic character and are naturally prone to absorbing water [23].

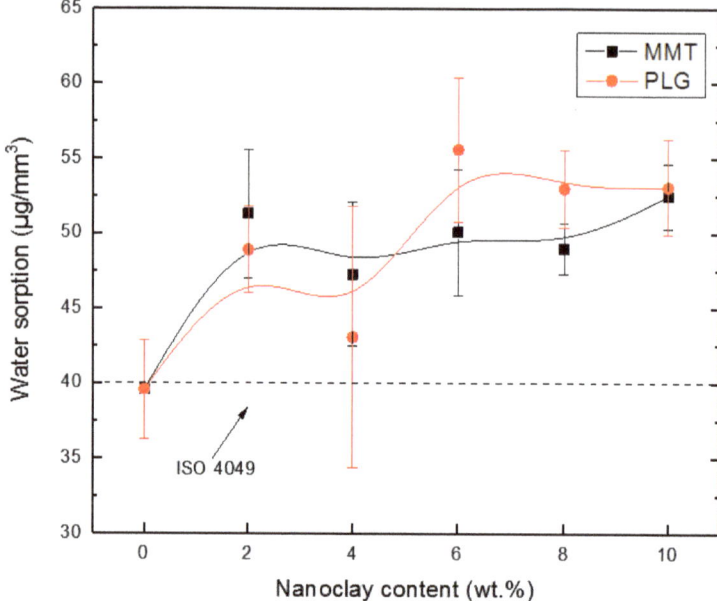

Figure 6. Water sorption of dental composites prepared with either MMT or PLG at several concentrations.

The hydrophilic behavior of composites depends on characteristics of constituents, i.e., organic matrix and inorganic filler. In this regard, unfilled resin exhibited a water absorption slightly below 40 mg/mm^3 (which is the maximum value for dental restorative materials stated by ISO 4049), whereas all nanocomposites exhibited water sorption values slightly higher.

Interestingly, composites containing PLG exhibited a similar water sorption behavior to that displayed by materials prepared with MMT, although the latter clay was modified organically by a cation exchange reaction between the silicate and methyl, tallow, bis-2-hydroxyethyl and quaternary ammonium chloride in order to reduce the clay hydrophilicity. This could be attributed to the presence of two hydroxyethyl groups in the organoclay as suggested by Mucci et al. [8].

Water uptake in dental resin composites occurs by diffusion of water molecules within a polymeric matrix and may cause hydrolytic degradation of the matrix and/or filler matrix interface [23], yielding leachable substances, which could be quantified in a solubility test.

Figure 7 shows the solubility results obtained from composites prepared with either MMT or PLG at several concentrations. It is clear that materials containing palygorskite exhibit a different behavior than that displayed by MMT composites. For instance, when PLG was added to dental resin, solubility decreased at lower nanoclay contents and then practically returned to the initial value at higher PLG concentrations; in contrast, solubility of composites containing MMT decreased monotonically with increasing nanoclay content. Regardless of the above fact, a statistically significant difference was only detected between nanocomposites containing 10 wt.% of clay. Further, it is interesting to note that solubility measurements for all composites (including an unfilled sample) remained below 7.5 mg/mm^3 as suggested by the ISO 4049 [17] standard for dentistry-polymer-based restorative materials.

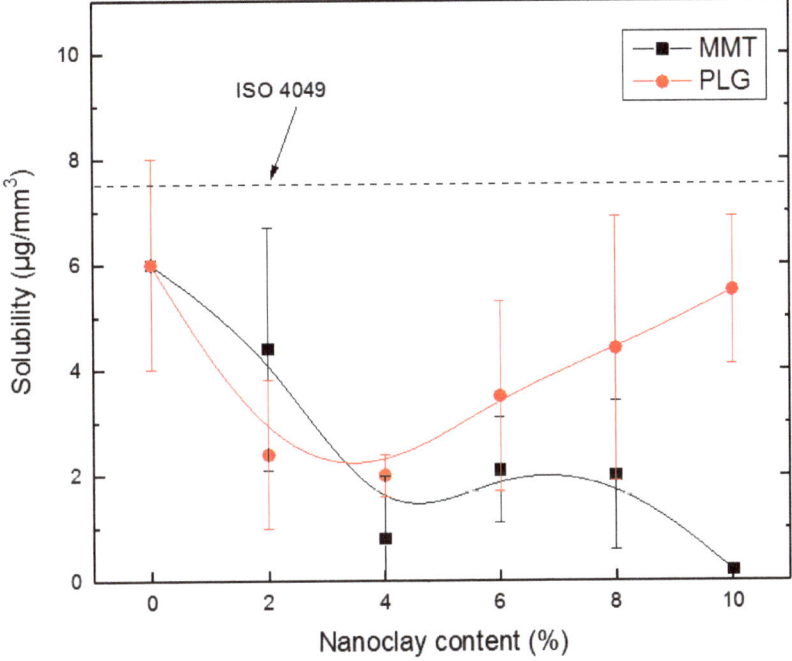

Figure 7. Solubility of dental composites prepared with either MMT or PLG at several concentrations.

4. Conclusions

Bis-GMA/TTEGDMA-based resin composites with two different types of nanoclays were successfully prepared and characterized. Results indicate that Tg, Td, depth of cure and water absorption were not greatly affected by the type of nanoclay, while the mechanical properties of dental resin composites depended on nanoclay type and concentration of inorganic filler. In general, MMT composites displayed higher mechanical strength than those shown by resins prepared with PLG, due to dispersion problems as revealed by SEM. Solubility of the composites was also dependent on nanoclay type and the mineral concentration. In general, dental composites prepared in this study fulfilled the minimum depth cure and solubility criteria set by the ISO 4049 standard. In contrast, the minimum bending strength (50 MPa) established by the international standard was only satisfied by dental resins containing MMT. Based on these results, composites containing either MMT or PLG (at low filler contents) are potentially suitable for use in dental restorative resins, although those prepared with MMT displayed better results.

Author Contributions: Conceptualization, J.M.C.-U., Y.V.-P. and J.J.E.-A.; Methodology, J.J.E.-A., Y.V.-P. and J.M.C.-U.; Validation, J.M.C.-U. and Y.V.-P.; Formal Analysis, J.J.E.-A. and J.M.C.-U; Investigation, J.J.E.-A.; Resources, J.M.C.-U., Y.V.-P., J.A.U.-C. and J.V.C.-R.; Writing-Original Draft Preparation, J.M.C.-U.; Writing-Review & Editing, J.J.E.-A., Y.V.-P., J.A.U.-C., J.V.C.-R.; Visualization, J.J.E.-A. and J.M.C.-U; Supervision, J.M.C.-U. and Y.V.-P.; Project Administration, J.M.C.-U.; Funding Acquisition, J.M.C.-U. All authors have read and agreed to the published version of the manuscript.

Funding: This work was supported by the Consejo Nacional de Ciencia y Tecnologia (CONACYT).

Acknowledgments: The authors thank Q.I. Santiago Duarte for the SEM micrographs, and W.H.K. for the physicochemical characterization.

Conflicts of Interest: The authors declare no conflict of interest. The funders had no role in the design of the study; in the collection, analyses, or interpretation of data; in the writing of the manuscript, or in the decision to publish the results.

References

1. Habib, E.; Wang, R.; Wang, Y.; Zhu, M.; Zhu, X.X. Inorganic Fillers for Dental Resin Composites: Present and Future. *ACS Biomater. Sci. Eng.* **2016**, *2*, 1–11. [CrossRef]
2. Moszner, N.; Salz, U. Recent Developments of New Components for Dental Adhesives and Composites. *Macromol. Mater. Eng.* **2007**, *292*, 245–271. [CrossRef]
3. Cramer, N.B.; Stansbury, J.W.; Bowman, C.N. Recent Advances and Developments in Composite Dental Restorative Materials. *J. Dent. Res.* **2011**, *90*, 402–416. [CrossRef]
4. Alsharif, S.O.; Akil, H.B.M.; El-Aziz, N.A.; Ahmad, Z.A. Effect of alumina particles loading on the mechanical properties of light-cured dental resin composites. *Mater. Des.* **2014**, *54*, 430–435. [CrossRef]
5. Campos, L.M.; Boaro, L.C.; Santos, T.M.; Marques, P.A.; Almeida, S.R.; Braga, R.R.; Parra, D.F. Evaluation of flexural modulus, flexural strength and degree of conversion in BISGMA/TEGDMA resin filled with montmorillonite nanoparticles. *J. Compos. Mater.* **2017**, *51*, 927–937. [CrossRef]
6. Discacciati, J.A.; Oréfice, R.L. Structural analysis on photopolymerized dental resins containing nanocomponents. *J. Mater. Sci.* **2007**, *42*, 3883–3893. [CrossRef]
7. Mahmoodian, M.; Pourabbas, B.; Arya, A.B. Preparation and Characterization of Bis-GMA/TEGDMA/Clay Nanocomposites at Low Filler Content Regimes. *J. Compos. Mater.* **2010**, *44*, 1379–1395. [CrossRef]
8. Mucci, V.; Pérez, J.; Vallo, C.I. Preparation and characterization of light-cured methacrylate/montmorillonite nanocomposites. *Polym. Int.* **2010**, *60*, 247–254. [CrossRef]
9. Terrin, M.M.; Poli, A.L.; Horn, M.A.; Neumann, M.G.; Cavalheiro, E.T.; Correa, I.C.; Schmitt, C.C. Effect of the loading of organomodified clays on the thermal and mechanical properties of a model dental resin. *Mat. Res.* **2016**, *19*, 40–44. [CrossRef]
10. Campos, L.M.; Boaro, L.C.; Ferreira, H.P.; Gomes dos Santos, L.K.; Ribeiro dos Santos, T.; Parra, D.F. Evaluation of polymerization shrinkage in dental restorative experimental composites based: BisGMA/TEGDMA, filled with MMT. *J. Appl. Polym. Sci.* **2016**, *133*, 1–10. [CrossRef]
11. Campos, L.M.; Lugao, A.B.; Vasconcelos, M.R.; Parra, D.F. Polymerization Shrinkage Evaluation on Nanoscale-Layered Silicates: Bis-GMA/TEGMA Nanocomposites, in Photo-Activated Polymeric Matrices. *J. Appl. Polym. Sci.* **2014**, *131*, 1–6.
12. Tian, M.; Gao, Y.; Liu, Y.; Liao, Y.; Hedin, N.E.; Fong, H. Fabrication and evaluation of Bis-GMA/TEGDMA dental resins/composites containing nano fibrillar silicate. *Dent. Mater.* **2008**, *24*, 235–243. [CrossRef] [PubMed]
13. Zhang, L.; Gao, Y.; Chen, Q.; Tian, M.; Fong, H. Bis-GMA/TEGDMA dental composites reinforced with nano-scaled single crystals of fibrillar silicate. *J. Mater. Sci.* **2010**, *45*, 2521–2524. [CrossRef]
14. Degrazia, F.W.; Leitune, V.C.B.; Takimi, A.S.; Collares, F.M.; Sauro, S. Physicochemical and bioactive properties of innovative resin-based materials containing functional halloysite-nanotubes fillers. *Dent. Mater.* **2016**, *32*, 1133–1143. [CrossRef]
15. Rüttermann, S.; Dluzhevskaya, I.; Großsteinbeck, C.; Raab, W.H.; Janda, R. Impact of replacing Bis-GMA and TEGDMA by other commercially available monomers on the properties of resin-based composites. *Dent. Mater.* **2010**, *26*, 353–359. [CrossRef]
16. Indrani, D.J.; Cook, W.D.; Televantos, F.; Tyas, M.J.; Harcourt, J.K. Fracture toughness of water-aged resin composite restorative materials. *Dent. Mater.* **1995**, *11*, 201–207. [CrossRef]
17. *ISO 4049:2009 Dentistry—Polymer-Based Restorative Materials*; ISO: Geneva, Switzerland, 2009.
18. *ASTM D 695—02a Standard Test Method for Compressive Properties of Rigid Plastic*; ASTM International: West Conshohocken, PA, USA, 2002.
19. Thorat, S.; Patra, N.; Ruffilli, R.; Diaspro, A.; Salerno, M. Preparation and characterization of a BisGMA-resin dental restorative composites with glass, silica and titania fillers. *Dent. Mater. J.* **2012**, *31*, 635–644. [CrossRef]
20. Collares, F.M.; Portella, F.F.; Leitune, V.C.; Samuel, S.M. Discrepancies in degree of conversión measurements by FTIR. *Braz. Oral Res.* **2013**, *28*, 453–454.
21. Teshima, W.; Nomura, Y.; Ikeda, A.; Kawahara, T.; Okazaki, M.; Nahara, Y. Thermal degradation of photo-polymerized BisGMA/TEGDMA-based dental resins. *Polym. Degrad. Stab.* **2004**, *84*, 167–172. [CrossRef]

22. Cervantes-Uc, J.M.; Cauich-Rodríguez, J.V.; Vázquez-Torres, H.; Garfias-Mesías, L.F.; Paul, D.R. Thermal degradation of commercially available organoclays studied by TGA–FTIR. *Thermochima Acta* **2007**, *457*, 92–102. [CrossRef]
23. Munhoz, T.; Fredholm, Y.; Rivory, P.; Balvay, S.; Hartmann, D.; Silva, P.; Chenal, J.M. Effect of nanoclay addition on physical, chemical, optical and biological properties of experimental dental resin composites. *Dent. Mater.* **2017**, *33*, 271–279. [CrossRef] [PubMed]
24. Menezes, L.R.; Silva, E.O. The Use of Montmorillonite Clays as Reinforcing Fillers for Dental Adhesives. *Mat. Res.* **2016**, *19*, 236–242. [CrossRef]
25. Menezes, L.R.; Silva, E.O.; Rocha, A.C.; Oliveira, D.C.; Campos, P.R. The applicability of organomodified nanoclays as new fillers for mechanical reinforcement of dental composites. *J. Compos. Mater.* **2018**, *52*, 963–970. [CrossRef]
26. Aydınoğlu, A.; Yoruç, A.B. Effects of silane-modified fillers on properties of dental composite resin. *Mater. Sci. Eng. C* **2017**, *79*, 382–389. [CrossRef] [PubMed]

© 2020 by the authors. Licensee MDPI, Basel, Switzerland. This article is an open access article distributed under the terms and conditions of the Creative Commons Attribution (CC BY) license (http://creativecommons.org/licenses/by/4.0/).

Article

Cartilage Tissue-Mimetic Pellets with Multifunctional Magnetic Hyaluronic Acid-Graft-Amphiphilic Gelatin Microcapsules for Chondrogenic Stimulation

Kai-Ting Hou [1], Ting-Yu Liu [2], Min-Yu Chiang [1], Chun-Yu Chen [3,4], Shwu-Jen Chang [3,*] and San-Yuan Chen [1,5,6,*]

1. Department of Materials Science and Engineering, National Chiao Tung University, Hsinchu City 30010, Taiwan; carolhou8@gmail.com (K.-T.H.); minyu28@gmail.com (M.-Y.C.)
2. Department of Materials Engineering, Ming Chi University of Technology, New Taipei City 24301, Taiwan; tyliu@mail.mcut.edu.tw
3. Department of Biomedical Engineering, I-Shou University, No.8, Yida Rd., Jiaosu Village, Yanchao District, Kaohsiung City 82445, Taiwan; iergy2000@gmail.com
4. Department of Orthopedics, Kaohsiung Veterans General Hospital, Kaohsiung City 81362, Taiwan
5. Frontier Research Centre on Fundamental and Applied Sciences of Matters, National Tsing Hua University, Hsinchu City 30013, Taiwan
6. School of Dentistry, College of Dental Medicine, Kaohsiung Medical University, Kaohsiung City 80708, Taiwan
* Correspondence: sjchang@isu.edu.tw (S.-J.C.); sanyuanchen@mail.nctu.edu.tw (S.-Y.C.); Tel.: +886-3573-1818 (S.-Y.C.)

Received: 31 January 2020; Accepted: 18 March 2020; Published: 2 April 2020

Abstract: Articular cartilage defect is a common disorder caused by sustained mechanical stress. Owing to its nature of avascular, cartilage had less reconstruction ability so there is always a need for other repair strategies. In this study, we proposed tissue-mimetic pellets composed of chondrocytes and hyaluronic acid-graft-amphiphilic gelatin microcapsules (HA-AGMCs) to serve as biomimetic chondrocyte extracellular matrix (ECM) environments. The multifunctional HA-AGMC with specific targeting on CD44 receptors provides excellent structural stability and demonstrates high cell viability even in the center of pellets after 14 days culture. Furthermore, with superparamagnetic iron oxide nanoparticles (SPIOs) in the microcapsule shell of HA-AGMCs, it not only showed sound cell guiding ability but also induced two physical stimulations of static magnetic field(S) and magnet-derived shear stress (MF) on chondrogenic regeneration. Cartilage tissue-specific gene expressions of Col II and SOX9 were upregulated in the present of HA-AGMC in the early stage, and HA-AGMC+MF+S held the highest chondrogenic commitments throughout the study. Additionally, cartilage tissue-mimetic pellets with magnetic stimulation can stimulate chondrogenesis and sGAG synthesis.

Keywords: cartilage tissue; amphiphilic gelatin microcapsules; tissue-mimetic pellets; magnetic stimulation; CD44 receptor

1. Introduction

Articular cartilage disorder most commonly occurs at the conjunction between bones, and the condition is progressively worsened by constant and unavoidable mechanical degeneration. The loss of the normal cartilage tissue will lead to more serious joint diseases like osteoarthritis [1]. Cartilage reconstruction has been a clinical issue for decades because of poor intrinsic ability to repair defects and lacking of specific diagnostic biomarkers. Several types of clinical treatments and pharmacologic therapies are available and effective in reducing pain and increasing mobility of patients [2–4].

Nevertheless, those treatments are merely a temporary relief and unable to restore the damaged tissue into its original function [5]. The long-term clinical solutions for cartilage repair are still in demand, and diverse regenerative therapies are consequently being brought to the table.

The hydrogel-based scaffolds have received a big interest as providing a temporary three-dimensional structure [6–9]. High functional, strong biomechanical properties and long-term biocompatible scaffolds can be made by regulating different materials to promote the cartilage tissue therapies [10,11]. Scaffolds which allow physical supporting often combines with primary chondrocytes therapies to overcome low cell proliferation or dedifferentiation problems. However, these 3D networks still struggle with several challenges such as cell viability, growth factor burst release or low oxygen content. On the other hand, particles serve as a superior medicine approach. Particles have been widely developed for drug delivery systems in tissue engineering applications due to their wide variety and highly regulated potential [12]. In particular, microscale particles were popular among delivering anti-inflammatory drugs or growth factors and hold promise in performing as building block in scaffold for cartilage repair [13,14]. Cruz et al. [15] aggregated gelatin microparticles and chondrocytes into 3D pellets and found higher cell viability over long periods.

As one of the major components in chondrocytes extracellular matrix (ECM), hyaluronic acid (HA) had a linear polysaccharide structure which functioned as a structural element, providing backbones for the component distribution and aggrecan aggregation. HA was also well known to interact with specific receptors including CD44 to regulate signal transduction, cell migration and differentiation [16,17]. Gelatin was a denatured protein from collagen and widely used in drug delivery systems. Because of gelatin's biocompatibility, low antigenicity, chemical modification possibility and low-cost [18], many studies had encapsulated growth factor in modified gelatin-based particles to stimulate chondrogenesis [8,19,20]. Moreover, gelatin-based microparticles as building blocks for three-dimensional (3D) structures have been wildly used in cartilage engineering [13].

In addition to chemical stimulations and highly cartilage-like materials, mechanical forces acting as additional tools were also applied to improve cartilage reconstruction in recent studies to mimic in vivo environments. The proper biophysical stimulations were able to increase proteoglycan synthesis and cartilage-specific gene expression [21,22]. Magnetic nanoparticles are thought of as excellent candidates to apply remote magnetic induced physical stimulation, which also holds the capability of targeting a specific site. Nathalie et al. [23] labeled mesenchymal stem cells (MSC) with magnetic nanoparticles to enhance seeding density and condensation in scaffold. Together with a dynamic bioreactor, MSC differentiation performance was markedly improved. Son et al. [24] exposed bone-marrow-derived human MSC (BM-hMSC) to static magnetic field and magnet-derived shear stress via magnetic nanoparticle. Without hypertrophic effect, biophysical stimulation led BM-hMSC to higher chondrogenic differentiation efficiency.

We developed a brand-new platform, cartilage tissue-mimetic pellets, to combine biochemical and biophysical treatments to mimic native cartilage tissue. As illustrated in Figure 1, cartilage tissue-mimetic pellets were composed by rabbit primary chondrocytes and hyaluronic acid-graft-amphiphilic gelatin microcapsules (HA-AGMCs). HA polymer chains on the microcapsules surface are expected to expand space between each microcapsule with its highly hydrophilic and polyanionic characteristics. In addition, HA can enhance chondrocytes attachment through CD44 receptors and act stabilize the pellets structure as ECM component at the beginning of formation. We encapsulated superparamagnetic iron oxide nanoparticles (SPIOs) in hydrophobic shells of HA-AGMCs to guide cells and serve physical stimulations by applying static magnetic field and magnet-derived shear stress. The inner hydrophilic space of microcapsules is capable of encapsulating growth factor or biomolecules for cell proliferation or repair. Such HA-AGMC approaches can be used to provide cartilage structure stability, rule biochemical and biophysical stimulations, and thereby promote faster and more complete cartilage reconstruction.

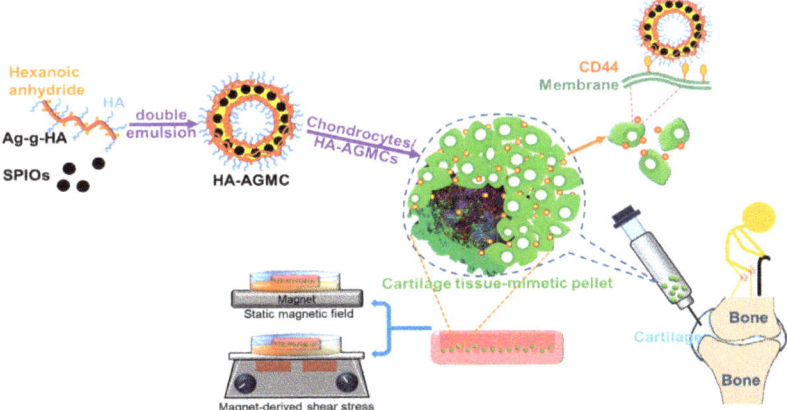

Figure 1. Synthesis of the hyaluronic acid-graft-amphiphilic gelatin microcapsules (HA-AGMCs) structure to fabricate cartilage tissue-mimetic pellets with combined biochemical and biophysical treatments.

2. Materials and Methods

2.1. Materials

Gelatin from porcine skin (type A 300 bloom), hexanoic anhydride, absolute ethanol (99.5%), sodium hydroxide, 2,4,6-Trinitrobenzene Sulfonic Acid (TNBS), 1,2-hexadecanediol (97%), oleic acid (90%), oleylamine (>70%), and iron(III) acetylacetonate (Fe(acac)3) were purchased from Sigma-Aldrich Co. (St. Louis, MO, USA) Hyaluronic acid (Hyalo-Oligo) was purchased from Tannmer Enterprise Co., Ltd. (New Taipei City, Taiwan). N-hydroxysuccinimide (NHS) and 1-ethyl-3-(3-dimethylaminopropyl) carbodiimide hydrochloride (EDC) were purchased from Echo Chemical Co., Ltd. (Miaoli, Taiwan). Platinum® PCR SuperMix and GScript First-Strand Synthesis Kit were purchased from Thermo Fisher Scientific (Waltham, MA, USA).

2.2. Synthesis of Hyaluronic Acid-Graft-Amphiphilic Gelatin (AG-g-HA)

Gelatin powder (3 g) was completely dissolved in deionized water (40 mL) at 70 °C with constant stirring for an hour. Eethanol (95%, 30 mL) and hexanoic anhydride (3 mL) were sequentially added dropwise in succession and stirred for 4 h. In this process, the amphiphilic gelatin (AG) form and was tuned to a pH value around 7. The final solution was dialyzed against a mixture of water and ethanol (3:4). AG was collected and dried at 60 °C and then ground into powder with grinding machine. Hyaluronic acid (HA) was grafted to the amino group of the gelatin via EDC/NHS method. HA (1 g) was first dissolved in phosphate buffer solution (40 mL) under stirring. 1-ethyl-3-(-3-dimethylaminopropyl) carbodiimide hydrochloride (EDC, 1 g) and N-hydroxysuccinimide (NHS, 1.9 g) were added to the solution and stirred for one hour at pH 5.5. After the pH was adjusted to 7 with NaOH (10 N), AG (1 g) was added directly and stirred for 2 h. After the reaction, AG-g-HA formed and dialyzed (Spectra/Por, MWCO = 20000) against water before dried by a freeze vacuum dryer. AG-g-HA was analyzed by nuclear magnetic resonance spectroscopy (VARIAN, UNIYTINOVA 500 NMR) with 600-MHz 1H-NMR. Each copolymer (10 wt %) was dissolved in D2 O to obtain 1 H NMR and 13 C NMR spectra.

2.3. Quantification of HA Grafting Rate

We used 2,4,6-Trinitrobenzene Sulfonic Acid (TNBS) reagent to determine the content of free amino groups. TNBS (0.025% w/v) was dissolved in sodium hydrogen carbonate solution (4%) serving as reaction buffer. Gelatin, AG, and AG-g-HA were dissolved in dH2O (1 mL), and mixed with reaction buffer (0.5 mL) separately. Calibration curve was made by mixing Lysine (2–20 μg in 1 mL dH2O) with reaction buffer (0.5 mL). After incubating at 37 °C for two hours, sodium dodecyl sulfate (SDS, 10% *w/v*) and 1 N HCl were added to each sample. An absorbance peak at 336 nm was measured with UV-vis spectroscopy (UV-vis, Evolution 300, Thermo, Waltham, MA, USA). The number of free amines contained in each polymer was quantified by correlating with the calibration curve.

2.4. Synthesis of Superparamagnetic Iron Oxide Nanoparticles (SPIOs)

The synthesis of SPIOs (6~8 nm) were prepared according to Sun et al [25]. In brief, Fe(acac)3 (2 mmol), 1,2–hexadecanediod (10 mmol), oleic acid (6 mmol), and olecylamine (6 mmol) were mixed in benzyl ether (20 mL) in three-necked bottle. Under nitrogen atmosphere, the mixture was refluxed at 100 °C for 30 min, and then sequentially heated to 200 °C for 1 h and 300 °C for 30 min. After cooling to room temperature, the product was collected by centrifugation at 6000 rpm for 5 min and washed with ethanol three times. The black-brown SPIOs were stored in ethanol at 4 °C.

2.5. Preparation of Hyaluronic Acid-Graft-Amphiphilic Gelatin Microcapsule (HA-AGMC)

The amphiphilicity and the self-assembly property of AG together with SPIOs have been investigated by Li et al. and Chiang et al. [19,26]. HA-AGMCs were prepared through a simple double emulsion process. First, AG-g-HA (80 mg) was added to deionized water (1.6 mL) and NaOH solution (1.6 mL, 0.1 N). The solution (0.6 mL) with hydrophobic SPIO (10 mg) in chloroform (1 mL) were emulsified to obtain a water-in-oil (W/O) emulsion. AG-g-HA solution (2.4 mL) was added to proceed the secondary emulsion to form the W/O/W emulsion. Afterward, the organic solvent was removed by rotary evaporator at 33 °C. The final HA-AGMC products were washed 3 times and re-dispersed with deionized water.

2.6. Characterization of HA-AGMCs

The size and zeta potential of HA-AGMCs were characterized by dynamic light scattering (Beckman Coulter Delsa™ Nano C particle analyzer, Beckman Coulter, Brea, CA, USA). The morphology of HA-AGMCs were characterized by scanning electron microscope (SEM, JEOL-6700, JEOL, Tokyo, Japan) and transmission electron microscope (TEM, JEM-2100, JEOL, Tokyo, Japan). The samples of SEM were prepared by placing HA-AGMCs solution on a silicon wafer and drying in a vacuum desiccator. TEM samples were prepared by laying HA-AGMCs solution on a carbon coated grid, followed by removing the excess liquid on the grid, and drying in a vacuum desiccator.

2.7. Loading Efficiency of SPIOs

SPIOs content was measured using UV-vis spectroscopy (UV-vis, Evolution 300, Thermo). Calibration curve was made by dissolving SPIOs (10 mg) and AG-g-HA (75 mg) in HCl (0.5 N). From UV-vis spectroscopy, the absorbance of the solution was measured at 363 nm. HA-AGMC was also diluted with HCl (0.5 N) to calculate the corresponding loaded SPIOs concentration.

2.8. Chondrocyte Isolation and Culture

Chondrocytes were isolated from the articular cartilage of New Zealand White rabbits (0.4–0.8 kg). All procedures conformed to the guidelines of the Institute of Animal Care and Use Committee of I-Shou University in Taiwan. All the surgical instruments were sterilized before use. After rinsing thighbones with phosphate-buffered saline (PBS) two times, cartilage tissue from the joint was dissected and cut into pieces of approximately 1×1 mm^2 samples. These samples were digested with protease (20 mg in 10 mL DMEM/F-12) for 2 h and transferred to collagenase (20 mg in 10 mL DMEM/F-12) for another 3 h. The cell suspension was centrifuged at 2000 rpm for 5 min and resuspended in DMEM/F12 medium with 10% FBS in 75T culture flask.

Chondrocytes were seeded in monolayer and used within two passages. The cells were cultured in DMEM/F12 medium with 10% FBS in a 5% CO_2 incubator at 37 °C. The medium was renewed every 2 days and cells were passaged once it reached confluence.

2.9. Cell cytotoxicity Test

Cell cytotoxicity was carried out using the MTS method. In brief, chondrocytes were harvested from culture flask and seeded in a 24-well tissue culture plates at a density of 5×10^4 cells/well. HA-AGMCs were added to the cell culture in different concentrations (0, 0.05, 0.11, 0.21, 0.43, 0.85, 1.7, 3.4, 6.8, 13.6 mg mL^{-1}). After 1, 2, or 4 days incubation, culture medium was replaced with DMEM/F-12 containing 10% MTS and reacted for 2 h. Each medium was collected and centrifuged at 6000 rpm for 5 min to remove HA-AGMCs. The absorbance of supernatant was monitored by ELISA reader at wavelength 490 nm. The experiment was done in triplicate.

2.10. 3D culture Methods

The experiment method of agarose hydrogels followed the protocol of 3D Petri dish molds [27]. Briefly, agarose powder (1 g) was dissolved in PBS and then pipetted into micro-mold without creating any bubble in proper temperature. After solidification, the agarose hydrogels were placed in culture medium to equilibrate for at least 15 min, and then transferred to fresh medium for further use or storage. All steps were in sterile conditions. Cells at a final density of 2.56×10^5 cells/190 µL were seeded in each agarose hydrogel placing in 6 well plate and waited 10 min to settle cells. Finally, we added additional medium to the plate (2.5 mL/well).

2.11. Cell proliferation and Cell Compatibility Analysis

MTS assay was performed to assess the proliferation of chondrocytes. In short, fresh media containing different concentrations of HA-AGMCs (42.5, 85, 170, 340, and 680 µg mL^{-1}) were mixed with chondrocytes before seeding in each agarose hydrogel placing in plate and waited 10 min to settle cells. Finally, additional medium was added to cover the hydrogel. After incubation for various time period, cell pellets were collected and counted before adding 10% MTS medium. Each medium was collected after 3 h reaction and the optical density was monitored by ELISA reader at wavelength 490 nm. The experiment was run in four times. Cell compatibility was also evaluated by Live/Dead assay. After days of incubation, cell pellets were collected and washed with PBS buffer. Staining reagent Calcein AM and Ethidium homodimer-1 in PBS covered the pellets at 37 °C for 30 min. The samples were observed under confocal microscopy. The experiment was done in triplicate. The fluorescence intensity in each pellet was measured by ImageJ and further analyzed by unpaired t-test.

2.12. HA-AGMCs Cellular Attachment Efficiency

To measure the precise amount of HA-AGMCs attached on chondrocytes, we examined the element content of Fe by inductively coupled plasma mass spectrometry (ICP-MS). To prepare the sample, we collected 1 mL medium in vials after chondrocytes mixing with various amounts of HA-AGMCs and seeded on hydrogel overnight. Hydrochloric acid (12 M), which can totally corrode the cell sample and redox iron in HA-AGMC, was mixed in all the samples. The content of Fe element was quantified by ICP-MS.

2.13. RNA Isolation, cDNA Synthesis and Quantitative PCR Analysis

The procedure of RNA isolation was bottomed on Molecular Research Center, Inc. In short, chondrocytes pellets were washed with PBS buffer and added 1mL of RNAzol® RT reagent to lyse cells. Subsequently, added DEPC-treated water (0.4 mL) to lysate and shook the mixture vigorously for 15 s. Centrifuged samples for 15 min at 13,500 rpm at 4 °C after storing for 15 min. The RNA remained soluble in the supernatant. Each of the supernatant (1 mL) was transferred to a new tube and precipitated by mixing with 0.4 mL of 75% ethanol. The samples were centrifuged for 10 min at 12000 rpm at 4 °C after stored for 15 min. RNA precipitated and formed a white pellet at the bottom of the tube. Next, we washed the RNA twice with 75% ethanol to remove the ethanol and dissolve the RNA precipitation pellet with DEPC-treated water.

Complementary DNA was performed for reverse transcription using the GScript First-Strand Synthesis Kit (GeneDireX). Specific cDNA was amplified by PCR using Platinum® Blue PCR SuperMix. The protocol of experiment method followed Invitrogen manufacturer. cDNA samples were mixed up with Platinum® Blue PCR SuperMix and forward/reverse primer. The PCR amplification condition was illustrated as follows: heating up to 95 °C for 5 min, denaturing the DNA at 95°C for 30 s, annealing with the primer at 55 °C for 30 s, extending the length of product at 72 °C for 30 s, and finally cycled for 35 times. PCR cycle ended at 72 °C for 3 min and cooled down to 4 °C. The samples were stored at −20 °C and electrophoresis experiment ran at voltage 80 V, 35 min. The experiment was also conducted in three times.

2.14. Alcian Blue Staining

Alcian blue solution was prepared beforehand by dissolving Alcian blue 8GX powder in 40% acetic acid and 60% ethanol mixture solvent in 1 wt % concentration, and the solution was stored at 4 °C. At the end of the culture period, cell pellets were washed with PBS buffer and fixed with 4% formaldehyde for 1 h. The fixed-cell samples were washed with PBS buffer and then incubated with Alcian blue solution overnight. The stained samples were washed with PBS buffer and mounted by mounting solution. The observation was using optical microscope. The experiment was run in triplicate and the results were statistically analyzed using two-way ANOVA Dunnett's multiple comparisons test comparing with the control group (* $P < 0.05$, ** $P < 0.01$, *** $P < 0.001$, and **** $P < 0.0001$).

2.15. Static Magnetic Field (MF) and Magnet-Derived Shear Stress (S) Stimulations

A cylinder-shaped neodymium magnet with a magnetic field of 0.22 T was used to static magnetic field induction [24,28]. On the other hand, a magnetic stirrer of 50 rpm was applied for magnet-derived shear stress stimulation on the cell-pellets. For combined stimulation, cell-pellets were first placed on the top of the magnet and then transferred to the magnetic stirrer on pellets in consecutive order following static magnetic field and/or magnet-derived shear stress for 1 h one day for five consecutive days

3. Results and Discussion

3.1. Synthesis and Characterization of AG-g-HA

Figure 2A illustrated the reaction scheme for the synthesis of Hyaluronic acid-grafted amphiphilic gelatin (AG-g-HA). As a widely used biopolymer, gelatin is composed of a series of amino acids, serving as hydrophilic backbone and modified by hexanoic anhydride to gain amphiphilic characteristic. HA was conjugated on gelatin's Arginine (Arg) part by the well-known EDC/NHS method. The chemical signals of the gelatin (Figure S1 and Table S1) from peak 1 to peak 11 assigned to the protons on the primitive gelatin macromolecules [29]. According to the NMR spectra in Figure 2B, chemical signals of amphiphilic gelatin (AG) at peak 3.5 ppm and peak 1.2 ppm were the protons from the hexanoyl group. The disappearance of peak at 2.9 ppm which was referred to the primary amino group of the gelatin confirmed that amino group, arginine, was partially substituted to hexanoyl group forming AG. The α-carbonyl group (COCαH3) at 1.9 ppm and glucose group between 3.2 and 3.9 ppm gave the cue of glycosylation (Figure 2C). TNBS is a well-known reagent specific for primary amino groups, which can be quantified and measured at absorbance 335 nm [30]. The substitution rate of the hexanoyl group was measured to be 58.2%. The second HA conjugation rate was 4.0% (Figure S2). The results demonstrated that there were finally 37.8% free amino groups remaining in the arginine and lysine of the gelatin where hexanoyl group and HA substitution were at 58.2% and 4.0%, respectively.

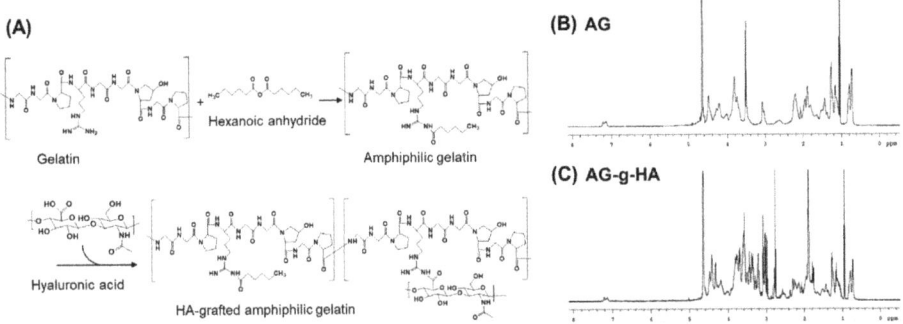

Figure 2. (**A**) The reaction scheme for the synthesis of hyaluronic acid-grafted amphiphilic gelatin (AG-g-HA); ^{1}H NMR spectrum of macromolecules in D2O for (**B**) amphiphilic gelatin (AG) and (**C**) AG-g-HA. As shown, the NMR spectra at 3.5, 1.2, 1.9 and around 3.5 ppm demonstrated the successful synthesis of AG-g-HA.

3.2. Characterization of HA-AGMCs and Cartilage Tissue-Mimetic Pellets

Hyaluronic acid-graft-amphiphilic gelatin microcapsule (HA-AGMC) was synthesized via double emulsification using SPIOs and AG-g-HA. The diameter and zeta potential of HA-AGMCs were measured by dynamic light scattering (DLS) (Table S2). HA-AGMC exhibited an average size of 1.24 ± 0.1 μm in diameter with an excellent monodispersity (Figure S3). The surface charge was approximately −16 mV caused by the replacement of positively charged amino groups to hexanoyl group and HA. HA-AGMCs demonstrated the appearance of deflated balloons structure in scanning electron microscope (SEM) images (Figure 3A). The round hollow morphology can be further confirmed by transmission electron microscopy (TEM) image in Figure 3B where the little dark spots of SPIOs were clearly observed in the shell of microcapsules. The loading efficiency of SPIOs in HA-AGMCs was 92.2% determined by ultraviolet-visible spectroscopy (UV-vis).

The biocompatibility of HA-AGMC with rabbit primary chondrocytes was investigated by MTS assay (Figure 3C). The cells were treated with different concentrations of HA-AGMC from 0.05 to 13.6 mg mL^{-1}. HA-AGMC illustrated outstanding biocompatibility at high concentration in which chondrocytes viability was higher than 84% after 24 h co-culture with HA-AGMC under 6.8 mg mL^{-1}.

In this study, hyaluronic acid (HA) on HA-AGMC surface was designed to help sticking chondrocytes and microcapsules together. The CD44 antigen was the main receptor for hyaluronic acid, which was responsible for cell proliferation, differentiation and migration [16,17,31]. Hyaluronic acid conjugated on the surface of microcapsule was capable of binding to the chondrocytes due to CD44/HA receptors and serving as a backbone to recruit proteoglycans and glycoproteins into extracellular matrix structures [32]. Attachment efficiency of HA-AGMCs to chondrocytes was used to evaluate the formation of cartilage tissue-mimetic pellets by co-culturing chondrocytes with HA-AGMCs on non-adhesive micro-molds (Figure 3D). The Fe amount was also quantified by using ICP-MS and increased with HA-AGMCs concentration. The attachment efficiency was over 90% in each HA-AGMCs group with the microcapsule concentration of 42.5 to 680 μg mL^{-1}. The result demonstrated that mostly HA-AGMCs can easily attach to chondrocytes and form pellets together.

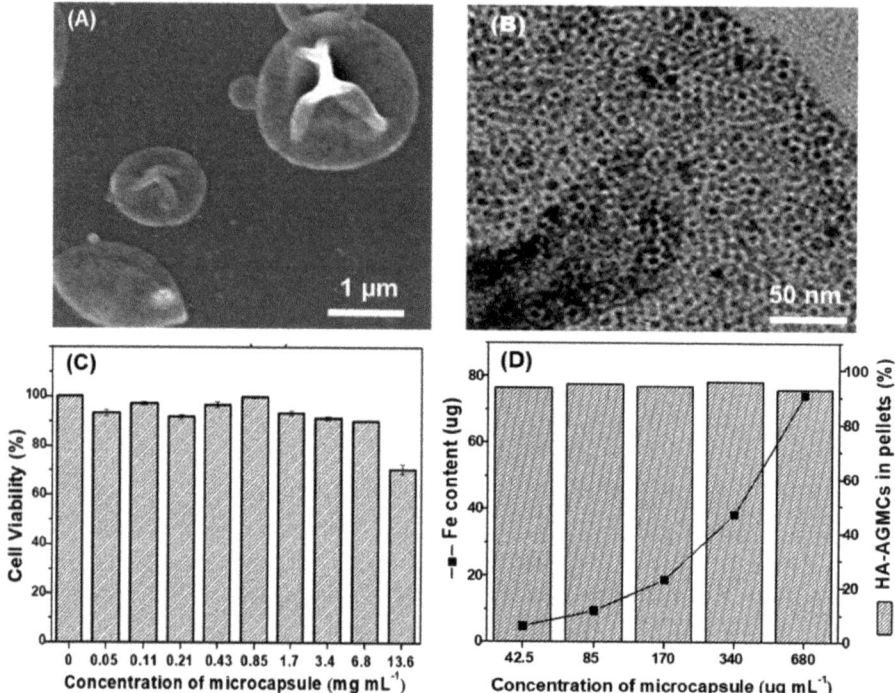

Figure 3. Morphological appearance and characterization of the HA-AGMCs. (**A**) SEM and (**B**) TEM images. (**C**) Biocompatibility of HA-AGMCs to chondrocytes after 24 h incubation (n = 3). (**D**) Attachment efficiency of HA-AGMCs to chondrocytes in different concentrations of HA-AGMC. The results demonstrated HA-AGMCs have a spherical-shaped bilayers structure with average diameter of 1.2 μm and negligible cytotoxicity to chondrocytes.

We fabricated cartilage tissue-mimetic pellets with different concentrations of microcapsule HA-AGMCs and also used macromolecule AG-g-HA as control group. The number of viable chondrocytes and the morphology of cartilage tissue-mimetic pellets were examined by MTS assay and optical microscope at day 7 and 14 respectively. HA-AGMCs and AG-g-HA showed little difference in optical density (O.D.) over the cultural period at different concentrations from 42.5 to 680 μg mL^{-1} (in AG-g-HA concentration) (Figure 4A,B). The results illustrated that the HA-AGMCs microcapsules containing SPIOs do not lower the number of chondrocytes. Of note, after 14 days of culture, HA-AGMCs at the concentration of 170 μg mL^{-1} had the relative highest cell number correlated to the optical microscope images (Figure 4A). The morphology of pellets in different HA-AGMCs concentrations after culturing 14 days was shown (Figure 4C). Pellets in each group showed uniform size and rounded shape, and more microcapsules made the pellets looked darker responding to more SPIOs. During the culture period, a scatter of cell debris was found in the group with 680 mL^{-1} HA-AGMCs, which resulted in a smaller size of pellet. In contrast, HA-AGMCs at the concentration of 170 μg mL^{-1} had the biggest pellet size of 200 μm. Therefore, the HA-AGMCs at 170 μg mL^{-1} was selected for the following all experiments.

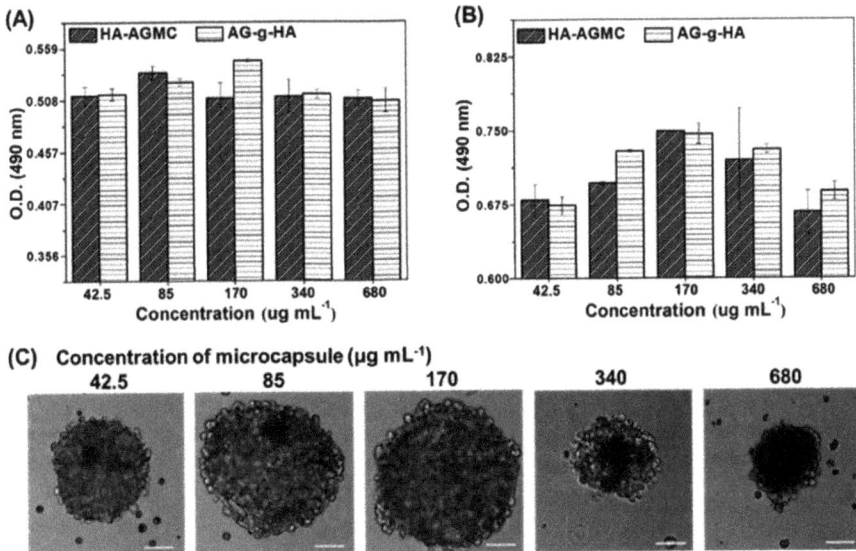

Figure 4. Cartilage tissue-mimetic pellets proliferation and morphology. The relative cell proliferation after culturing chondrocytes with HA-AGMCs microcapsule or AG-g-HA at different concentrations (from 42.5 to 680 μg mL^{-1}) at (**A**) 7 days and (**B**) 14 days (n = 4). (**C**) The cell pellet morphologies after culturing 14 days showed that HA-AGMCs at the concentration of 170 μg mL^{-1} had the biggest pellet size of 200 μm. Scale bar = 50 μm.

3.3. Cartilage Tissue-Mimetic Pellets Live/Dead Assay

We tested long-term cell viability with Live/Dead assay to confirm cell condition, and to investigate whether HA-AGMCs had any influence on the cartilage tissue-mimetic pellets. In 3D culture, the supplement of oxygen was mainly controlled by diffusion, consequently resulting in oxygen gradient. Oxygen tension was often found in the center of the 3D structure and also dependent on the number of cells. The nutrient and oxygen supply in the central of pellets would be increasingly inadequate during the culture period [33,34]. In this study, fluorescence images showed after 14 days, the vast majority of cells remain viable in both pure cells control group and HA-AGMCs group (Figure 5A,B). The mean fluorescence intensity in each HA-AGMCs pellet showed more cells and higher viability compared to cell only group as shown in Figure 5C. In comparison to clear boundary between each cell in control group (Figure 5A), HA-AGMC group showed blurry edge, indicating cell pellets in HA-AGMC group had stronger connection to ECM (Figure 5B). This further confirmed that HA-AGMC are able to promote the connection between cells and construct chondrocytes into a compact pellet. To observe the localization of HA-AGMC, quantum dots were dissolved in chloroform with SPIOs and encapsulated in the shell of HA-AGMC. The confocal fluorescent images showed cooperative formation of HA-AGMC and chondrocytes at 14 days (Figure 5D), HA-AGMCs still remained in the pellets and scattered uniformly in the ball-shaped pellets. Our studies demonstrated that microcapsule HA-AGMCs can maintain great cell compatibility and viability in cartilage tissue-mimetic pellets.

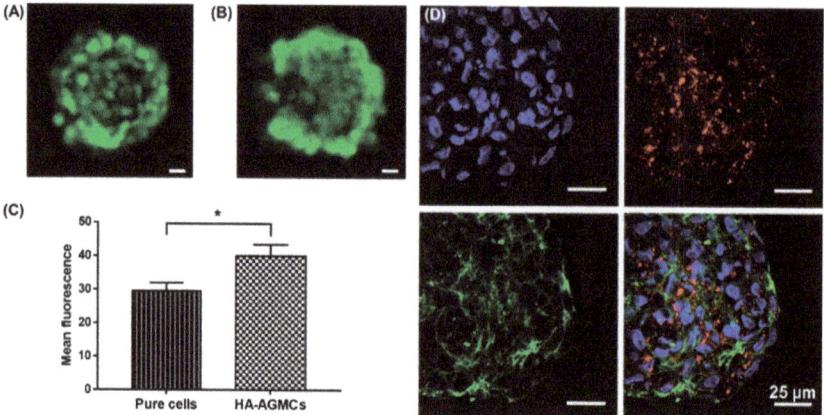

Figure 5. Confocal fluorescent image of cartilage tissue-mimetic pellets after 14 days culture. Live/Dead fluorescent image of (**A**) pure cells control group and (**B**) cells with HA-AGMCs group. Calcein AM (green), Ethidium homodimer-1 (red). (**C**) The fluorescence intensity in each pellet was measured by ImageJ and further analyzed by unpaired t-test ($P = 0.0123$, n = 3). (**D**) Quantum dots were encapsulated in the shell of HA-AGMCs and used to visualize the localization of cartilage tissue-mimetic pellets consisting of DAPI (blue), actin (green), HA-AGMC (red quantum dots). As shown, chondrocytes remained viable in HA-AGMCs group and had stronger connection to ECM. Scale bar = 25 m.

3.4. Gene expression of Cartilage Tissue-Mimetic Pellets under Physical Stimulations

HA-AGMC was designed as a multifunctional platform to guide the cartilage tissue-mimetic pellets and serve physical stimulations. Static magnetic field (MF) and magnet-derived shear stress (S) were applied on the HA-AGMCs one hour each day for five consecutive days. In addition to mimic cartilage ECM environment, joint movements mechanical forces also played a crucial part in growth and development of articular cartilage tissue. It was found that stem cell performed better properties when cells grew in scaffold under external mechanical stimulations [23,35,36]. More recently, magnetic nanoparticles captured more attention to bioreactor as they were capable of providing different mechanical forces and were more suitable for applying loading to improve cell condensation and scaffold seeding efficiency [37–39]. We performed biophysical stimulations via SPIOs and analyzed the gene expression of Aggrecan (Agg), collagen type I (Col I), collagen type II (Col II), and Sox9 with polymerase chain reaction (PCR) after 7 and 21 days of culture. The primer sequences were shown in Table S3 and housekeeping genes GAPDH were used in comparison with the samples. After 7 days culture period, the presence of HA-AGMCs significantly increased Col I, Col II and Sox9 gene expression compared to control group. HA-AGMC+MF+S and control group showed similar Agg gene expression (Figure 6A). The gene expression of Col II and Agg gave us an indication of chondrocytes' functionality as they attributed to the secretion of ECM to stabilize the structure of ECM [40]. It is worth noting that Sox9 gene regulated chondrogenesis and chondrocyte proliferation [41], and the expressions of Sox9 were dramatically upregulated 2-fold in all HA-AGMCs-added groups after 7 days. Taking up-regulation of Agg, Col II, and Sox9 contrasting to Col I gene expression altogether, HA-AGMC-treated group demonstrated a slightly better chondrocyte expansion. The HA-AGMC with physical stimulation-treated group showed significant Agg gene expression compared to control group after 21 days culture period (Figure 6B). Furthermore, gene expression of both Col I and Col II revealed significant difference between HA-AGMC+MF+S and control group. Taking the results together, the application of HA-AGMCs and biophysical stimulation did not generate dedifferentiation effect on chondrocytes due to the relatively lower expression of Col I than any other group during culture period. Furthermore, HA-AGMC+MF+S group performed the highest Col II gene expression level throughout the culture period, despite of HA-AGMC+S with sound Col II gene expression at the early stage. These data indicated that HA-AGMC can help chondrocytes present in cartilage tissue-specific gene at the beginning of pellets formation, and somewhat maintain functional gene expression better after applying both static magnetic field and magnet-derived shear stress. However, further studies are needed to optimize the mimic native cartilage environment in terms of biophysical parameters.

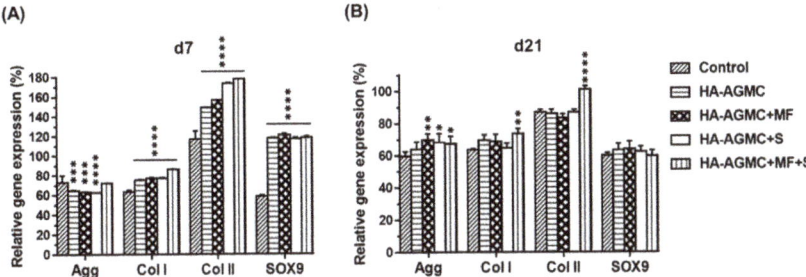

Figure 6. Gene expressions of cartilage tissue-mimetic pellets under static magnetic field (MF) and magnet-derived shear stress (S) stimulations at (**A**) day 7 and (**B**) day 21. The results were statistically analyzed using two-way ANOVA Dunnett's multiple comparisons test comparing with the control group (* $P < 0.05$, ** $P < 0.01$, *** $P < 0.001$, and **** $P < 0.0001$). HA-AGMC helped chondrocytes maintain cartilage tissue-specific gene at the beginning of pellets forming and applying both MF and S hold the gene expression better.

3.5. Synthesis of Sulfated Glycosaminoglycan under Physical Stimulations

The synthesis of sulfated glycosaminoglycan (sGAG) is one important index of chondrocyte biochemical function [41,42], and the diminishing presence of sGAG indicated a tendency to de-differentiate into fibrochondrocytes [43]. The cartilage tissue-mimetic pellets productivity of sGAG under physical simulations was examined by Blyscan assay (Figure 7A). The sGAG production increased with incubation time. Physical stimulations further enhanced cartilage tissue-specific ECM production, and HA-AGMC+MF+S group exhibited the greatest sGAG secretion in medium after 21 days culture (Figure 7B). Alcian blue staining results further demonstrated that both and HA-AGMCs group had high content of sGAG excretion in comparison with pure cells control group in Figure 7C where, staining results were not further quantified because of the SPIOs interference.

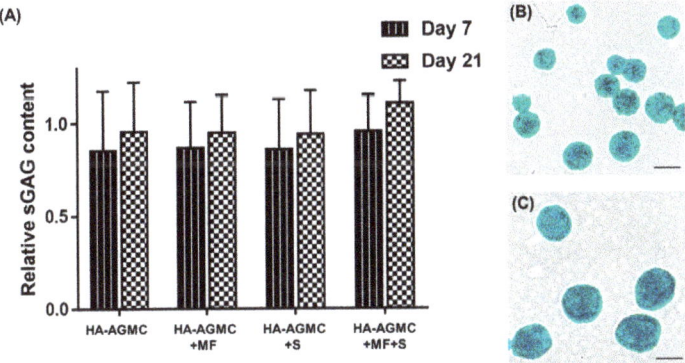

Figure 7. The cartilage tissue-mimetic pellets productivity of sGAG. (**A**) Blyscan assay detected sGAG content in culture medium under different stimulations at 7 and 21 day (n = 3). Alcian blue staining demonstrated the secretion of sGAG after 21 days culture of (**B**) control and (**C**) HA-AGMCs. scale bars = 200 μm.

3.6. Histological Analysis

The cell pellets with HA-AGMCs at the concentration of 170 µg mL^{-1} was used for animal implant experiments to make in vivo chondrogenic analysis in an osteoarthritic mode. Osteoarthritis was surgically induced by anterior cruciate ligament transection (ACLT) and partial medial meniscectomy on one knee of male New Zealand rabbits. Half of the rabbit's femoral head are implanted with a magnet as a source of magnetic force as shown in Figure 8A. In this in-vivo experiment, the HA-AGMCs are pre-cultured with chondrocytes before implantation into rabbit OA knee. Subsequently, after 6 weeks of surgery, the rabbits received implants of pellets containing either only cells or HA-AGMCs with cells or nothing as control groups.

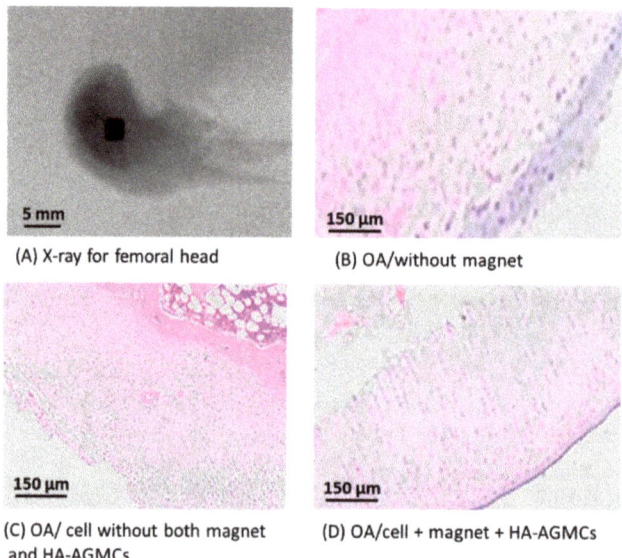

(A) X-ray for femoral head
(B) OA/without magnet
(C) OA/ cell without both magnet and HA-AGMCs
(D) OA/cell + magnet + HA-AGMCs

Figure 8. (A) Half of the rabbit's femoral head implanted with a magnet as a source of magnetic force. H&E stain of the cartilage tissue treated with various groups of cells/magnet/HA-AGMCs for 4 weeks in (**B–D**).

After 4 weeks of implantation, the preliminary results revealed that control groups without HA-AGMCs displayed irregular layer of cartilage surface and degeneration of the cartilage tissue (as illustrated in Figure 8B). In addition, the abnormal growth and disorder of cartilage tissues were observed in pure cell without magnet and HA-AGMCs group shown in Figure 8C. More importantly, a newly formed tissue is evident in the group treated with a combination of chondrocytes with A-AGMCs and magnet in Figure 8D, which demonstrated that the cells treated with HA-AGMCs and magnetic force can improve the retention and biofunctionality of transplanted chondrocytes to form ordering arrangement in cartilage matrix, which is very important for cartilage repair. However, the comprehensive investigation about chondrogenic analysis and cartilage repair is in subsequent progress, which will be reported in the future.

4. Conclusions

In summary, cartilage tissue-mimetic pellets have considerable advantages in over articular cartilage disorder obstacles. In this study, Am-HA-gelatin has been synthesized by modifying primary amino group on gelatin with hexanoic anhydride to obtain the amphiphilic property and then grafting hyaluronic acid on the amphiphilic gelatin which is capable to form simple core-shell hollow structure. The multifunctional HA-AGMC with specific targeting on CD44 receptors provided cartilage tissue-mimetic pellets with high structure stability and remained high cell viability even in the center of pellets after 14 days culture. Packing with superparamagnetic iron oxide nanoparticles (SPIOs) in the HA-AGMC microcapsule, the pellets with the magnetic HA-AGMCs demonstrated the combination of static magnetic field and magnet-derived shear stress can exhibit the highest cartilage tissue-specific gene expression. Our preliminary results have showed that the cells treated with HA-AGMCs and magnetic force exhibit the better growth and ordering of chondrocytes.

Supplementary Materials: The following are available online at http://www.mdpi.com/2073-4360/12/4/785/s1, Figure S1: ^{1}H NMR spectrum of molecules in D2O for gelatin., Figure S2: Quantification of gelatin-based molecules with TNBS. (A) Absorption spectra of primary amino group of the gelatin reacted with TNBS at different concentrations. (B) Characterization of different degrees of substitution for the gelatin-based molecules., Figure S3: Particle size distribution profile of HA-AGMCs, Table S1: The protons on the primitive gelatin exhibited chemical shift signals from peak 1 to peak 12., Table S2: Characterization of the HA-AGMCs, Table S3: List of genes and the primers used for PCR in chondrocytes.

Author Contributions: Conceptualization, S.-J.C. and S.-Y.C.; methodology, K.-T.H. and M.-Y.C; software, K.-T.H.; validation, T.-Y.L. and M.-Y.C.; formal analysis, K.-T.H.; investigation, M.-Y.C.; resources, C.-Y.C.; data curation, T.-Y.L. and C.-Y.C.; writing—original draft preparation, K.-T.H., S.-J.C. and S.-Y.C.; writing—review and editing, K.-T.H., T.-Y.L., C.-Y.C., S.-J.C. and S.-Y.C.; supervision, S.-J.C. and S.-Y.C.; project administration, S.-J.C. and S.-Y.C.; funding acquisition, S.-J.C. and S.-Y.C. All authors have read and agreed to the published version of the manuscript.

Funding: This research was funded by Ministry of Science and Technology, Taiwan, grant number MOST 107-2221-E-214-012-MY3, MOST 108-2218-E-080-001, MOST 106-2221-E-009-065-MY3.

Conflicts of Interest: The authors declare no conflict of interest.

References

1. Johnstone, B.; Alini, M.; Cucchiarini, M.; Dodge, G.R.; Eglin, D.; Guilak, F.; Madry, H.; Mata, A.; Mauck, R.L.; Semino, C.E.; et al. Tissue engineering for articular cartilage repair—The state of the art. *Eur. Cells Mater.* **2013**, *25*, 248–267. [CrossRef] [PubMed]
2. Zhang, W.; Ouyang, H.; Dass, C.R.; Xu, J. Current research on pharmacologic and regenerative therapies for osteoarthritis. *Bone Res.* **2016**, *4*, 15040. [CrossRef] [PubMed]
3. Kwon, H.; Brown, W.E.; Lee, C.A.; Wang, D.; Paschos, N.; Hu, J.C.; Athanasiou, K.A. Surgical and tissue engineering strategies for articular cartilage and meniscus repair. *Nat. Rev. Rheumatol.* **2019**, *15*, 550–570. [CrossRef]
4. Grässel, S.; Lorenz, J. Tissue-engineering strategies to repair chondral and osteochondral tissue in osteoarthritis: Use of mesenchymal stem cells. *Curr. Rheumatol. Rep.* **2014**, *16*, 452. [CrossRef]
5. Mithoefer, K.; McAdams, T.; Williams, R.J.; Kreuz, P.C.; Mandelbaum, B.R. Clinical efficacy of the microfracture technique for articular cartilage repair in the knee: An evidence-based systematic analysis. *Am. J. Sports Med.* **2009**, *37*, 2053–2063. [CrossRef]
6. Sophia Fox, A.J.; Bedi, A.; Rodeo, S.A. The basic science of articular cartilage: Structure, composition, and function. *Sports Health* **2009**, *1*, 461–468. [CrossRef]
7. Gao, Y.; Liu, S.; Huang, J.; Guo, W.; Chen, J.; Zhang, L.; Zhao, B.; Peng, J.; Wang, A.; Wang, Y.; et al. The ECM-cell interaction of cartilage extracellular matrix on chondrocytes. *Biomed. Res. Int.* **2014**, *2014*, 648459. [CrossRef]
8. Goude, M.C.; McDevitt, T.C.; Temenoff, J.S. Chondroitin sulfate microparticles modulate transforming growth factor-beta1-induced chondrogenesis of human mesenchymal stem cell spheroids. *Cells Tissues Organs* **2014**, *199*, 117–130. [CrossRef]
9. Li, J.; Chen, G.; Xu, X.; Abdou, P.; Jiang, Q.; Shi, D.; Gu, Z. Advances of injectable hydrogel-based scaffolds for cartilage regeneration. *Regen. Biomater.* **2019**, *6*, 129–140. [CrossRef]
10. Lim, J.J.; Hammoudi, T.M.; Bratt-Leal, A.M.; Hamilton, S.K.; Kepple, K.L.; Bloodworth, N.C.; McDevitt, T.C.; Temenoff, J.S. Development of nano- and microscale chondroitin sulfate particles for controlled growth factor delivery. *Acta Biomater.* **2011**, *7*, 986–995. [CrossRef]

11. Levett, P.A.; Melchels, F.P.; Schrobback, K.; Hutmacher, D.W.; Malda, J.; Klein, T.J. A biomimetic extracellular matrix for cartilage tissue engineering centered on photocurable gelatin, hyaluronic acid and chondroitin sulfate. *Acta Biomater.* **2014**, *10*, 214–223. [CrossRef] [PubMed]
12. Tan, M.L.; Choong, P.F.; Dass, C.R. Recent developments in liposomes, microparticles and nanoparticles for protein and peptide drug delivery. *Peptides* **2010**, *31*, 184–193. [CrossRef] [PubMed]
13. Oliveira, M.B.; Mano, J.F. Polymer-based microparticles in tissue engineering and regenerative medicine. *Biotechnol. Prog.* **2011**, *27*, 897–912. [CrossRef] [PubMed]
14. Lam, J.; Lu, S.; Kasper, F.K.; Mikos, A.G. Strategies for controlled delivery of biologics for cartilage repair. *Adv. Drug Deliv. Rev.* **2015**, *84*, 123–134. [CrossRef]
15. Cruz, D.G.; Sardinha, V.; Ivirico, J.E.; Mano, J.F.; Ribelles, J.G. Gelatin microparticles aggregates as three-dimensional scaffolding system in cartilage engineering. *J. Mater. Sci. Mater. Med.* **2013**, *24*, 503–513. [CrossRef]
16. Knudson, C.B. Hyaluronan and CD44: Strategic players for cell–matrix interactions during chondrogenesis and matrix assembly. *Birth Defects Res. Part C Embryo Today Rev.* **2003**, *69*, 174–196. [CrossRef]
17. Akmal, M.; Singh, A.; Anand, A.; Kesani, A.; Aslam, N.; Goodship, A.; Bentley, G. The effects of hyaluronic acid on articular chondrocytes. *J. Bone Joint Surg.* **2005**, *87*, 1143–1149. [CrossRef]
18. Elzoghby, A.O. Gelatin-based nanoparticles as drug and gene delivery systems: Reviewing three decades of research. *J. Control. Release* **2013**, *172*, 1075–1091. [CrossRef]
19. Chiang, C.S.; Chen, J.Y.; Chiang, M.Y.; Hou, K.T.; Li, W.M.; Chang, S.J.; Chen, S.Y. Using the interplay of magnetic guidance and controlled TGF-beta release from protein-based nanocapsules to stimulate chondrogenesis. *Int. J. Nanomed.* **2018**, *13*, 3177–3188. [CrossRef]
20. Nguyen, A.H.; McKinney, J.; Miller, T.; Bongiorno, T.; McDevitt, T.C. Gelatin methacrylate microspheres for controlled growth factor release. *Acta Biomater.* **2015**, *13*, 101–110. [CrossRef]
21. Vaca-González, J.J.; Guevara, J.M.; Moncayo, M.A.; Castro-Abril, H.; Hata, Y.; Garzón-Alvarado, D.A. Biophysical Stimuli: A Review of Electrical and Mechanical Stimulation in Hyaline Cartilage. *Cartilage* **2017**, *10*, 157–172. [CrossRef] [PubMed]
22. Kamei, G.; Kobayashi, T.; Ohkawa, S.; Kongcharoensombat, W.; Adachi, N.; Takazawa, K.; Shibuys, H.; Deie, M.; Hattori, K.; Goldberg, J.L.; et al. Articular cartilage repair with magnetic mesenchymal stem cells. *Am. J. Sports Med.* **2013**, *41*, 1255–1264. [CrossRef] [PubMed]
23. Luciani, N.; Du, V.; Gazeau, F.; Richert, A.; Letourneur, D.; Le Visage, C.; Wilhelm, C. Successful chondrogenesis within scaffolds, using magnetic stem cell confinement and bioreactor maturation. *Acta Biomater.* **2016**, *37*, 101–110. [CrossRef] [PubMed]
24. Son, B.; Kim, H.D.; Kim, M.; Kim, J.A.; Lee, J.; Shin, H.; Hwang, N.S.; Park, T.H. Physical Stimuli-Induced Chondrogenic Differentiation of Mesenchymal Stem Cells Using Magnetic Nanoparticles. *Adv. Healthc. Mater.* **2015**, *4*, 1339–1347. [CrossRef] [PubMed]
25. Sun, S.; Zeng, H.; Robinson, D.B.; Raoux, S.; Rice, P.M.; Wang, S.X.; Li, G. Monodisperse MFe2O4 (M = Fe, Co, Mn) Nanoparticles. *J. Am. Chem. Soc.* **2004**, *126*, 273–279. [CrossRef]
26. Li, W.M.; Liu, D.M.; Chen, S.Y. Amphiphilically-modified gelatin nanoparticles: Self-assembly behavior, controlled biodegradability, and rapid cellular uptake for intracellular drug delivery. *J. Mater. Chem.* **2011**, *21*, 12381–12388. [CrossRef]
27. Napolitano, A.P.; Dean, D.M.; Man, A.J.; Youssef, J.; Ho, D.N.; Rago, A.P.; Lech, M.P.; Morgn, J.R. Scaffold-free three-dimensional cell culture utilizing micromolded nonadhesive hydrogels. *Biotechniques* **2007**, *43*, 496–500. [CrossRef]
28. Amin, H.D.; Brady, M.A.; St-Pierre, J.P.; Stevens, M.M.; Overby, D.R.; Ethier, C.R. Stimulation of Chondrogenic Differentiation of Adult Human Bone Marrow-Derived Stromal Cells by a Moderate-Strength Static Magnetic Field. *Tissue Eng. Part A* **2014**, *20*, 1612. [CrossRef]
29. Rodin, V.V.; Izmailova, V.N. NMR method in the study of the interfacial adsorption layer of gelatin. *Colloids Surf. A Physicochem. Eng. Asp.* **1996**, *106*, 95–102. [CrossRef]
30. Grotzky, A.; Manaka, Y.; Fornera, S.; Willeke, M.; Walde, P. Quantification of α-polylysine: A comparison of four UV/Vis spectrophotometric methods. *Anal. Methods* **2010**, *2*, 1448. [CrossRef]
31. Misra, S.; Hascall, V.C.; Markwald, R.R.; Ghatak, S. Markwald and Shibnath Ghatak, Interactions between Hyaluronan and Its Receptors (CD44, RHAMM) Regulate the Activities of Inflammation and Cancer. *Front. Immunol.* **2015**, *6*, 201. [CrossRef] [PubMed]

32. Bhattacharya, D.S.; Svechkarev, D.; Souchek, J.J.; Hill, T.K.; Taylor, M.A.; Natarajan, A.; Mohs, A.M. Impact of structurally modifying hyaluronic acid on CD44 interaction. *J. Mater. Chem. B* **2017**, *5*, 8183–8192. [CrossRef] [PubMed]
33. Buckley, C.T.; Meyer, E.G.; Kelly, D.J. The influence of construct scale on the composition and functional properties of cartilaginous tissues engineered using bone marrow-derived mesenchymal stem cells. *Tissue Eng. Part A* **2012**, *18*, 382–396. [CrossRef] [PubMed]
34. Ong, S.Y.; Dai, H.; Leong, K.W. Inducing hepatic differentiation of human mesenchymal stem cells in pellet culture. *Biomaterials* **2006**, *27*, 4087–4097. [CrossRef] [PubMed]
35. Sun, M.; Lv, D.; Zhang, C.; Zhu, L. Culturing functional cartilage tissue under a novel bionic mechanical condition. *Med. Hypotheses* **2010**, *75*, 657–659. [CrossRef]
36. Fahy, N.; Alini, M.; Stoddart, M.J. Mechanical stimulation of mesenchymal stem cells: Implications for cartilage tissue engineering. *J. Orthop. Res.* **2018**, *36*, 52–63. [CrossRef]
37. Zhang, N.; Lock, J.; Sallee, A.; Liu, H. Magnetic Nanocomposite Hydrogel for Potential Cartilage Tissue Engineering: Synthesis, Characterization, and Cytocompatibility with Bone Marrow Derived Mesenchymal Stem Cells. *ACS Appl. Mater. Interfaces* **2015**, *7*, 20987–20998. [CrossRef]
38. Shimizu, K.; Ito, A.; Honda, H. Mag-seeding of rat bone marrow stromal cells into porous hydroxyapatite scaffolds for bone tissue engineering. *J. Biosci. Bioeng.* **2007**, *104*, 171–177. [CrossRef]
39. Ghosh, S.; Kumar, S.R.P.; Puri, I.K.; Elankumaran, S. Magnetic assembly of 3D cell clusters: Visualizing the formation of an engineered tissue. *Cell Prolif.* **2016**, *49*, 134–144. [CrossRef]
40. Bosnakovski, D.; Mizuno, M.; Kim, G.; Takagi, S.; Okumura, M.; Fujinaga, T. Chondrogenic differentiation of bovine bone marrow mesenchymal stem cells (MSCs) in different hydrogels: Influence of collagen type II extracellular matrix on MSC chondrogenesis. *Biotechnol. Bioeng.* **2006**, *93*, 1152–1163. [CrossRef]
41. Caron, M.M.; Emans, P.J.; Coolsen, M.M.; Voss, L.; Surtel, D.A.; Cremers, A.; van Rhijn, L.W.; Welting, T.J.M. Redifferentiation of dedifferentiated human articular chondrocytes: Comparison of 2D and 3D cultures. *Osteoarthr. Cartil.* **2012**, *20*, 1170–1178. [CrossRef] [PubMed]
42. Pfeiffer, E.; Vickers, S.M.; Frank, E.; Grodzinsky, A.J.; Spector, M. The effects of glycosaminoglycan content on the compressive modulus of cartilage engineered in type II collagen scaffolds. *Osteoarthr. Cartil.* **2008**, *16*, 1237–1244. [CrossRef] [PubMed]
43. McCorry, M.C.; Puetzer, J.L.; Bonassar, L.J. Characterization of mesenchymal stem cells and fibrochondrocytes in three-dimensional co-culture: Analysis of cell shape, matrix production, and mechanical performance. *Stem Cell Res. Ther.* **2016**, *7*, 39. [CrossRef] [PubMed]

 © 2020 by the authors. Licensee MDPI, Basel, Switzerland. This article is an open access article distributed under the terms and conditions of the Creative Commons Attribution (CC BY) license (http://creativecommons.org/licenses/by/4.0/).

Article

The Ophthalmic Performance of Hydrogel Contact Lenses Loaded with Silicone Nanoparticles

Nguyen-Phuong-Dung Tran and Ming-Chien Yang *

Department of Materials Science and Engineering, National Taiwan University of Science and Technology, Taipei 10607, Taiwan; thaonguyeng89@gmail.com
* Correspondence: myang@mail.ntust.edu.tw; Tel.: +886-2-2737-6528; Fax: +886-2-2737-6544

Received: 13 April 2020; Accepted: 11 May 2020; Published: 14 May 2020

Abstract: In this study, silicone nanoparticles (SiNPs) were prepared from polydimethylsiloxane (PDMS) and tetraethyl orthosilicate (TEOS) via the sol-gel process. The resultant SiNPs were characterized by dynamic light scattering (DLS), transmission electron microscope (TEM), and scanning electron microscope (SEM). These SiNPs were then blended with 2-hydroxyethylmethacrylate (HEMA) and 1-vinyl-2-pyrrolidinone (NVP) before polymerizing into hydrogel contact lenses. All hydrogels were subject to characterization, including equilibrium water content (EWC), contact angle, and oxygen permeability (Dk). The average diameter of SiNPs was 330 nm. The results indicated that, with the increase of SiNPs content, the oxygen permeability increased, while the EWC was affected insignificantly. The maximum oxygen permeability attained was 71 barrer for HEMA-NVP lens containing 1.2 wt% of SiNPs with an EWC of 73%. These results demonstrate that by loading a small amount of SiNPs, the Dk of conventional hydrogel lenses can be improved greatly. This approach would be a new method to produce oxygen-permeable contact lenses.

Keywords: silicone nanoparticles; PDMS; TEOS; hydrogels; soft contact lenses

1. Introduction

Contact lenses are employed for correcting eye vision. The global market for contact lenses is about US$8.1 billion in 2018. Two major classes of soft contact lenses are silicone hydrogel and conventional hydrogel lenses. The market shares for the former is 69% while the latter takes 19% in 2018 [1]. Conventional hydrogel contact lenses are synthesized from hydrophilic monomers such as 2-hydroxyethyl methacrylate (HEMA), offering the wearer comfort due to their hydrophilicity. However, this class of contact lenses exhibits low oxygen permeability that may cause red-eye syndrome for long-term wearing [2,3]. To cope with this problem, silicon-containing polymers such as 3-(methacryloyloxy) propyltris(trimethylsiloxy) silane (TRIS) or polydimethylsiloxane (PDMS) are incorporated into the hydrogel to increase the oxygen permeability, leading to the inception of silicone hydrogel lenses [4–11].

Polydimethylsiloxane exhibits a hydrophobic nature and is low cost, simple to fabricate, and shows good biocompatibility, flexibility, thermal and oxidative stability, high optical transparency, and especially high oxygen permeability [12–18]. The main drawback of PDMS is the restriction of water absorptivity, surface wettability, and higher lipid deposition because of its inherent hydrophobicity [19–21]. These limitations can be improved through the combination with hydrophilic materials. The incorporation of hydrophobic and hydrophilic monomers usually takes two oppositional tendencies. The first trend is to increase equilibrium water content. However, this will decrease the oxygen permeability because of hydrophilic monomers. On the other hand, a higher PDMS content will enhance oxygen permeability but reduce the water uptake ability [11,22,23]. In silicone hydrogel contact lenses, water is a main factor to restrict the oxygen permeability [24–26]. However, the second trend may improve both wettability and oxygen transmissibility when hydrophobic and hydrophilic monomers are cooperated at a proper ratio.

In this work, a novel approach is adopted to improve the oxygen permeability of HEMA-based hydrogels. Instead of incorporating PDMS into the polymer chain, poly(dimethylsiloxane) dialkanol (PDMS diol) was reacted with tetraethyl orthosilicate (TEOS) through hydrolysis and condensation to form silicone nanoparticles (SiNPs) via the sol-gel process [27–29]. Thereafter, SiNPs were loaded into hydrogel synthesized from HEMA and 1-vinyl-2-pyrrolidinone (NVP). All the resultant hydrogels were subject to characterization, including Fourier transform infrared spectroscopy (FTIR), Raman, scanning electron microscope (SEM), transmission electron microscope (TEM), dynamic light scattering (DLS), equilibrium water content (EWC), oxygen permeability (Dk), optical transparency, mechanical strength, and contact angle measurements. We think this novel approach of loading silicone nanoparticles would improve the oxygen permeability without reducing the hydrophilicity and wettability of HEMA hydrogels.

2. Materials and Methods

2.1. Materials

Poly(dimethylsiloxane) dialkanol (PDMS-diol, KF-6001) was purchased from Shin-Etsu Chemical Co. Ltd., Tokyo, Japan. 1-Vinyl-2-pyrrolidinone (NVP), tetraethyl orthosilicate (TEOS), and 2-hydroxy-2-methylbenzene acetone (D-1173) were purchased from Sigma-Aldrich (St. Louis, Mo USA). Further, 2-hydroxyethylmethacrylate (HEMA) and ethylene glycol dimethacrylate (EGDMA) were obtained from Acros Organics (NJ, USA). Phosphate buffered saline solution (PBS, 0.1 M, pH 7.4) was prepared in our laboratory.

2.2. Preparation of Silicone Nanoparticles

Silicone nanoparticles (SiNPs) were synthesized by cross-linking PDMS-diol with TEOS through the sol-gel process shown in Figure 1. The reaction was conducted in a solution containing 0.1 mL of HCl (37%), 4 mL of water, and 2 mL of ethanol (95%). Subsequently, 1 mL of TEOS was added into the solution and then stirred for 1 h at room temperature, followed by adding 4 mL of PDMS-diol dropwise into the reacting solution. The reaction was stirred for 24 h at room temperature in the dark after completion of the addition. After condensing in a vacuum oven at 80 °C for 24 h, SiNPs were harvested and purified with ethanol before sonication and centrifugation. Finally, SiNPs were stored after drying in an oven at 80 °C for 12 h.

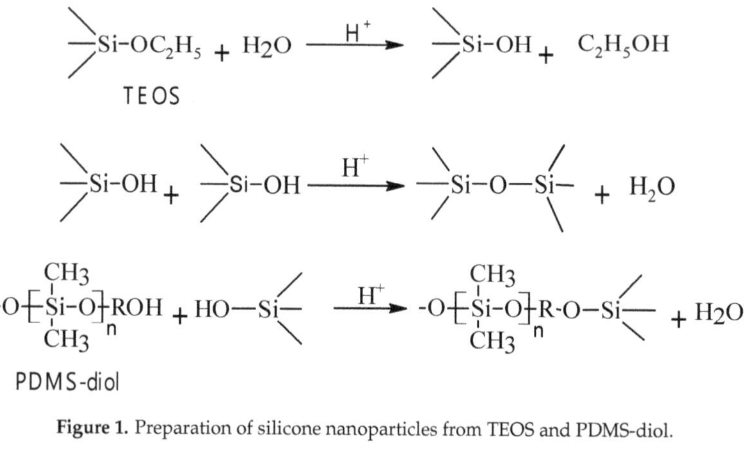

Figure 1. Preparation of silicone nanoparticles from TEOS and PDMS-diol.

Figure 1 shows the synthetic reactions for silicone nanoparticles (SiNPs) from PDMS and TEOS. Firstly, the ethoxy groups of TEOS were hydrolyzed into the hydroxyl groups. In the subsequent condensation, silica and silicone were formed by removing hydroxyl groups of Si-OH [27,29,30].

2.3. Preparation of SiNPs-Loaded Hydrogel Composites

The hydrogels were polymerized from NVP and HEMA in the presence of SiNPs, cross-linking agent EGDMA and photo initiator D-1173 as shown in Table 1 and Figure 2. For all formulations, the mixture contained 0.5 wt% of EGDMA and D-1173. Then, the mixture was stirred in the dark at room temperature for 5 h. Afterward, the mixture was transferred to polypropylene molds and cured under UV light (365 nm) for 40 min. After demolding, lenses were soaked in 50% ethanol for 20 h at 50 °C to remove un-reacted monomers and photo initiator. Then, the lenses were immersed in distilled water for 4 h at 50 °C to wash out ethanol. Finally, the lenses were preserved in PBS (pH 7.4) at room temperature.

Table 1. Formulation of soft lenses including HEMA, NVP, and silicone nanoparticles (SiNPs).

Sample	Gel Composition (mg)		
	HEMA	NVP	SiNPs
HS0	100	0	0
HS04	100	0	0.4
HS08	100	0	0.8
HS12	100	0	1.2
HN2S0	80	20	0
HN2S04	80	20	0.4
HN2S08	80	20	0.8
HN2S12	80	20	1.2
HN5S0	50	50	0
HN5S04	50	50	0.4
HN5S08	50	50	0.8
HN5S12	50	50	1.2

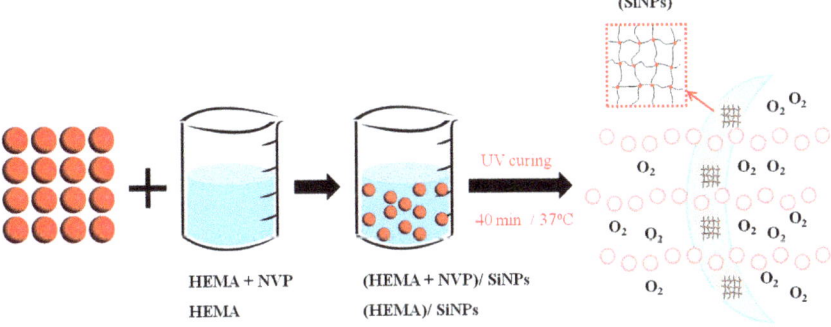

Figure 2. Preparation of SiNPs-loaded hydrogel lenses.

2.4. Elemental Analysis and Size of Particles

The elemental composition of dry particles was determined using a field emission scanning electron microscope (FE-SEM/EDS, JSM-6500F, JEOL, Tokyo, Japan). The particle size was analyzed using dynamic light scattering (DLS-DKSH, Malvern Instruments Ltd., Malvern, UK), and transmission electron microscopy (TEM, JEM-2000FXII, JEOL, Japan).

2.5. Equilibrium Water Content

The equilibrium water content (EWC) of the hydrogel was calculated as follows:

$$\text{EWC (\%)} = \frac{W_2 - W_1}{W_2} \times 100 \tag{1}$$

where W_1 and W_2 are the weights of the dry lens and the rehydrated lens in distilled water for one day at room temperature, respectively.

2.6. Optical Transparency

After swelling in PBS solution, the lens was adhered on the surface of cuvette containing 2 mL distilled water. The optical transparency was determined in a wavelength range of 400–700 nm using a UV-Vis spectrophotometer (Cary 300, Agilent Technologies, Santa Clara, CA, USA).

2.7. Surface Characterization

The contact angle of the contact lens was measured using a contact angle goniometer (DSA 100, Krüss GmbH, Hamburg, Germany) at room temperature. The contact angle was an average of three repetitions.

2.8. Chemical structure

The structure of SiNPs and contact lenses were examined using Raman Spectroscopy and FTIR. The FTIR of SiNPs (Nicolet 170 SX, Thermo Fisher Scientific, Madison, WI, USA) were performed in the wavenumber range of 600–4000 cm^{-1}. SiNPs were pelletized with potassium bromide (KBr) before being scanned over 32 times by an infrared ray. The lenses were detected over 32 scans based on FTIR-ATR. The Raman spectroscopy (iHR550, Horiba Scientific, Kyoto, Japan) was determined in the wavenumber range of 400–4000 cm^{-1}.

2.9. Mechanical Properties

The mechanical properties of hydrogel specimens were determined by modulus and tensile strength. Samples were cut as dog bone shape after hydrated in DI water. Modulus and tensile strength of specimens were measured based on a tensile tester (MTS 810, Material Test System, Eden Prairie, MN, USA) via ASTM D1708 standard at a crosshead speed of 50 mm/min.

2.10. Oxygen Permeability

The oxygen permeability (Dk, barrer) of the lens was determined according to ISO18369-4:2006 which is based on polarographic method using an oxygen permeometer (201 T O2 permeometer, Createch, Chesterfield Twp, MI, USA). Polarography measures the oxygen permeation through a sample by measuring the current produced in a cell by reducing oxygen at a noble metal electrode. Before testing, guard ring polarographic cell (8.6 mm radius, CreaTech/Rehder Development Co., Chesterfield Twp, MI, USA), buffer solutions and the lenses were placed in a temperature and humidity-controlled box at 37 °C and 98% relative humidity till the temperature equilibrium. After fully hydrated, the lenses were stacked to measure the electronic current at a various number of lenses to correct the boundary effect. The linear plot of t/Dk versus thickness was drawn and determined Dk/t from the slope [31–33].

3. Results and Discussion

3.1. Size and Elemental Composition of Particles

The size of SiNPs was determined based on DLS measurement as shown in Figure 3. The average diameter of SiNPs was 330 ± 100 nm. The TEM image in Figure 4 shows similar SiNPs sizes.

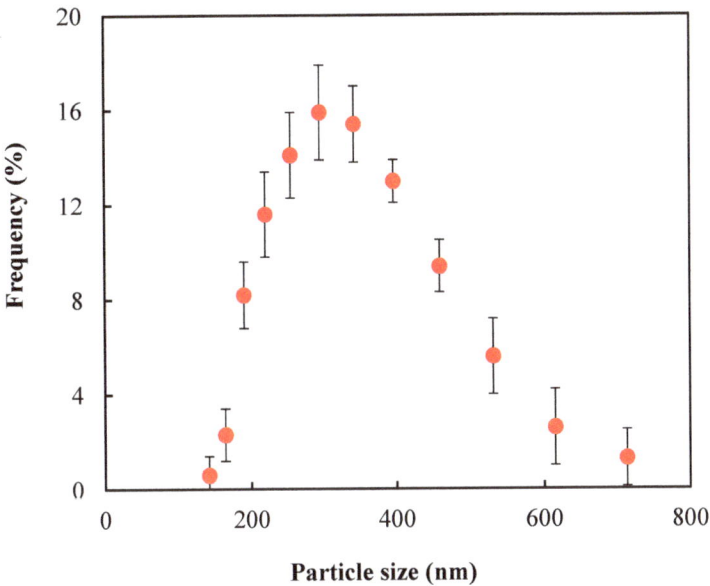

Figure 3. Size distribution of silicone nanoparticles obtained from DLS.

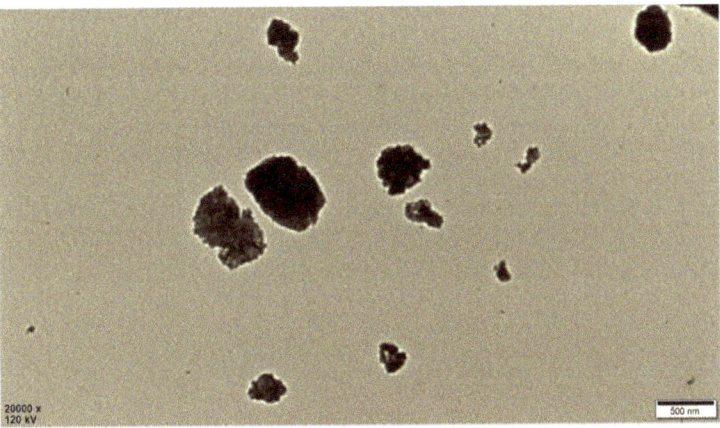

Figure 4. TEM image of SiNPs.

The hydrolysis and condensation reaction of TEOS would produce silica particles based on the sol-gel process [27–29,34]. In this work, the condensation of TEOS occurred in the presence of PDMS-diol to result in silicone nanoparticles. The EDX results in Figure 5 showed that the elemental composition of SiNPs was consisted of 28.6% Si, 40.8% O, and 30.6% C. Considering that the atomic ratio of Si:O in PDMS is 1:1 while that in SiO_2 is 1:2, the mole fractions of PDMS and SiO_2 in SiNP can be estimated by solving the following equations

$$\begin{cases} x + y = 0.286 & \text{for Si} \\ x + 2y = 0.408 & \text{for O} \end{cases}$$
$$\therefore y = 0.122, \ x = 0.164$$

Thus, the mole fraction of silicone is $\frac{x}{x+y} = \frac{0.164}{0.286} = 57.4\%$ and that of silica is $\frac{y}{x+y} = 42.6\%$. In addition, the presence of carbon in SiNPs indicated that PDMS did incorporated into SiNPs.

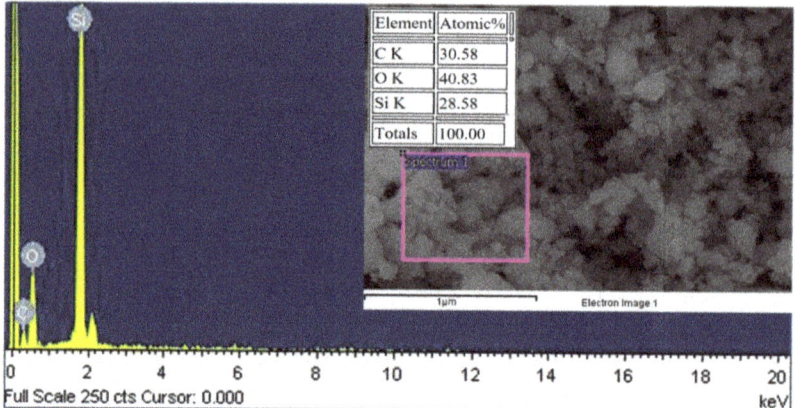

Figure 5. Micrograph of SiNPs from SEM-EDS.

Raman and FTIR were employed to examine the chemical structure of SiNPs synthesized from PDMS and TEOS. Figure 6 shows that the Raman spectrum of SiNPs exhibited strong PDMSs peaks appearing at 2909–2968 cm^{-1}, 1409–1459 cm^{-1} (C-H groups), and 706–782 cm^{-1} (Si-C groups). These results were also found in the literature [35–37]. Furthermore, the peak at 483 cm^{-1} were attributed to Si-O-Si [38]. These characteristic peaks indicated that the reaction of PDMS and TEOS through the sol-gel process was successful. Comparing the spectra of TEOS, PDMS, and SiNPs in Figure 7, the peaks of Si-O-Si (1074 cm^{-1}, 808 cm^{-1}, 796 cm^{-1}, 786 cm^{-1}, 495 cm^{-1}, 480 cm^{-1}, 460 cm^{-1}) and SiOH (958 cm^{-1}) were observed. The stretching vibrations of C-H occurred at 2792 cm^{-1} of TEOS, 2956 cm^{-1} of PDMS, and 2966 cm^{-1} of SiNPs while the peaks of Si-C (1251 cm^{-1} and 1263 cm^{-1}) were respectively observed in the spectra of PDMS and SiNPs [24].

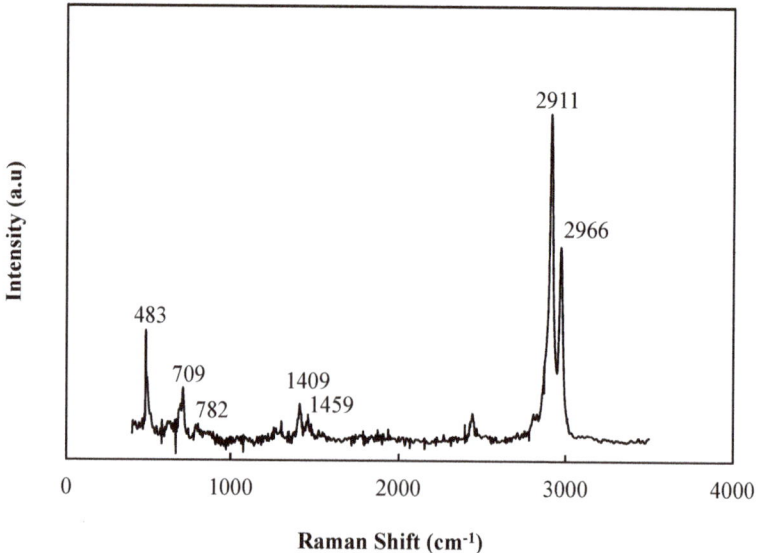

Figure 6. Raman spectrum of SiNPs.

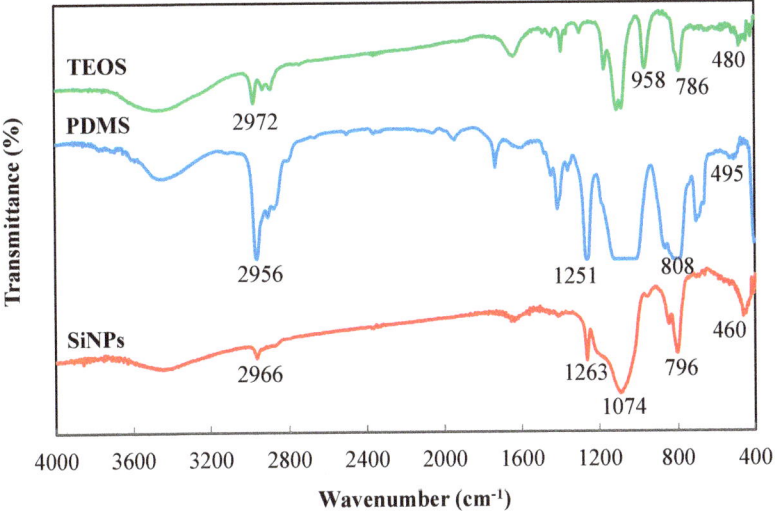

Figure 7. FTIR spectra of TEOS, PDMS, and SiNPs.

3.2. Optical Transparency

Figure 8 shows the photos of hydrated soft lenses including poly(HEMA-co-NVP) and poly(HEMA-co-NVP)-SiNPs. These lenses appeared transparent.

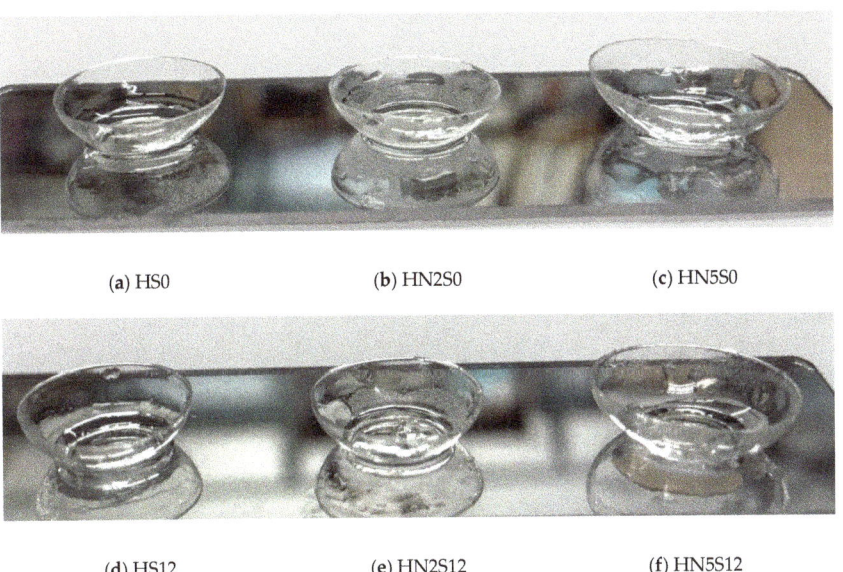

Figure 8. Photos of hydrated contact lenses.

Figure 9 shows that the light transmittance (T%) of the contact lens decreased with the increase of SiNPs content. The reduction in transparency can be attributed to the size distribution of SiNPs (see Figure 4) that caused light scattering. At a content of 1.2 wt%, the transmittance dropped to 90%. The light transmittance of a contact lens is preferred to be above 90% [39]. A higher SiNPs content would further decrease the transparency of the contact lens below 90%. Thus, in this study, the maximum content of SiNPs was limited to 1.2 wt%.

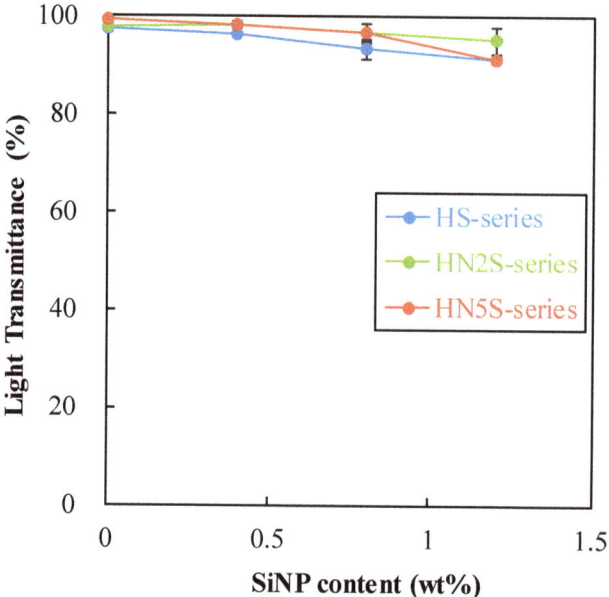

Figure 9. Light transmittance of SiNPs samples.

3.3. Equilibrium Water Content

Equilibrium water content (EWC) is an important index conferring comfortable wearing for patients because of the softness and wettability as well as the limitation of dry corneal eye [40,41]. In this present work, three series of soft lenses were prepared with different EWC values. The basic ingredient of these hydrogels was HEMA, which is a well-known monomer for contact lens. The other ingredient, NVP, is to increase the hydrophilicity of the hydrogel. All these formulations were crosslinked with 0.5 wt% of EGDMA.

Figure 10 shows that the presence of SiNPs in hydrogel network did not significantly affect the EWC of soft lenses. The values of EWC for the HS-, HN2-, and HN5- series varied around 34%, 42%, and 73%, respectively, regardless of the content of SiNPs. This is because that the content of SiNPs was low, and that these nanoparticles interacted little with the matrix of HEMA and NVP. In other words, a small number of nanoparticles were simply dispersed in the hydrogel matrix. On the contrary, for commercialized silicone soft lenses, hydrophobic ingredients such as TRIS, SiMA, and PDMS were incorporated into the main chains of the hydrogel and caused the reduction in EWC of contact lenses [19,22,42,43].

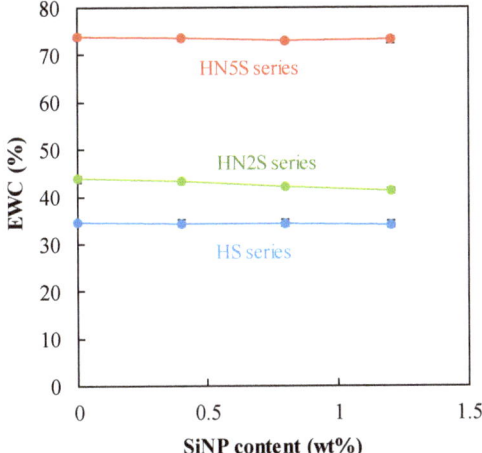

Figure 10. EWC of SiNPs samples.

3.4. Contact Angle

Figure 11 shows that the contact angle of the soft lenses increased slightly with the content of SiNPs. This may be attributed to the inherent hydrophobic PDMS in SiNPs embedded in hydrogel matrix [24,44,45]. However, the difference was less than 5°, and all the contact angles were below 70°. Thus, the addition of SiNPs did little to change the wettability of these lenses. The slight increase may be attributed to a small quantity of hydrophobic PDMS (from SiNPs) exposed on the surface [23].

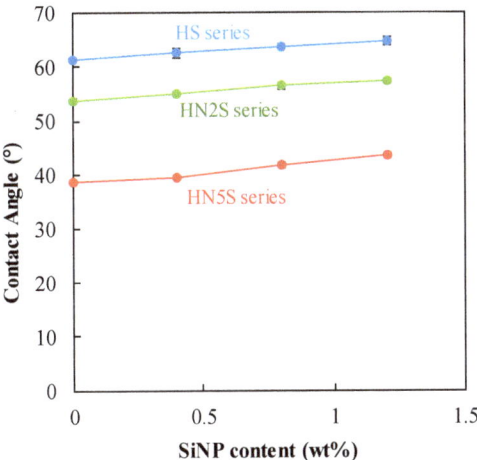

Figure 11. Contact angle of SiNPs samples.

3.5. IR Spectra of Contact Lenses

Figure 12 shows that the FTIR spectra of HS-series lenses differed little as the contents of SiNPs increased from 0 wt% to 1.2 wt%. This phenomenon was observed for HN2S- and HN5S- series as well (not shown). In the spectra of HS-series lenses, peaks of OH groups appeared at 3350 cm^{-1} while peaks of Si-O-Si groups were detected at 1074 cm^{-1}. Further, in the spectrum of HS-series, the presence of Si-C and C=O were respectively observed at the peaks of 1253 cm^{-1} and 1706 cm^{-1} [24]. In the spectra of HN2S- and HN5S- series, the peaks of OH (3338, 3338 cm^{-1}) and Si-O-Si (1072, 1080 cm^{-1}), Si-C (1257 cm^{-1}, 1257 cm^{-1}) and C=O (1643 cm^{-1}, 1631 cm^{-1}) were also detected.

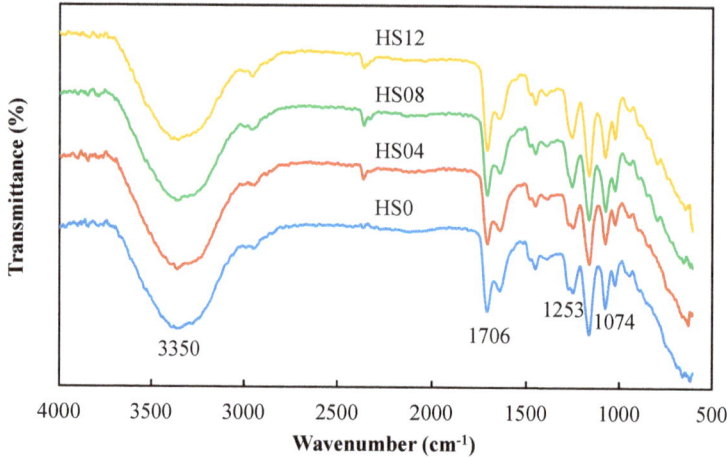

Figure 12. The FTIR spectra of HS- series SiNPs-containing contact lenses.

3.6. Mechanical Properties

Figure 13 presents the modulus and tensile strength of all HS, HN2S, and HN5S series. The moduli of all HN5S samples were not significantly affected by the SiNPs content increasing from 0 wt% to 1.2 wt%. Particularly, the moduli of all HN5S lenses were approximately 0.48 MPa for all SiNPs contents. This tendency was also found for HS and HN2S series samples. The moduli of HS and HN2S samples were approximately 0.65 MPa and 0.56 MPa, respectively. In this research, the moduli of all HN5S, HN2S, and HS series were around 0.48–0.65 MPa, which were similar to commercial lenses such as Acuvue Oasys, Acuvue Advance, and Biomedics 38 [43,46]. Hence, these SiNPs-contained contact lenses exhibited mechanical properties comparable to those commercial contact lenses.

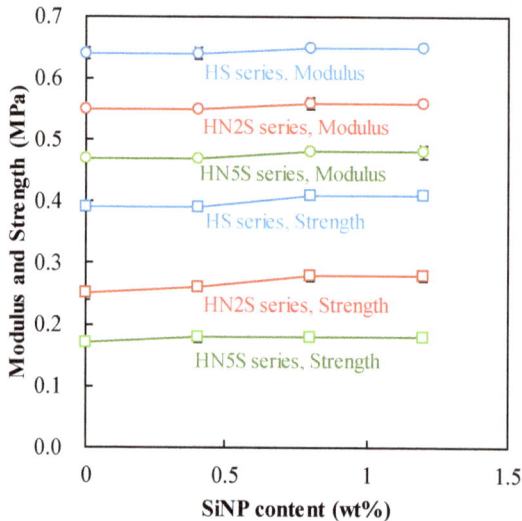

Figure 13. The modulus and strength of SiNPs samples.

3.7. Oxygen Permeability

Figure 14 shows that Dk is linearly depending on the SiNPs content. Furthermore, the slop increased with the hydrophilicity. Although the Dk increases with the EWC for conventional non-silicone hydrogels, the loading of SiNPs did accelerate the permeation of oxygen in these composite hydrogels. For the HS series, the Dk increased from 9 to 29 barrer as the content of SiNPs increased to 1.2 wt%. For a more hydrophilic series, Dk of HN5 soft lenses increased rapidly from 39 to 71 barrer when the content of SiNPs increased from 0 to 1.2 wt%. Additionally, all formulations containing a higher concentration of NVP monomers exhibited higher oxygen transmissibility than others. As a result, the oxygen permeation of three formulation series including p(HEMA-co-NVP) and pure HEMA polymers was affected by the SiNPs content.

It is well-known that hydrophobic PDMS can improve oxygen transmissibility of soft lenses based on its siloxane groups (-Si(CH$_3$)$_2$-O-), especially silicon-oxygen bond [7,9,47]. However, the main weakness of the hydrophobic component is to impair the water absorbability of the lens, [16,24,48] thus EWC and Dk usually follow an inverse correlation. Accordingly, water is usually the limiting factor of oxygen transport for silicone hydrogel lenses [26,49]. On the contrary, for the SiNPs-loaded hydrogel, the addition of SiNPs did not affect the water uptake ability while increasing the oxygen permeability of lenses, as shown in Table 1. This phenomenon broke the reverse relationship of EWC and Dk in comparison to other non SiNPs lenses.

For hydrogels loaded with SiNPs, the content of these nanoparticles was less than 1.2 wt% of the matrix. Although SiNPs were hydrophobic, the effect on the water absorbability was low due to this small content. Furthermore, hydrophilic matrix of the hydrogel interacted little with these hydrophobic nanoparticles. Thus, the mobility of these nanoparticles would be higher than in the case of silicone hydrogels where PDMS chains were integrated into the matrix. Apparently, higher mobility would facilitate the permeation of oxygen through the lens. Although in this study the highest Dk was 70 barrer, we believe that higher Dk could be attained for higher SiNPs content once the particle size was reduced through improving the synthesizing process of SiNPs, thereby breaking through the transparency limitation.

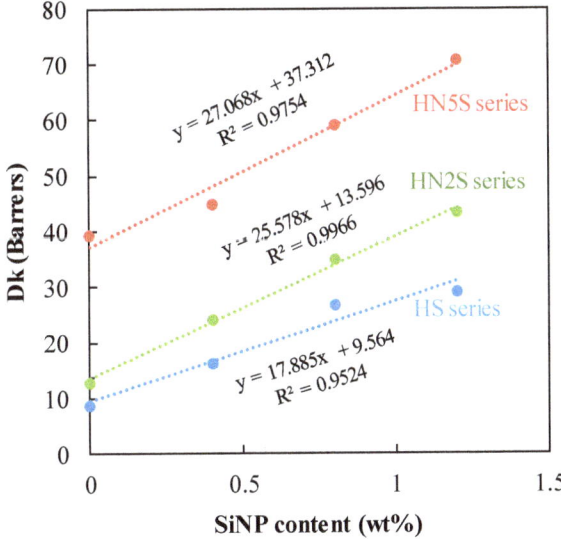

Figure 14. The effect of SiNPs content on Dk.

4. Conclusions

As demonstrated in this study, silicone nanoparticles (SiNPs) were synthesized from TEOS and PDMS-diol. The existence of PDMS and TEOS in SiNPs was verified through Raman and FTIR spectroscopy. The resultant nanoparticles exhibited a diameter of 330 ± 100 nm and composed of 57% of silicone and 43% of silica. These nanoparticles were further entrapped in hydrogels polymerized from HEMA and NVP. The resultant SiNPs-loaded hydrogel lenses exhibited an unusual correlation between the oxygen permeability (Dk) and the equilibrium water content (EWC): the Dk increased with the content of silicone nanoparticles while the EWC changed insignificantly. Moreover, based on the result of the contact angle and Young's modulus, the loading of SiNPs slightly influenced the wetting surface and mechanical properties. The transparency was reduced to 91% when the content of SiNPs was 1.2 wt%, probably due to the light scattering from the nanoparticles. Further effort to reduce the particle size is underway in our lab. With this work, we demonstrate a novel approach to improve the oxygen permeability without impairing the hydrophilicity of soft contact lenses. These results would be beneficial to the development of soft contact lenses.

Author Contributions: N.-P.-D.T. prepared experiments, as well as wrote the original draft. M.-C.Y. supervised the research project and finalized the manuscripts. All authors have read and agreed to the published version of the manuscript.

Acknowledgments: This work was supported by the Ministry of Science and Technology, Taiwan through Grant No. MOST 106-2622-E-011-004-CC2.

Conflicts of Interest: The authors declare no conflict of interest.

References

1. Nichols, J.J.; Fisher, D. Overview of general market trends. *Contact Lens Spectrum* **2019**, *34*, 18–23.
2. Calossi, A.; Fossetti, A.; Lupelli, L.; Rossetti, A. The effects of short-term contact lens-induced hypoxia tried on ourselves. *Cont. Lens Anterior Eye* **2015**, *38*, 16. [CrossRef]
3. Duench, S.; Sorbara, L.; Keir, N.; Simpson, T.; Jones, L. Impact of silicone hydrogel lenses and solutions on corneal epithelial permeability. *Optom. Vis. Sci.* **2013**, *90*, 546–556. [CrossRef] [PubMed]
4. Awasthi, A.; Meng, F.; Künzler, J.; Linhardt, J.; Papagelis, P.; Oltean, G.; Myers, S. Ethylenically unsaturated polycarbosiloxanes for novel silicone hydrogels: Synthesis, end-group analysis, contact lens formulations, and structure–property correlations. *Polym. Adv. Technol.* **2013**, *24*, 557–567. [CrossRef]
5. Chekina, N.; Pavlyuchenko, V.; Danilichev, V.; Ushakov, N.; Novikov, S.; Ivanchev, S. A new polymeric silicone hydrogel for medical applications: Synthesis and properties. *Polym. Adv. Technol.* **2006**, *17*, 872–877. [CrossRef]
6. Wu, J.; He, C.; He, H.; Cheng, C.; Zhu, J.; Xiao, Z.; Zhang, H.; Li, X.; Zheng, J.; Xiao, J. Importance of zwitterionic incorporation into polymethacrylate-based hydrogels for simultaneously improving optical transparency, oxygen permeability, and antifouling properties. *J. Mater. Chem. B* **2017**, *5*, 4595–4606. [CrossRef]
7. Saini, A.; Rapuano, C.J.; Laibson, P.R.; Cohen, E.J.; Hammersmith, K.M. Episodes of microbial keratitis with therapeutic silicone hydrogel bandage soft contact lenses. *Eye Contact Lens* **2013**, *39*, 324–328. [CrossRef]
8. Seitz, M.E.; Wiseman, M.E.; Hilker, I.; Loos, J.; Tian, M.; Li, J.; Goswami, M.; Litvinov, V.M.; Curtin, S.; Bulters, M. Influence of silicone distribution and mobility on the oxygen permeability of model silicone hydrogels. *Polymer* **2017**, *118*, 150–162. [CrossRef]
9. Szczotka-Flynn, L.; Jiang, Y.; Raghupathy, S.; Bielefeld, R.A.; Garvey, M.T.; Jacobs, M.R.; Kern, J.; Debanne, S.M. Corneal inflammatory events with daily silicone hydrogel lens wear. *Optom. Vis. Sci.* **2014**, *91*, 3–12. [CrossRef]
10. Tran, N.P.D.; Yang, M.C. Synthesis and Characterization of Silicone Contact Lenses Based on TRIS-DMA-NVP-HEMA Hydrogels. *Polymers* **2019**, *11*, 944. [CrossRef]
11. Wang, J.J.; Li, X.S. Improved oxygen permeability and mechanical strength of silicone hydrogels with interpenetrating network structure. *Chin. J. Polym. Sci.* **2010**, *28*, 849–857. [CrossRef]

12. Chen, D.; Chen, F.; Hu, X.; Zhang, H.; Yin, X.; Zhou, Y. Thermal stability, mechanical and optical properties of novel addition cured PDMS composites with nano-silica sol and MQ silicone resin. *Compos. Sci. Technol.* **2015**, *117*, 307–314. [CrossRef]
13. Mi, H.Y.; Jing, X.; Huang, H.X.; Turng, L.S. Novel polydimethylsiloxane (PDMS) composites reinforced with three-dimensional continuous silica fibers. *Mater. Lett.* **2018**, *210*, 173–176. [CrossRef]
14. Abbasi, F.; Mirzadeh, H.; Katbab, A.A. Modification of polysiloxane polymers for biomedical applications: A review. *Polym. Int.* **2001**, *50*, 1279–1287. [CrossRef]
15. Rudy, A.; Kuliasha, C.; Uruena, J.; Rex, J.; Schulze, K.D.; Stewart, D.; Angelini, T.; Sawyer, W.; Perry, S.S. Lubricous hydrogel surface coatings on polydimethylsiloxane (PDMS). *Tribol. Lett.* **2017**, *65*, 3. [CrossRef]
16. Ghoreishi, S.; Abbasi, F.; Jalili, K. Hydrophilicity improvement of silicone rubber by interpenetrating polymer network formation in the proximal layer of polymer surface. *J. Polym. Res.* **2016**, *23*, 115. [CrossRef]
17. Abbasi, F.; Mirzadeh, H.; Simjoo, M. Hydrophilic interpenetrating polymer networks of poly (dimethyl siloxane)(PDMS) as biomaterial for cochlear implants. *J. Biomater. Sci. Polym. Ed.* **2006**, *17*, 341–355. [CrossRef]
18. Lin, C.H.; Lin, W.C.; Yang, M.C. Fabrication and characterization of ophthalmically compatible hydrogels composed of poly (dimethyl siloxane-urethane)/Pluronic F127. *Colloids Surf. B* **2009**, *71*, 36–44. [CrossRef]
19. Chen, J.S.; Liu, T.Y.; Tsou, H.M.; Ting, Y.S.; Tseng, Y.Q.; Wang, C.H. Biopolymer brushes grown on PDMS contact lenses by in situ atmospheric plasma-induced polymerization. *J. Polym. Res.* **2017**, *24*, 69. [CrossRef]
20. Jones, L.; Senchyna, M.; Glasier, M.A.; Schickler, J.; Forbes, I.; Louie, D.; May, C. Lysozyme and lipid deposition on silicone hydrogel contact lens materials. *Eye Contact Lens* **2003**, *29*, 75–79. [CrossRef]
21. Lin, C.H.; Cho, H.L.; Yeh, Y.H.; Yang, M.C. Improvement of the surface wettability of silicone hydrogel contact lenses via layer-by-layer self-assembly technique. *Colloids Surf. B* **2015**, *136*, 735–743. [CrossRef] [PubMed]
22. Van Beek, M.; Weeks, A.; Jones, L.; Sheardown, H. Immobilized hyaluronic acid containing model silicone hydrogels reduce protein adsorption. *J. Biomater. Sci. Polym. Ed.* **2008**, *19*, 1425–1436. [CrossRef] [PubMed]
23. Zhao, Z.B.; An, S.S.; Xie, H.J.; Han, X.L.; Wang, F.H.; Jiang, Y. The relationship between the hydrophilicity and surface chemical composition microphase separation structure of multicomponent silicone hydrogels. *J. Phys. Chem. B* **2015**, *119*, 9780–9786. [CrossRef] [PubMed]
24. Lin, C.H.; Yeh, Y.H.; Lin, W.C.; Yang, M.C. Novel silicone hydrogel based on PDMS and PEGMA for contact lens application. *Colloids Surf. B* **2014**, *123*, 986–994. [CrossRef] [PubMed]
25. Pozuelo, J.; Compañ, V.; González-Méijome, J.M.; González, M.; Mollá, S. Oxygen and ionic transport in hydrogel and silicone-hydrogel contact lens materials: An experimental and theoretical study. *J. Membr. Sci.* **2014**, *452*, 62–72. [CrossRef]
26. Zhao, Z.; Xie, H.; An, S.; Jiang, Y. The relationship between oxygen permeability and phase separation morphology of the multicomponent silicone hydrogels. *J. Phys. Chem. B* **2014**, *118*, 14640–14647. [CrossRef]
27. Kim, G.D.; Lee, D.A.; Moon, J.W.; Kim, J.D.; Park, J.A. Synthesis and applications of TEOS/PDMS hybrid material by the sol–gel process. *Appl. Organomet. Chem.* **1999**, *13*, 361–372. [CrossRef]
28. Kohjiya, S.; Ochiai, K.; Yamashita, S. Preparation of inorganic/organic hybrid gels by the sol-gel process. *J. Non Cryst. Solids* **1990**, *119*, 132–135. [CrossRef]
29. Huang, H.H.; Orler, B.; Wilkes, G.L. Structure-property behavior of new hybrid materials incorporating oligomeric species into sol-gel glasses. 3. Effect of acid content, tetraethoxysilane content, and molecular weight of poly (dimethylsiloxane). *Macromolecules* **1987**, *20*, 1322–1330. [CrossRef]
30. Loccufier, E.; Geltmeyer, J.; Daelemans, L.; D'hooge, D.R.; De Buysser, K.; De Clerck, K. Silica nanofibrous membranes for the separation of heterogeneous azeotropes. *Adv. Funct. Mater.* **2018**, *28*, 1804138. [CrossRef]
31. Compan, V.; Andrio, A.; Lopez-Alemany, A.; Riande, E.; Refojo, M. Oxygen permeability of hydrogel contact lenses with organosilicon moieties. *Biomaterials* **2002**, *23*, 2767–2772. [CrossRef]
32. Compan, V.; Guzman, J.; Riande, E. A potentiostatic study of oxygen transmissibility and permeability through hydrogel membranes. *Biomaterials* **1998**, *19*, 2139–2145. [CrossRef]
33. Compan, V.; Villar, M.; Valles, E.; Riande, E. Permeability and diffusional studies on silicone polymer networks with controlled dangling chains. *Polymer* **1996**, *37*, 101–107. [CrossRef]
34. De, G.; Karmakar, B.; Ganguli, D. Hydrolysis–condensation reactions of TEOS in the presence of acetic acid leading to the generation of glass-like silica microspheres in solution at room temperature. *J. Mater. Chem.* **2000**, *10*, 2289–2293. [CrossRef]

35. Chrimes, A.F.; Khoshmanesh, K.; Stoddart, P.R.; Mitchell, A.; Kalantar-zadeh, K. Microfluidics and Raman microscopy: Current applications and future challenges. *Chem. Soc. Rev.* **2013**, *42*, 5880–5906. [CrossRef]
36. González-Vázquez, M.; Hautefeuille, M. Controlled Solvent-Free formation of embedded PDMS-derived carbon nanodomains with tunable fluorescence using selective laser ablation with a Low-Power CD laser. *Micromachines* **2017**, *8*, 307. [CrossRef]
37. Warner, J.; Polkinghorne, J.; Gonerka, J.; Meyer, S.; Luo, B.; Frethem, C.; Haugstad, G. Strain-induced crack formations in PDMS/DXA drug collars. *Acta. Biomater.* **2013**, *9*, 7335–7342. [CrossRef]
38. Wang, D.; Yang, P.; Hou, P.; Zhang, L.; Zhou, Z.; Cheng, X. Effect of SiO2 oligomers on water absorption of cementitious materials. *Cem. Concr. Res.* **2016**, *87*, 22–30. [CrossRef]
39. Korogiannaki, M.; Guidi, G.; Jones, L.; Sheardown, H. Timolol maleate release from hyaluronic acid-containing model silicone hydrogel contact lens materials. *J. Biomater. Appl.* **2015**, *30*, 361–376. [CrossRef]
40. Jacob, J.T. Biocompatibility in the development of silicone-hydrogel lenses. *Eye Contact Lens* **2013**, *39*, 13–19. [CrossRef]
41. Maldonado-Codina, C.; Efron, N. Hydrogel lenses-material and manufacture: A review. *Optom. Pract.* **2003**, *4*, 101–115.
42. Abbasi, F.; Mirzadeh, H. Properties of poly (dimethylsiloxane)/hydrogel multicomponent systems. *J. Polym. Sci. B Polym. Phys.* **2003**, *41*, 2145–2156. [CrossRef]
43. Wang, J.; Li, X. Preparation and characterization of interpenetrating polymer network silicone hydrogels with high oxygen permeability. *J. Appl. Polym. Sci.* **2010**, *116*, 2749–2757. [CrossRef]
44. Maldonado-Codina, C.; Morgan, P.B.; Efron, N.; Canry, J.-C. Characterization of the surface of conventional hydrogel and silicone hydrogel contact lenses by time-of-flight secondary ion mass spectrometry. *Optom. Vis. Sci.* **2004**, *81*, 455–460. [CrossRef]
45. French, K. Contact lens material properties part 1: Wettability. *Optician* **2005**, *230*, 20–28.
46. Green-Church, K.B.; Nichols, K.K.; Kleinholz, N.M.; Zhang, L.; Nichols, J.J. Investigation of the human tear film proteome using multiple proteomic approaches. *Mol. Vis.* **2008**, *14*, 456.
47. Tran, N.P.D.; Yang, M.C. Synthesis and characterization of soft contact lens based on the combination of silicone nanoparticles with hydrophobic and hydrophilic monomers. *J. Polym. Res.* **2019**, *26*, 143. [CrossRef]
48. Song, M.; Shin, Y.H.; Kwon, Y. Synthesis and properties of siloxane-containing hybrid hydrogels: Optical transmittance, oxygen permeability and equilibrium water content. *J. Nanosci. Nanotechnol.* **2010**, *10*, 6934–6938. [CrossRef]
49. Gavara, R.; Compañ, V. Oxygen, water, and sodium chloride transport in soft contact lenses materials. *J. Biomed. Mater. Res. B* **2017**, *105*, 2218–2231. [CrossRef]

© 2020 by the authors. Licensee MDPI, Basel, Switzerland. This article is an open access article distributed under the terms and conditions of the Creative Commons Attribution (CC BY) license (http://creativecommons.org/licenses/by/4.0/).

Article

Transdermal Composite Microneedle Composed of Mesoporous Iron Oxide Nanoraspberry and PVA for Androgenetic Alopecia Treatment

Jen-Hung Fang [1], Che-Hau Liu [1], Ru-Siou Hsu [1], Yin-Yu Chen [1], Wen-Hsuan Chiang [2], Hui-Min David Wang [3] and Shang-Hsiu Hu [1,*]

1. Department of Biomedical Engineering and Environmental Sciences, National Tsing Hua University, Hsinchu 300, Taiwan; s102012803@m102.nthu.edu.tw (J.-H.F.); u9912052@gms.ndhu.edu.tw (C.-H.L.); hsu.ru.siou@gmail.com (R.-S.H.); anny99943@gmail.com (Y.-Y.C.)
2. Department of Chemical Engineering, National Chung Hsing University, Taichung 402, Taiwan; whchiang@dragon.nchu.edu.tw
3. Graduate Institute of Biomedical Engineering, National Chung Hsing University, Taichung 402, Taiwan; davidw@dagon.nchu.edu.tw
* Correspondence: shhu@mx.nthu.edu.tw

Received: 29 May 2020; Accepted: 19 June 2020; Published: 22 June 2020

Abstract: The transdermal delivery of therapeutic agents amplifying a local concentration of active molecules have received considerable attention in wide biomedical applications, especially in vaccine development and medical beauty. Unlike oral or subcutaneous injections, this approach can not only avoid the loss of efficacy of oral drugs due to the liver's first-pass effect but also reduce the risk of infection by subcutaneous injection. In this study, a magneto-responsive transdermal composite microneedle (MNs) with a mesoporous iron oxide nanoraspberry (MIO), that can improve the drug delivery efficiency, was fabricated by using a 3D printing-molding method. With loading of Minoxidil (Mx, a medication commonly used to slow the progression of hair loss and speed the process of hair regrowth), MNs can break the barrier of the stratum corneum through the puncture ability, and control the delivery dose for treating androgenetic alopecia (AGA). By 3D printing process, the sizes and morphologies of MNs is able to be, easily, architected. The MIOs were embedded into the tip of MNs which can deliver Mx as well as generate mild heating for hair growth, which is potentially attributed by the expansion of hair follicle and drug penetration. Compared to the mice without any treatments, the hair density of mice exhibited an 800% improvement after being treated by MNs with MF at 10-days post-treatment.

Keywords: 3D printing process; mesoporous iron oxide; microneedles; minoxidil

1. Introduction

For male baldness, this genetic disease occurs at age 30–60, which is estimated that over 50% of the general population suffer from hair loss [1]. Traditionally in the treatment of hereditary male baldness, the drug was mixed into the inactive component which can soften the stratum corneum, such as glycerol and propylene glycol, facilitating the active ingredient effective for percutaneous absorption [2]. However, the stratum corneum softening takes a long time, with side effects including redness, swelling, rash, and subjective feeling of burning sensation was caused in some users. Additionally, it may lead to hair growth on the undesired local with inaccurate using [3].

As a new type of painless administration, microneedles (MNs) with the advantages of convenient use, minimally invasive procedure, and relatively inexpensive cost, are easy to overcome stratum corneum to facilitate pharmaceuticals entering epidermal as well as dermal layers [4]. MNs are applied in many kinds of researches nowadays. For example, Li et al., last year, designed biodegradable MNs with an air bubble inside the center for rapidly separating chemicals to skin and caused sustained release [5]. However, the conventional fabrications of MNs are rigid, expensive, and inconvenient by directly purchasing commercial products, and using laser ablation, photolithograph, and magneto liquid forming, which prodigiously limits the applications of MNs [6–12].

Magnetic mesoporous nanoparticles consisted of silica as well as iron oxide, and are broadly employed in many fields such as catalyst, cancer therapy, antibacterial composites, energy storage, and drug delivery system [13–19]. These nano-systems provide promising and novel approaches for improving the efficacy with their large pore surface, which can also be easily controlled by adjusting the composition and concentration of surfactant. Especially, iron oxide nanoparticles allow them to be manipulated with externally applied magnetic field, which possess on-demand release behaviors from their cargos without any lag in response. By means of employing magnetic field, iron oxide can slightly generate heat to cause cutaneous blood flow increases as well as vasodilation, leading to the promotion of hair growth [20–22]. Furthermore, iron is one of the most important nutritional factors in human body, as it can help oxygen delivery as well as carbon dioxide removal from animal tissue, and has the composition of serum ferritin [23]. Many studies also show that iron-deficiency is also common with hair loss [24,25].

In this study, the concept of a physical promotion method for patching increases the efficiency of percutaneous absorption and reduces the chance of inducing side effects. A highly biocompatible polyvinyl alcohol was used to make a dissolvable microneedle patch, reaching the smart device in biomedical applications. Digital light processing (DLP) 3D printing technology was used to fabricate the MNs master with high resolution, inexpensive cost, flexible manufacturing, customization, and fast forming prototype compared with conventional process [26]. A mesoporous iron oxide nanoraspberry (MIO) inside MNs (MIOs@MNs) encapsulated Minoxdil (Mx), a therapeutic drug for hair regrowth, which can be triggered by external magnetic field (MF), resulting in local temperature increases as well as controlled release (Figure 1). We further evaluated the biocompatibility and mechanical properties of the MNs, suggesting a potential application in biomedical engineering filed.

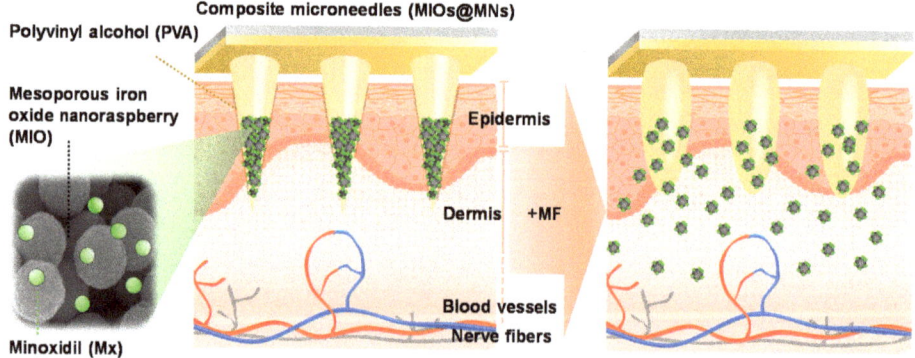

Figure 1. Schematic mechanism of microneedle patch system applying on the surface of skin. Polyvinyl alcohol (PVA) was used to prepare microneedles (MNs) with dissolvable properties. A mesoporous iron oxide nanoraspberry (MIO) encapsulating Minoxdil (Mx) inside the MNs were triggered by external magnetic field (MF), leading to local temperature increases as well as controlled release.

2. Materials and Methods

2.1. Materials

$FeCl_3 \cdot 6H_2O$, ethylene glycol, sodium acetate (NaAc), oleylamine, Mx, and phosphate buffered saline were purchased from Sigma-Aldrich, St. Louis, Missouri, USA. CdSe/ZnS core/shell Quantum Dot

Solid (QDs) were purchased form Ocean Nano Tech, San Diego, California, USA. 3-(4,5-dimethylthiazol-2-yl)-2,5-diphenyltetrazolium bromide (MTT reagent) were bought from Amresco, Solon, Ohio, USA. Ham's F-12 medium, fetal bovine serum (FBS) and penicillin/streptomycin were purchased from Gibco, Thermo Fisher Scientific, Waltham, Massachusetts, USA. Trypsin (2.21 mM EDTA Solution) was purchased from Corning, Steuben County, New York, USA. C57BL/6 mice obtained from National Laboratory Animal Center, NLAC, Taiwan.

2.2. Synthesis of MIOs

The synthesis process of MIOs follows the hydrothermal method of our previous studies [27]. Briefly, 5.4 g of $FeCl_3 \cdot 6H_2O$ were dissolved in 40 mL of ethylene glycol with vigorous stirring till the color of the solution changed to a clear yellow. Then, anhydrous NaAc (3.29 g) was added to the mixture in gentle stirring thereafter to the extent of homogeneity overnight. After the color of the mixtures changed to black, 0.6 mL of oleylamine was added under ultrasonication for effective mixing. Subsequently, 20 mL of the solution was transferred into a teflon-lined stainless steel autoclave (25 mL capacity) for hydrothermal reactions at 220 °C for 5, 15, and 25 h, respectively. After the Teflon-lined stainless steel autoclave was allowed to cool down to room temperature, the precipitate was washed by ethanol, three times, and stored in 4 °C, this was called MIOs. MIOs were dried under vacuum at room temperature before characterization and applications later. For Mx encapsulation, Mx was loaded to MIOs-5h by mixing 1% Mx with MIOs in dimethyl sulfoxide (DMSO) for 12 h. Then, the mixture was placed into vacuum at 50 °C for 24 h to evaporate DMSO. The drug concentration was determined by ultraviolet–visible (UV-vis) spectroscopy (Metertech, Taipei, Taiwan).

2.3. Fabrication Process of MNs

We used 3D printing technologies to prepare MNs master so that we can freely adjust the specification of MNs including height, width, and length of MNs. Each needle was 300 μm in a round base diameter and 600 μm in height. Needles were arranged with 600 μm tip-tip spacing. Polydimethylsiloxane (PDMS) was placed over the 3D printed MNs master to make female mold, followed by curing in the oven at 60 °C for 8 h, and the female mold was finally finished by carefully peeling the 3D printed MNs. For preparation polyvinyl alcohol (PVA)-MNs patch, the mixture, PVA and Mx-MIOs, were loaded onto PDMS female mold, and placed in a centrifuge at 4500 rpm for 1 h, so that the mixed solution would fill the pyramidal cavity of the mold. The back of MNs was filled with pure PVA solution, and centrifuged at 4500 rpm and 25 °C for 1 h. Subsequently, the device was allowed to dry at room temperature for 12 h, and then the patch was carefully detached from PDMS mold.

2.4. Characterizations of MIOs and MNs

The morphology of MIOs are analyzed by field emission scanning electron microscope (JEOL Ltd., Tokyo, Japan), and MIOs were dried up on silicon wafers. The wafers were anchored to SEM specimen mounts using double-sided carbon tape, and sputter deposited with platinum in 10 mA for 120 s. An X-ray diffractometer (D8 Advance X-ray Diffractometer, Bruker, Billerica, MA, USA) was used for identification of crystal dimensions. A thermogravimetric analyzer (Seiko Instruments Inc., Tokyo, Japan) provided a proportion of each compositions based on differential melting temperature with physical and chemical properties. Dynamics laser scattering (DLS) analysis, by using a particle sizer (Nano-ZS, Malvern, Worcestershire, UK), determined the average size and size distribution of MIOs diluted with deionized water. Field-dependent magnetization curves were evaluated by a superconducting quantum interference device (San Diego, CA, USA) from −10,000 to 10,000 Oe.

2.5. Controllable Properties of MIOs

For heat generation studies, MIOs-5h, MIOs-15h, and MIOs-25h were inserted to the center of the coils, and MIOs were sequentially conducted with a MF (President Honor Industries, Taipei, Taiwan) at a power of 3.2 kW and frequency with 1 MHz. The temperature of solution increased and was detected at various times.

Mx was used to evaluate the drug loading capacity and release properties of MIOs. Then, 1% Mx were mixed with MIOs in DMSO for 12 h, and then the mixture was placed into vacuum at 50 °C for 24 h to evaporate the DMSO. After the organic solvent was evaporated, the mixture was washed by DI water through being centrifuged twice. Mx was quantified by a UV–vis spectrometer with the wavelength in 230 nm. The loading capacity of Mx in MIOs were calculated by the equation: EE% = (A − B)/A × 100, where A is the total amount of Mx, and B is the amount of Mx remaining in the supernatant. To estimate the drug release profile, Mx containing MIOs were dispersed in phosphate buffered saline (PBS). The solution was centrifuged at 4000 rpm and taken at various times for detection.

For magnetically triggered release, MF was applied to activate and induce heat of MIOs at a power of 3.2 kW and frequency with 1 MHz for 1 min at the beginning. Mx-MIOs were dispersed in water and placed into tubes for detection at various times.

2.6. In Vitro Experiments

HIG-82 was obtained from the American Type Culture Collection (ATCC), and were maintained in F12 medium containing with 10% (v/v) FBS and 1% (v/v) penicillin-streptomycin at 37 °C and in 5% CO_2. The culture medium was replaced every three days with a fresh one. For biocompatibility, we followed "Biological evaluation of medical devices—Part 5: Tests for in vitro cytotoxicity" (ISO 10993-5) to investigate the MNs. Briefly, we extracted the substance released from MNs in culture medium with serum following ISO 10993-12. After seeding HIG-82 cells in 96-well (10,000 cells per well) with 100 µL of medium for 24 h, we added the extraction into each well for another 24 h incubation. Finally, each well was added with 100 µL of MTT agent (1 mg/mL) for 4 h, and then dissolved in DMSO to make purple color liquid. The absorbance value was read with a microplate reader (Synergy™ HT Multi-detection microplate reader, BioTek Instruments, Inc., Winooski, VT, USA) with 570 nm in wavelength and 650 nm as reference. Cell viability was calculated by comparison with untreated cells and calculated according to the following: Cell viability (%) = absorbance of experimental group/absorbance of control group.

2.7. In Vivo Experiments

All C57BL/6 mice aged 6 weeks were purchase from National Laboratory Animal Center, NLAC, Taiwan as the animal model of hair growth experiment. All mice were maintained under conditions at 22 ± 2 °C on a 12 h of dark-day cycle with access water and food. All surgical procedures were performed in accordance with the protocol approved by the Institutional Animal Care and Use Committee (IACUC), National Tsing Hua University, Hsinchu, Taiwan (IACUC protocol and approval number is 10704). Mice with 8 weeks of age were shaved and divided into 4 groups, including drug-containing control group (PBS), Mx-loaded MNs (Mx@MNs), Mx-MIOs@MNs, and Mx-MIOs@MNs+MF. Mice was treated with patches at day one after shaving. For control group, the shaved back of mice was covered with PBS solution. For Mx@MNs, Mx-MIOs@MNs, and Mx-MIOs@MNs+MF, the patch was pressed firmly for the first 30 s to penetrate through the epidermis and pressed softly for additional 2 min. The patch base was peeled at 10 min postinsertion into the skin, leaving the MNs settled in the skin for further sustained drug release. For Mx-MIOs@MNs+MF group, the mice were further treated under MF (power cube 32/900, President Honor Industries, Taiwan) at a power of 3.2 kW and frequency with 1 MHz. Photos of the mice back were recorded for the quantitation of the hair density by Zen Desk software.

2.8. Statistical Analysis

As an analysis of the variance, both one- and two-way ANOVA were used, and statistical divergences were assessed by ANOVA with Tukey's multiple comparison test. The Graph Pad-Prism was used with a significance level of alpha 0.05 to evaluate the statistical significance.

3. Results and Discussions

3.1. Synthesis and Characterization of MIOs

The synthesized process of MIOs was illustrated in Figure S1a. MIOs (porous iron oxide particles) were prepared by using ligand-aided synthetic approach, in which oleic amine (OA) and sodium citrate as coordinating agents was applied in a hydrothermal reaction [27]. Several methods, such as thermal decomposition, coprecipitation, or electrochemical, were able to fabricate the iron oxide based particles, the ligand-aided synthetic iron oxide that can manipulate the surface property, and the crystal growth rate to form the porous structures in one-step [28]. The magnetic particles serving as a drug delivery system are promising for on-demand drug release at a specific place since the external magnetic field was able to manipulate them remotely for spatial guiding and magneto-thermal conversion by a unique alternating MF [29]. After various reaction times, scanning electron microscopy (SEM) was applied to evaluate these MIOs as shown in Figure 2a. Based on the reaction time, the MIOs hereafter were termed as MIOs-5h, MIOs-15h, and MIOs-25h. Once the reaction was 5 h, the uniform MIOs-5h with clear pores could be observed, and it had a mean size of ~150 nm in diameter. At a closer observation (lower panel of Figure 2a), the MIOs-5h were composed by several iron oxide domains with 20 nm and the pores were constructed by these IO domains. Furthermore, iron oxide domains of MIOs were also observed by TEM (Figure S1b).

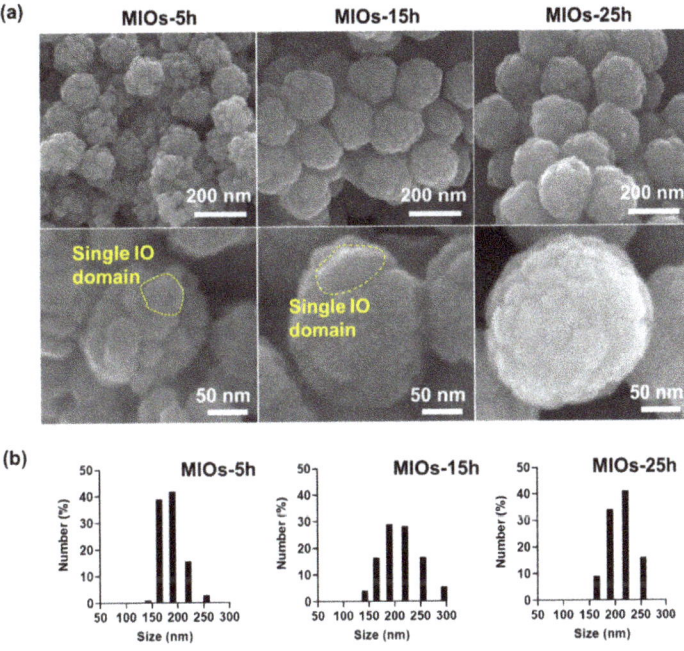

Figure 2. Morphologies and size distribution of MIOs. (**a**) scanning electron microscopy (SEM) images of different synthesis time (5 h, 15 h, and 25 h) of MIOs. (**b**) the size distribution of MIOs with different synthesis time (5 h, 15 h, and 25 h) measured by dynamic light scattering (DLS).

The size and surface roughness of MIOs were affected by the reaction time. With reaction time ranged from 5 to 25 h, an increase of particle sizes was monitored from 150 to 200 nm, respectively. Moreover, the IO domains were also grown with time, causing the lower pore size and smoother surface for MIOs-15h and MIOs-25h. Similar size distribution was also detected by dynamic light scattering (DLS) as shown in Figure 2b. The plausible mechanism of fabrication for hydrothermal reaction has also been documented as an oriented gather and subsequent local Ostwald ripening [30], where the nucleation and growth of iron salts in solvent via the oriented growth of iron crystal and OA restricted the growth of crystal to form the pores. Furthermore, to evaluate the surface area, the Brunauer–Emmett–Teller (BET) method was applied to measure gas absorption isotherms (N_2) at 77 K. The particles were treated at 80 °C under vacuum to remove the surface adsorption and were degassed at 180 °C for 4 h before BET analysis. The results revealed that each MIO-5h and MIO-10h had a surface area of ~265 and ~78 m^2g^{-1}, respectively (Figure S1c). Compared to MIO-5h, the lower surface area of MIO-10h was potentially attributed by the longer time of growth of iron oxide crystals. Moreover, such a OA ligand also adjusted the surface characteristics, efficiently improving the loading capacity of hydrophobic pharmaceuticals [30].

In Figure 3a, X-ray diffraction (XRD) analyses of MIOs exhibited the major diffraction planes at (200), (311), (400), (422), (511), and (440), which are the characteristic of the Fe_3O_4 crystal planes [31]. OA with primary amine, can adsorb onto building blocks as a coating agent, and also like a steric hindrance agent as a stabilizer during a hydrothermal reaction [32]. MIOs with different reactions represented identical characteristic peaks, and MIOs-5h has the smaller half width at (311) plane than MIOs-15h and MIOs-25h, suggesting the smaller grain size of IO particles. The results were also consistent to SEM images. Furthermore, a SQUID was utilized to investigate the magnetic properties of MIOs (Figure 3b). MIOs-5h, MIOs-15h, and MIOs-25h had similar magnetization-field patterns with negligible hysteresis but different magnetization saturations (Ms). The Ms of MIOs-5h was approximately 62 emu/g lower than MIOs-15h and MIOs-25h. As expected, while increasing the size of MIOs, an enhancing in Ms was observed. The difference was also reflected in the particle and grain sizes [33].

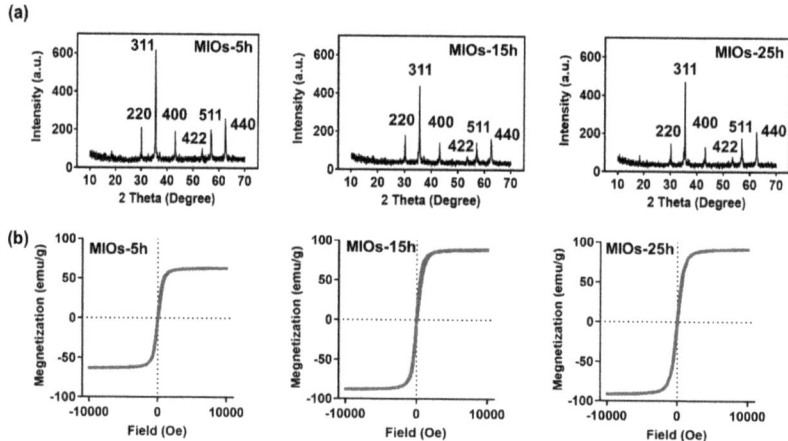

Figure 3. Characterization of MIOs. (a) X-ray diffraction (XRD). (b) superconducting quantum interference device (SQUID).

While 10 µg/mL of MIOs were subjected to a high frequency MF at a strength of 3.2 kW with a frequency of 1 MHz, an obvious temperature increase could be monitored (Figure 4a). Within 6 min of MF treatment, the temperature of solution with MIOs-25h increased to 86 °C. The mechanism of MF generating heat through magnetic matters can be understood by the energy relaxation and dissipation, known as Brown and Néel relaxations [34,35]. Brownian relaxation is due to rotational diffusion of the whole particle in the carrier liquid, and Néel relaxation is caused by the reorientation of the magnetization vector inside the magnetic core against an energy barrier. The energy was induced by the friction between particle–particle and particles–solvent under the alternating magnetic moments. On the other hand, the different IO domains with distinctive magnetic direction in one particle can also cause intense internal energy through the energy relaxation and dissipation. Therefore, both particle sizes and domains with various magnetic property affected the heating rates. After 6 min of MF application, the temperature of MIOs-5h was heated to 61 °C. The heating rate was affected by the size of iron oxide domain. The smaller iron oxide crystal usually offered the lower heating rate due to the weaker saturation of magnetization [36–38]. Therefore, MIOs-5h possessing smaller IO crystal sizes had a weaker heating rate in solution under MF treatment. For superparamagnetic iron oxide nanoparticles, the most influential analytic solution was derived using a linear response theory. The suitable magneto–thermal conversion was convenient for remotely controlled drug release and tissue treatment [39].

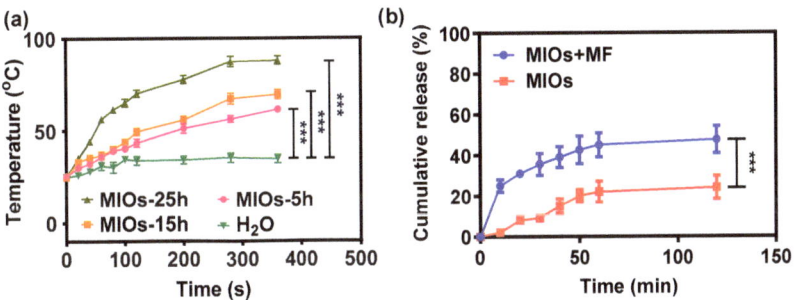

Figure 4. Controllable properties of MIOs. (**a**) temperature changes of MIOs solution under MF at a power of 3.2 kW and a frequency of 1 MHz. (**b**) cumulative release profile of Mx from MIOs with and without external MF treatment. Data points represent mean ± SD (N = 3, *** $p < 0.005$).

Considering the porosity and mild heating rate, MIOs-5h was used in the further studies. Figure 4b displayed the drug release profiles of MIOs-5h under MF. Before the release experiments, Mx was loaded to MIOs-5h by mixing 1% Mx with MIOs in DMSO for 12 h. Then, the mixture was placed into a vacuum at 50 °C for 24 h to evaporate DMSO. Estimated by the un-loaded Mx in supernatant, the loading efficiency of MIOs was about 88%, where loading was potentially attributed by the affinity between Mx and the large hydrophobic surface of MIOs-5h [40]. In the drug loading process of porous particles, oleic acid on the surface was usually utilized to offer hydrophobic–hydrophobic interaction for the physical adsorption of guest agents [30]. While coating hydrophobic ligands on the pore structure, the drug delivery system would efficiently improve the loading capacity of hydrophobic agents [40]. In the Mx loading, it was dissolved in organic solvent (DMSO) and mixed with MIOs. During the evaporation process for drug loading, the capillary motion of Mx would also potentially improve the loading efficiency.

The Mx release from MIOs was monitored in PBS at 37 °C for 120 min (Figure 4b). Without treated by MF, a sustained released profile was detected, in which about 20% of Mx was released after 120 min. The slow release also reflected the affinity between Mx and MIOs. Once applying 1 min of MF to MIOs at the beginning, a faster release of Mx was monitored, and the treatment caused about 23% of the Mx to release within 10 min. Even the MF was turned off, the release was sustained. For 120 min, more than 40% of Mx can be released. The release was driven by the heat energy generated by MIOs, which was able to increase the molecules kinetics to overcome the adsorption on MIOs. Overall, the results indicated the on-demand Mx release from MIOs can be achieved by a MF.

3.2. Fabrication and Characterization of MNs

To prepare the MNs, PDMS served as the negative mold was reprinted according to the microneedle template printed in advance by 3D printing (Figure 5a). Due to the convenience of 3D printing, the different sizes and shapes of MNs can be designed and printed by digital light processing (Figures S2 and S3). In this study, for androgenetic alopecia treatment, the height of needle was designed as 600 µm with 11 by 11 array to reach the hair follicles. Using PDMS as a mold, 10 wt% of PVA with a molecular weight of 10,000 g/mol was filled, centrifuged, and dried in the mode. After solvent completely evaporated, the PVA MNs can be formed and separated from mold. The resulting PVA MNs was shown in Figure 5b. It provided the sharp and clear needle shapes with strong mechanical property.

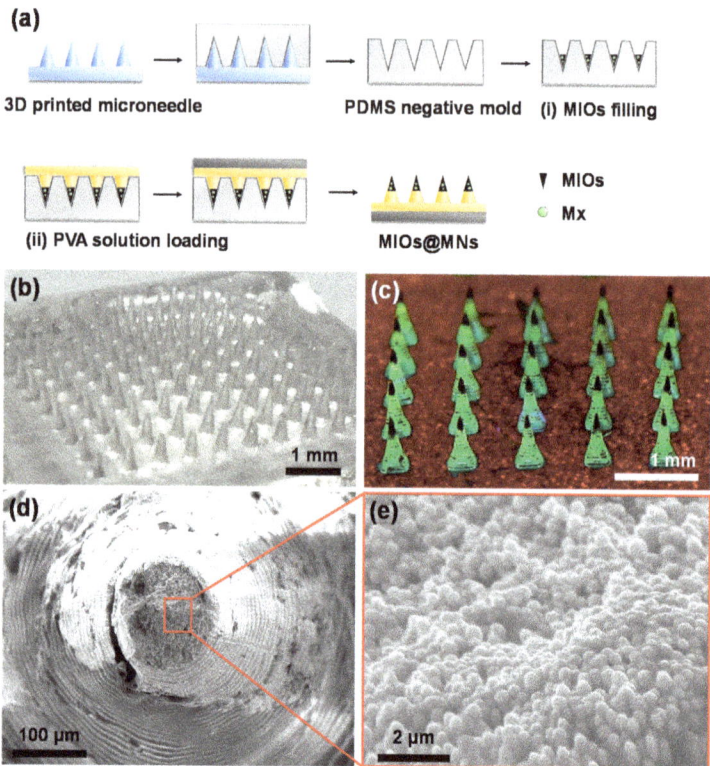

Figure 5. (a) Synthesis process of composite MNs with digital light processing (DLP) 3D printing and polydimethylsiloxane (PDMS) molding. Two steps of filling were carried out: (i) MIOs and (ii) PVA filling. The images of (b) PVA MNs and (c) composite MNs after mold preparation. MNs with MIOs (at needle tips) and green fluorescence PVA. SEM image of composite MNs in (d) low and (e) high magnification.

To form the composite MNs, a two step filling process was applied. First, the Mx-loaded MIOs (Mx-MIOs) were filled to PDMS modes. Followed by centrifugation and drying, the similar processes of PVA MNs were used to construct the MNs in the second step. The resulting composite MNs was shown in Figure 5c, in which MIOs as dark parts in MNs were observed in each tip. Once the green fluorescence dye was dissolved in PVA, for second layer in advance, the resulting fluorescence MNs could also be exhibited. The third layer labeled by red fluorescence could also easily formed. The results suggested that each layers of MNs can be established through the reliable step by step filling processes. Under SEM observation, the high resolution of composite MNs was investigated (Figure 5d). The ring-like patterns of MNs was caused by the DLS printing building the structures layer by layer. In the tip of MNs, the high density of MIOs was displayed, consistent to the previous results (Figure 5e).

Have demonstrated the structures of composite MNs, the mechanical property of composite MNs was estimated to confirm the ability to break through the stratum corneum barrier. The setup of the experiment was illustrated in Figure 6a, and the pressure versus displacement map obtained by applying pressure under the CT3 Texture (Figure 6b). The composite MNs were able to break the stratum corneum, where the minimum pressure required to break the stratum corneum was 0.17 N/needle [41]. Both PVA and composite MNs had sufficient mechanical strength to break the stratum. Furthermore, the thermogravimetric analyzer (TGA) was used to determine the weight ratio of MIOs and PVA in a single needle (Figure 6c). The result revealed the MIOs contributed 10% weight ratio in composite MNs, and 60% weight ratio of composite MNs was offered from organic materials such as PVA.

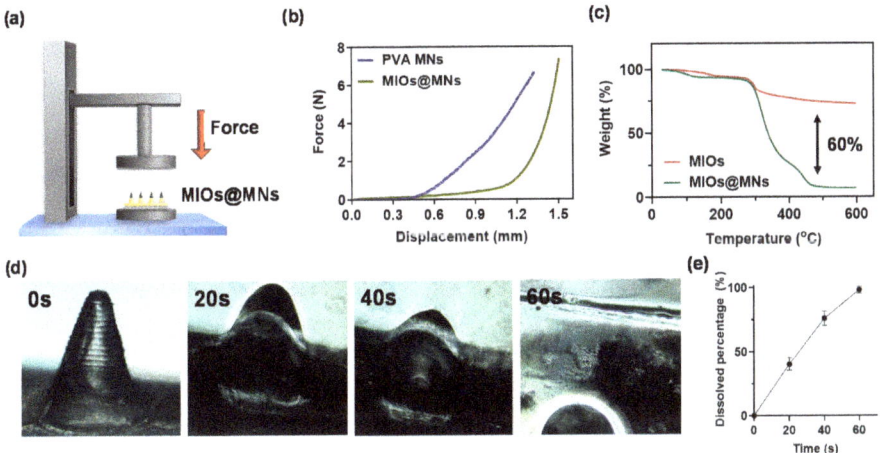

Figure 6. (a) schematic illustration of mechanical strength test for PVA and composite MNs. (b) compression displacement curve of PVA and composite MNs. (c) thermogravimetric analyzer (TGA) analysis of PVA and composite MNs. (d) optical images of dissolvable composite MNs in water at various times. (e) quantitation of dissolving rate.

In addition, we performed an in vitro microneedle swelling test to simulate the in vivo condition. PVA MNs swelled quickly within a minute (Figure 6d,e). This microneedle patch was proved to be soluble within a few seconds in water, and the fast dissolution characteristics have great potential for the delivery of the drug, followed by decomposing the patch through skin interstitial fluid. The experimental results show that the microneedle patch can effectively damage the stratum corneum and dissolve in the skin, and the transmission efficiency should be improved for the drug in the transdermal route.

3.3. In Vitro Experiments

The MNs were applied to a pork skin to investigate the insertion (Figure S4a). With 10 min of MF applications, the skin surface was cleaned to remove the excess MNs. Under the images, the clear MNs patterns can be detected on the skin surfaces, indicating the MN was able to insert into the skin. The histological evaluation also further confirmed the MNs insertion (Figure S4b), in which the barrier of the stratum corneum was broken.

Having exhibited the insertion of MNs, the toxicity of PVA and MIOs@MNs were evaluated in Figure 7. The in vitro cell compatibility of MNs were determined using fibroblast cells (HIG-82). Briefly, HIG-82 cells were plated on 96 well microplates at a density of 1×10^4 cells/mL. Then, the culture medium was changed, and cells were exposed to serial dilutions of tested dissolved MNs. Cell viability was assessed after 72 h by means of MTT assay by measuring absorbance at 570 nm wavelength using ELISA Reader. In Figure 7a, the cell viability decreased with increasing PVA concentrations, but the cell viability can measure more than 90%, suggesting the low toxicity of PVA MNs. The cell viability of MIOs in Figure 7b also performed similar behaviors and weak influences in cell viability even at the Mx-MIOs@MNs at concentration of 200 mg/mL. While applying 1 min of MF, the higher cell toxicity was observed in composite MNs. The cell death was potentially caused by the heat. Although the temperature was less than 40 °C with 1 min treatment, the local energy may affect the cell proliferation at high concentration. However, the cell viability can still keep at 75% at the concentration of 200 mg/mL.

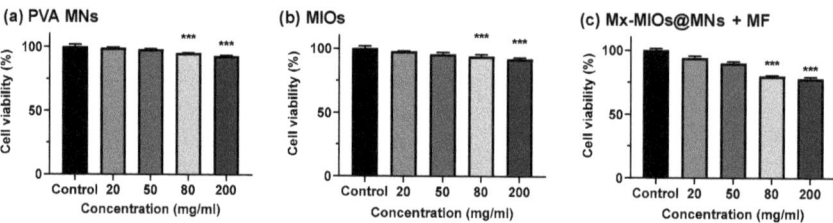

Figure 7. The cell viability of HIG-82 cells incubated with (**a**) PVA MNs, (**b**) MIOs, and (**c**) Mx-MIOs MNs+MF. Data points represent mean ± SD (N = 5, *** $p < 0.005$).

3.4. In Vivo Animal Hair Growth Experiments

The in vivo animal hair growth studies were carried by 8 weeks-old C57BL/6 mice. Before treating by MNs, the hairs were removed by shaving and hair removal cream. Then, C57BL/6 mice were divided into four groups, including control, Mx-loaded MNs (Mx@MNs), Mx-MIOs@MNs, and Mx-MIOs@MNs+MF (Figure 8a). A microneedle patch (11 by 11) prepared by the above-mentioned experimental optimization formula was administered at the 8th week, and the administration area was 1 cm². The hair was applied to the back of the hairless mouse and peeled off after 10 min of attachment. The same dosage of Mx was used in the four groups. After photographing the back of the mouse, a fixed range was tracked for 16 days and taken to quantify the percentage of the selected range of hair by the computer software ZEN zeiss at various times (Figure S5). The results of hair growth were exhibited in Figure 8b. At the first time point, the bright intensity was normalized as baseline. At the 3rd and 7th days of treatment, the obvious difference could be observed, where the mice treated by Mx-MIOs@MNs demonstrated 25% and 40% of hair re-growth, respectively, while no hair was found regrowth beyond the treated region. However, in the sharp contrast, the control group generated an inferior therapeutic effect, showing less than 12% recovery of hair within 1 week. No obvious hair regrowth was found in the mouse without simultaneous MF treatment and Mx during the test period, indicating that the hair follicles were still in the telogen phase of the hair cycle [42]. Once applying MF to Mx-MIOs@MNs, the faster hair growth rate can be monitored within two weeks. At the 10th day, the hair density of Mx-MIOs@MNs+MF was eight-fold greater than that of control group, indicating the improvement of thermal-derived Mx release and penetration. Furthermore, both liver and kidney functions were evaluated to understand the toxicity of treatments. Six indices of liver and kidney functions including albumin (ALB), alanine aminotransferase (ALT), blood urea nitrogen (BUN), creatinine (CRE), globulin (GLOB), and total protein (TP) were also evaluated at 3-day posttreatment (Figure 9). The results revealed that the PVA and composite MNs with MF treatment caused negligible toxicity.

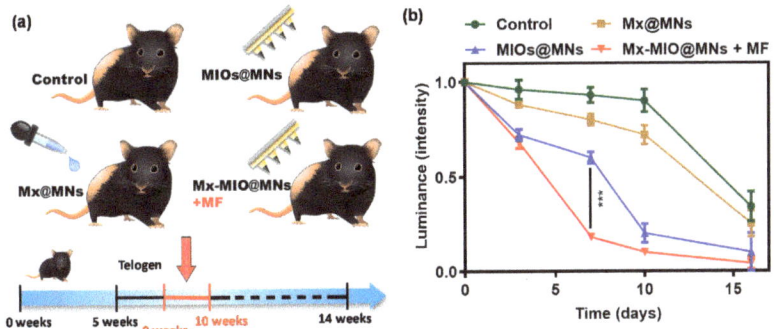

Figure 8. In vivo evaluation of dissolvable MNs for hair loss treatment. (**a**) schematic illustration represents control, Mx@MNs, Mx-MIOs@MNs, and Mx-MIOs@MNs+MF groups for hair loss treatment for 16 days. (**b**) the hair intensity treated by control, Mx@MNs, Mx-MIOs@MNs, and Mx-MIOs@MNs+MF groups for hair growth on mice. Data points represent mean ± SD (N = 3, *** $p < 0.005$).

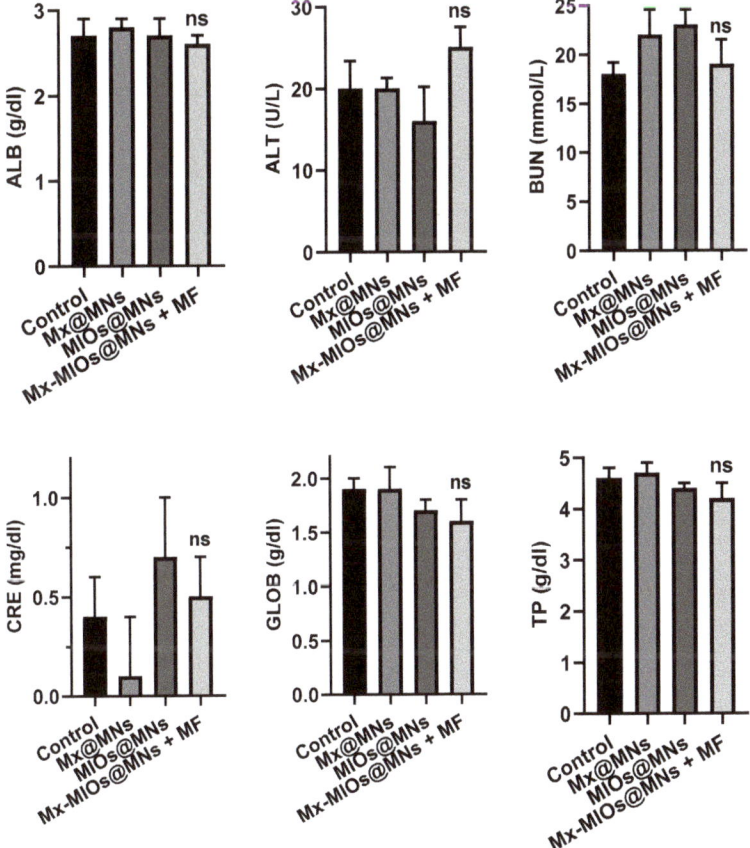

Figure 9. The toxicity of liver and kidney. Kidney functions of albumin (ALB), alanine aminotransferase (ALT), blood urea nitrogen (BUN), creatinine (CRE), globulin (GLOB), and total protein (TP) of mice treated by Mx@MNs, Mx-MIOs@MNs, and Mx-MIOs@MNs+MF (n = 3, mean ± SD, ns: No significant difference).

4. Conclusions

In summary, magneto-responsive transdermal composite microneedles with a mesoporous iron oxide nanoraspberry were developed to improve androgenetic alopecia (AGA) treatment. Unlike traditional administration, this approach can reduce side effect from treating on the undesired local with inaccurate use. MIOs with porous structure and hydrophobic surface can capture hydrophobic pharmaceuticals, which also are the most important nutrition avoiding hair loss in human body. Besides, under the non-contact trigger, MF, MIOs, would generate wild heating for aiding increases of cutaneous blood flow as well as vasodilation. Minoxidil is a drug for hair regrowth with high efficiency. By using 3D printed technology, it can build various morphologies microneedle molds for further uses according to the geometry of tissue anatomies. Through filling Mx-loaded MIOs and PVA subsequently to mold to MNs, the layer-by-layer composite MNs is constructed with high mechanical property for overcoming the barrier of the stratum corneum via the punctuation, and dissolved, later, after administration. The successful delivering the Mx-loaded MIOs to skin and integrating the magneto–thermal drug release achieved eight-fold greater improvements of hair growth with low toxicity. This composite MNs provides a new synergistic delivery strategy for transdermal drug delivery and potential for wide clinical applications.

Supplementary Materials: The following are available online at http://www.mdpi.com/2073-4360/12/6/1392/s1, Figure S1: Synthesis process of MIOs and hydrothermal reaction; Figure S2: Schematic of fabrication of MNs by 3D printing; Figure S3: Image of microneedle surface puncture; Figure S4: Skin insertion of MNs; Figure S5: Comparison in hair growth in an alopecia model of C57BL/6 mice applied with test compound topically for over two weeks.

Author Contributions: Conceptualization, S.-H.H.; methodology, J.-H.F., C.-H.L., Y.-Y.C., W.-H.C. and H.-M.D.W.; software, C.-H.L.; validation, J.-H.F., C.-H.L., Y.-Y.C., W.-H.C. and H.-M.D.W.; formal analysis, J.-H.F.; investigation, J.-H.F., C.-H.L., Y.-Y.C., W.-H.C. and H.-M.D.W.; resources, R.-S.H.; data curation, J.-H.F. and C.-H.L.; writing—original draft preparation, J.-H.F. and C.-H.L.; writing—review and editing, J.-H.F.; visualization, C.-H.L.; supervision, S.-H.H.; project administration, S.-H.H.; funding acquisition, S.-H.H. All authors have read and agreed to the published version of the manuscript.

Funding: This research was funded by the Ministry of Science and Technology of the Republic of China, Taiwan under contracts MOST 108-2636-E-007-001 and MOST 106-2628-E-007-003-MY3 and by National Tsing Hua University (107Q2512E1) in Taiwan.

Conflicts of Interest: The authors declare no conflict of interest.

References

1. Yang, G.; Chen, Q.; Wen, D.; Chen, Z.; Wang, J.; Chen, G.; Wang, Z.; Zhang, X.; Zhang, Y.; Hu, Q.; et al. A therapeutic microneedle patch made from hair-derived keratin for promoting hair regrowth. *ACS Nano* **2019**, *13*, 4354–4360. [CrossRef] [PubMed]
2. Liao, A.H.; Lu, Y.J.; Lin, Y.C.; Chen, H.K.; Sytwu, H.K.; Wang, C.H. Effectiveness of a layer-by-layer microbubbles-based delivery system for applying minoxidil to enhance hair growth. *Theranostics* **2016**, *6*, 817–827. [CrossRef] [PubMed]
3. Sfouq Aleanizy, F.; Yahya Alqahtani, F.; Alkahtani, H.M.; Alquadeib, B.; Eltayeb, E.K.; Aldarwesh, A.; Abdelhady, H.G.; Alsarra, I.A. Colored polymeric nanofiber loaded with minoxidil sulphate as beauty coverage and restoring hair loss. *Sci. Rep.* **2020**, *10*, 4084. [CrossRef] [PubMed]
4. Limcharoen, B.; Toprangkobsin, P.; Kroger, M.; Darvin, M.E.; Sansureerungsikul, T.; Rujwaree, T.; Wanichwecharungruang, S.; Banlunara, W.; Lademann, J.; Patzelt, A. Microneedle-facilitated intradermal proretinal nanoparticle delivery. *Nanomaterials* **2020**, *10*, 368. [CrossRef]
5. Li, W.; Terry, R.N.; Tang, J.; Feng, M.R.; Schwendeman, S.P.; Prausnitz, M.R. Rapidly separable microneedle patch for the sustained release of a contraceptive. *Nat. Biomed. Eng.* **2019**, *3*, 220–229. [CrossRef]
6. Lee, H.; Choi, T.K.; Lee, Y.B.; Cho, H.R.; Ghaffari, R.; Wang, L.; Choi, H.J.; Chung, T.D.; Lu, N.; Hyeon, T.; et al. A graphene-based electrochemical device with thermoresponsive microneedles for diabetes monitoring and therapy. *Nat. Nanotechnol.* **2016**, *11*, 566–572. [CrossRef]

7. Chen, S.; Matsumoto, H.; Moro-oka, Y.; Tanaka, M.; Miyahara, Y.; Suganami, T.; Matsumoto, A. Microneedle-array patch fabricated with enzyme-free polymeric components capable of on-demand insulin delivery. *Adv. Funct. Mater.* **2019**, *29*, 1807369. [CrossRef]
8. Boopathy, A.V.; Mandal, A.; Kulp, D.W.; Menis, S.; Bennett, N.R.; Watkins, H.C.; Wang, W.; Martin, J.T.; Thai, N.T.; He, Y.; et al. Enhancing humoral immunity via sustained-release implantable microneedle patch vaccination. *Proc. Natl. Acad. Sci. USA* **2019**, *116*, 16473–16478. [CrossRef]
9. McHugh, K.J.; Jing, L.; Severt, S.Y.; Cruz, M.; Sarmadi, M.; Jayawardena, H.S.N.; Perkinson, C.F.; Larusson, F.; Rose, S.; Tomasic, S.; et al. Biocompatible near-infrared quantum dots delivered to the skin by microneedle patches record vaccination. *Sci. Transl. Med.* **2019**, *11*, 7162. [CrossRef]
10. Bae, W.G.; Ko, H.; So, J.Y.; Yi, H.; Lee, C.H.; Lee, D.H.; Ahn, Y.; Lee, S.H.; Lee, K.; Jun, J.; et al. Snake fang-inspired stamping patch for transdermal delivery of liquid formulations. *Sci. Transl. Med.* **2019**, *11*, 3392. [CrossRef]
11. Li, Y.; Chen, Z.; Zheng, G.; Zhong, W.; Jiang, L.; Yang, Y.; Jiang, L.; Chen, Y.; Wong, C.-P. A magnetized microneedle-array based flexible triboelectric-electromagnetic hybrid generator for human motion monitoring. *Nano Energy* **2020**, *69*, 104415. [CrossRef]
12. Zhang, X.; Chen, G.; Bian, F.; Cai, L.; Zhao, Y. Encoded microneedle arrays for detection of skin interstitial fluid biomarkers. *Adv. Mater.* **2019**, *31*, e1902825. [CrossRef]
13. Yang, K.; Liu, Y.; Deng, J.; Zhao, X.; Yang, J.; Han, Z.; Hou, Z.; Dai, H. Three-dimensionally ordered mesoporous iron oxide-supported single-atom platinum: Highly active catalysts for benzene combustion. *Appl. Catal. B* **2019**, *244*, 650–659. [CrossRef]
14. Kim, T.; Fu, X.; Warther, D.; Sailor, M.J. Size-controlled Pd nanoparticle catalysts prepared by galvanic displacement into a porous Si-iron oxide nanoparticle Host. *ACS Nano* **2017**, *11*, 2773–2784. [CrossRef]
15. Fan, W.; Lu, N.; Shen, Z.; Tang, W.; Shen, B.; Cui, Z.; Shan, L.; Yang, Z.; Wang, Z.; Jacobson, O.; et al. Generic synthesis of small-sized hollow mesoporous organosilica nanoparticles for oxygen-independent X-ray-activated synergistic therapy. *Nat. Commun.* **2019**, *10*, 1241. [CrossRef]
16. Zhao, T.; Chen, L.; Wang, P.; Li, B.; Lin, R.; Abdulkareem Al-Khalaf, A.; Hozzein, W.N.; Zhang, F.; Li, X.; Zhao, D. Surface-kinetics mediated mesoporous multipods for enhanced bacterial adhesion and inhibition. *Nat. Commun.* **2019**, *10*, 4387. [CrossRef] [PubMed]
17. Sun, Z.; Cai, X.; Feng, D.-Y.; Huang, Z.-H.; Song, Y.; Liu, X.-X. Hybrid iron oxide on three-dimensional exfoliated graphite electrode with ultrahigh capacitance for energy storage applications. *ChemElectroChem* **2018**, *5*, 1501–1508. [CrossRef]
18. Pradhan, L.; Thakur, B.; Srivastava, R.; Ray, P.; Bahadur, D. Assessing therapeutic potential of magnetic mesoporous nanoassemblies for chemo-resistant tumors. *Theranostics* **2016**, *6*, 1557–1572. [CrossRef]
19. Chiang, M.-R.; Su, Y.-L.; Chang, C.-Y.; Chang, C.-W.; Hu, S.-H. Lung metastasis-targeted donut-shaped nanostructures shuttled by the margination effect for the PolyDox generation-mediated penetrative delivery into deep tumors. *Mater. Horiz.* **2020**, *7*, 1051–1061. [CrossRef]
20. Cockrem, F. Effect of a sympathomimetic agent (methoxamine hydrochloride) on growth of hair in the house mouse (mus musculus). *Nature* **1959**, *183*, 614–615. [CrossRef] [PubMed]
21. Choi, H.-I.; Kim, D.Y.; Choi, S.J.; Shin, C.Y.; Hwang, S.T.; Kim, K.H.; Kwon, O. The effect of cilostazol, a phosphodiesterase 3 (PDE3) inhibitor, on human hair growth with the dual promoting mechanisms. *J. Dermatol. Sci.* **2018**, *91*, 60–68. [CrossRef] [PubMed]
22. Woo, Y.M.; Kim, O.J.; Jo, E.S.; Jo, M.Y.; Ahn, M.Y.; Lee, Y.-H.; Li, C.-r.; Lee, S.-H.; Choi, J.-S.; Ha, J.M.; et al. The effect of Lactobacillus plantarum hydrolysates promoting VEGF production on vascular growth and hair growth of C57BL/6 mice. *J. Anal. Sci. Technol.* **2019**, *10*, 18. [CrossRef]
23. Andrews, N.C. Iron homeostasis: Insights from genetics and animal models. *Nat. Rev. Genet.* **2000**, *1*, 208–217. [CrossRef]
24. Rushton, D.H. Nutritional factors and hair loss. *Clin. Exp. Dermatol.* **2002**, *27*, 396–404. [CrossRef] [PubMed]
25. Du, X.; She, E.; Gelbart, T.; Truksa, J.; Lee, P.; Xia, Y.; Khovananth, K.; Mudd, S.; Mann, N.; Moresco, E.M.; et al. The serine protease TMPRSS6 is required to sense iron deficiency. *Science* **2008**, *320*, 1088–1092. [CrossRef]
26. Zhu, W.; Tringale, K.R.; Woller, S.A.; You, S.; Johnson, S.; Shen, H.; Schimelman, J.; Whitney, M.; Steinauer, J.; Xu, W.; et al. Rapid continuous 3D printing of customizable peripheral nerve guidance conduits. *Mater. Today* **2018**, *21*, 951–959. [CrossRef]

27. Hsu, R.S.; Fang, J.H.; Shen, W.-T.; Sheu, Y.C.; Su, C.K.; Chiang, W.H.; Hu, S.H. Injectable DNA-architected nanoraspberry depot-mediated on-demand programmable refilling and release drug delivery. *Nanoscale* **2020**, *12*, 11153. [CrossRef]
28. Laurent, S.; Forge, D.; Port, M.; Roch, A.; Robic, C.; Vander Elst, L.; Muller, R.N. Magnetic iron oxide nanoparticles: Synthesis, stabilization, vectorization, physicochemical characterizations, and biological applications. *Chem. Rev.* **2008**, *108*, 2064–2110. [CrossRef]
29. Wahajuddin; Arora, S. Superparamagnetic iron oxide nanoparticles: Magnetic nanoplatforms as drug carriers. *Int. J. Nanomed.* **2012**, *7*, 3445–3471.
30. Lu, B.Q.; Zhu, Y.J.; Ao, H.Y.; Qi, C.; Chen, F. Synthesis and characterization of magnetic iron oxide/calcium silicate mesoporous nanocomposites as a promising vehicle for drug delivery. *ACS Appl. Mater. Interfaces* **2012**, *4*, 6969–6974. [CrossRef]
31. Sun, Y.-k.; Ma, M.; Zhang, Y.; Gu, N. Synthesis of nanometer-size maghemite particles from magnetite. *Colloids Surf. A* **2004**, *245*, 15–19. [CrossRef]
32. Cushing, B.L.; Kolesnichenko, V.L.; O'Connor, C.J. Recent advances in the liquid-phase syntheses of inorganic nanoparticles. *Chem. Rev.* **2004**, *104*, 3893–3946. [CrossRef] [PubMed]
33. Rong, C.B.; Li, D.; Nandwana, V.; Poudyal, N.; Ding, Y.; Wang, Z.L.; Zeng, H.; Liu, J.P. Size-dependent chemical and magnetic ordering in L 10-FePt nanoparticles. *Adv. Mater.* **2006**, *18*, 2984–2988. [CrossRef]
34. Hu, S.H.; Liao, B.J.; Chiang, C.S.; Chen, P.J.; Chen, I.W.; Chen, S.Y. Core-shell nanocapsules stabilized by single-component polymer and nanoparticles for magneto-chemotherapy/hyperthermia with multiple drugs. *Adv. Mater.* **2012**, *24*, 3627–3632. [CrossRef] [PubMed]
35. Guardia, P.; Corato, R.D.; Lartigue, L.; Wilhelm, C.; Espinosa, A.; Garcia-Hernandez, M.; Gazeau, F.; Manna, L.; Pellegrino, T. Water-soluble iron oxide nanocubes with high values of specific absorption rate for cancer cell hyperthermia treatment. *ACS Nano* **2012**, *6*, 3080–3091. [CrossRef] [PubMed]
36. Tong, S.; Quinto, C.A.; Zhang, L.; Mohindra, P.; Bao, G. Size-dependent heating of magnetic iron oxide nanoparticles. *ACS Nano* **2017**, *11*, 6808–6816. [CrossRef]
37. Lartigue, L.; Innocenti, C.; Kalaivani, T.; Awwad, A.; Duque, M.M.S.D.; Guari, Y.; Larionova, J.; Guérin, C.; Montero, J.L.G.; Barragan-Montero, V.; et al. Water-dispersible sugar-coated iron oxide nanoparticles. An evaluation of their relaxometric and magnetic hyperthermia properties. *J. Am. Chem. Soc.* **2011**, *133*, 10459–10472. [CrossRef] [PubMed]
38. Fortin, J.P.; Wilhelm, C.; Servais, J.; Ménager, C.; Bacri, J.C.; Gazeau, F. Size-sorted anionic iron oxide nanomagnets as colloidal mediators for magnetic hyperthermia. *J. Am. Chem. Soc.* **2007**, *129*, 2628–2635. [CrossRef]
39. Chopra, R.; Shaikh, S.; Chatzinoff, Y.; Munaweera, I.; Cheng, B.; Daly, S.M.; Xi, Y.; Bing, C.; Burns, D.; Greenberg, D.E. Employing high-frequency alternating magnetic fields for the non-invasive treatment of prosthetic joint infections. *Sci. Rep.* **2017**, *7*, 7520. [CrossRef]
40. Su, Y.L.; Fang, J.H.; Liao, C.Y.; Lin, C.T.; Li, Y.T.; Hu, S.H. Targeted mesoporous iron oxide nanoparticles-encapsulated perfluorohexane and a hydrophobic drug for deep tumor penetration and therapy. *Theranostics* **2015**, *5*, 1233–1248. [CrossRef]
41. Lee, J.W.; Park, J.H.; Prausnitz, M.R. Dissolving microneedles for transdermal drug delivery. *Biomaterials* **2008**, *29*, 2113–2124. [CrossRef] [PubMed]
42. Flores, A.; Schell, J.; Krall, A.S.; Jelinek, D.; Miranda, M.; Grigorian, M.; Braas, D.; White, A.C.; Zhou, J.L.; Graham, N.A.; et al. Lactate dehydrogenase activity drives hair follicle stem cell activation. *Nat. Cell Biol.* **2017**, *19*, 1017–1026. [CrossRef] [PubMed]

© 2020 by the authors. Licensee MDPI, Basel, Switzerland. This article is an open access article distributed under the terms and conditions of the Creative Commons Attribution (CC BY) license (http://creativecommons.org/licenses/by/4.0/).

Article

Anti-Bacterial and Anti-Fouling Capabilities of Poly(3,4-Ethylenedioxythiophene) Derivative Nanohybrid Coatings on SUS316L Stainless Steel by Electrochemical Polymerization

Chuan-Chih Hsu [1], Yu-Wei Cheng [2,*], Che-Chun Liu [2], Xin-Yao Peng [2], Ming-Chi Yung [3,*] and Ting-Yu Liu [2,*]

1. Division of Cardiovascular Surgery, Department of Surgery, Taipei Medical University Hospital, Taipei Heart Institute, Taipei Medical University, Taipei 11031, Taiwan; cchsu1967@hotmail.com
2. Department of Materials Engineering, Ming Chi University of Technology, New Taipei City 24301, Taiwan; liu820201@gmail.com (C.-C.L.); asd5997591@gmail.com (X.-Y.P.)
3. Department of Cardiovascular Surgery, Taiwan Adventist Hospital, and School of Medicine, National Yang Ming University, Taipei 105, Taiwan
* Correspondence: ywcheng@mail.mcut.edu.tw (Y.-W.C.); mcyung52@hotmail.com (M.-C.Y.); tyliu0322@gmail.com (T.-Y.L.)

Received: 30 May 2020; Accepted: 28 June 2020; Published: 30 June 2020

Abstract: We have successfully fabricated poly(3,4-ethylenedioxythiophene) (PEDOT) derivative nanohybrid coatings on flexible SUS316L stainless steel by electrochemical polymerization, which can offer anti-fouling and anti-bacterial capabilities. PEDOT derivative nanohybrids were prepared from polystyrene sulfonates (PSS) and graphene oxide (GO) incorporated into a conducting polymer of PEDOT. Additionally, the negative charge of the PEDOT/GO substrate was further modified by poly-diallyldimethylammonium chloride (PDDA) to form a positively charged surface. These PEDOT derivative nanohybrid coatings could provide a straightforward means of controlling the surface energy, roughness, and charges with the addition of various derivatives in the electrochemical polymerization and electrostatically absorbed process. The characteristics of the PEDOT derivative nanohybrid coatings were evaluated by Raman spectroscopy, scanning electron microscopy (SEM), X-ray photoelectron spectroscopy (XPS), atomic force microscopy (AFM), water contact angle, and surface potential (zeta potential). The results show that PEDOT/PSS and PEDOT/GO nanohybrid coatings exhibit excellent anti-fouling capability. Only 0.1% of bacteria can be adhered on the surface due to the lower surface roughness and negative charge surface by PEDOT/PSS and PEDOT/GO modification. Furthermore, the anti-bacterial capability (7 mm of inhibition zone) was observed after adding PDDA on the PEDOT/GO substrates, suggesting that the positive charge of the PEDOT/GO/PDDA substrate can effectively kill bacteria (*Staphylococcus aureus*). Given their anti-fouling and anti-bacterial capabilities, PEDOT derivative nanohybrid coatings have the potential to be applied to biomedical devices such as cardiovascular stents and surgical apparatus.

Keywords: electrochemical polymerization; PEDOT; graphene oxide; anti-fouling capability; anti-bacterial capability

---

1. Introduction

In recent decades, the development of human medical care has been paid increasing attention. In relation to this, films with anti-bacterial coatings have also been widely studied, including antibiotic materials, silver nanoparticles, copper nanoparticles, metal ions, and carbon nanotubes [1–4]. It is well-known that anti-bacterial materials use surface charges to improve anti-bacterial activities through the interaction with the different surface charge of the bacterial cell wall [5]. Generally, the wall of a bacterium with negative charge can be captured by anti-bacterial materials with positive charge, which can provide an excellent anti-bacterial effect [6]. Thus, the anti-bacterial activities of anti-bacterial materials can be improved by surface modifications of materials [7].

Poly(3,4-ethylenedioxythiophene) (PEDOT), a conducting conjugated polymer, has been widely used in various applications due to its excellent properties, including flexibility, chemical stability, high optical transparency, and good conductivity [8]. In addition, conducting conjugated PEDOT can be polymerized from EDOT monomers using different polymerization methods, such as chemical oxidative polymerization, vapor phase polymerization, and electrochemical polymerization [9–11]. Furthermore, conductive PEDOT-based derivatives, such as electrode films doped with poly(4-styrenesulfonate) (PSS) or different hydrophilic materials have attracted wide attention [12–14]. Due to good chemical and physical stability, the PEDOT/PSS can be used in a wide variety of applications such as organic solar cells, supercapacitors, antistatic surfaces, thermoelectric devices, and printed electronics [15,16]. Although oxidized PEDOT has excellent conductivity, it is unstable in water and solvents [17]. Therefore, the negative charge of PSS can be doped with the positive charge of conducting conjugated PEDOT to address this issue and improve its applicability.

In 2004, graphene nanosheets were successfully exfoliated from graphite using the process of repeatedly peeling flakes of Scotch tape [18]. Graphene of two-dimensional nanosheets with an atomic layer of carbon atoms has attracted significant scientific attention and technical interest due to its flexibility, chemical stability, high carrier mobility, and excellent electrical conductivity [19]. In addition, graphene oxide (GO) nanosheets were prepared by chemical modifications in which the surface of the GO exhibited a large number of functional groups (epoxide, carboxyl, carbonyl groups) containing oxygen. This promotes a negative charge of the surface to improve its dispersion in water [20]. These properties of GO-based nanosheets have provided great potential for wide applications in the biosensor, anti-bacterial, and optoelectronic devices [6,21,22]. Moreover, the surface negative charge of GO nanosheets can also be modified by the electrostatic absorbed method of positive charged poly-diallyldimethylammonium chloride (PDDA) to induce anti-bacterial capability [23]. In addition, the poor conductivity of GO nanosheets was improved to fabricate a novel conducting composite by a combination of conducting polymers [24–26]. Therefore, the PEDOT of conducting polymers has been incorporated in GO nanosheets to produce novel nanohybrid materials via self-assembly or chemical processes for electrochemical sensors and anti-fouling coatings [24,27,28].

In this study, PEDOT derivative nanohybrid coatings were fabricated using electrochemical polymerization on a flexible SUS316L stainless steel substrate. These PEDOT derivative nanohybrids were prepared by incorporation of a PSS dopant and GO nanosheets. In addition, the negatively charged PEDOT-GO nanohybrid coating was modified to form a positively charged surface of anti-bacterial film by electrostatically absorbed PDDA. The characteristics of the PEDOT derivative nanohybrid coatings were evaluated by Raman spectroscopy, scanning electron microscopy (SEM), X-ray photoelectron spectroscopy (XPS), atomic force microscopy (AFM), water contact angle, and surface potential (zeta potential). The anti-bacterial and anti-fouling capabilities of PEDOT derivative nanohybrid coatings were investigated by an inhibition zone and anti-fouling test. We anticipated that the PEDOT derivative nanohybrid coatings composed of GO nanosheets and PDDA would demonstrate the excellent anti-bacterial and anti-fouling capabilities, which can be applied to bio-interface coatings and biomedical devices.

2. Materials and Methods

2.1. Materials

SUS316L stainless steel substrate was purchased from Sinkang Industries Co., Ltd., (New Taipei City, Taiwan). Graphite powder was purchased from Allightec Co., (Taichung City, Taiwan).

3,4-Ethylenedioxythiophene (EDOT), PSS solution (Mw 75,000 g/mol), hydrogen peroxide (H_2O_2), hydrochloric acid (HCl), sulfuric acid (H_2SO_4), nitric acid (HNO_3), sodium chloride (NaCl), alcohol, potassium permanganate ($KMnO_4$), and PDDA solution were purchased from Sigma-Aldrich Corp. (St. Louis, MO, USA) and Acors Organics (Ozaukee County, WI, USA).

2.2. Preparation of GO Nanosheets

GO nanosheets were prepared from graphite powder by a modified Hummers-Offeman method [29]. First, 1 g of graphite powder was dispersed into 36 mL of H_2SO_4 solution for 30 min in an ice bath. Then, 12 mL of HNO_3 solution and 5 g of $KMnO_4$ were slowly added to the resulting solution, and the mixtures were stirred for 40 min in an ice bath. Subsequently, 120 mL of deionized (DI) water was gently added and stirred for 2 h. Then, 6 mL of H_2O_2 was added to the stirred solution for 2 h and settled overnight. After removing the supernatant, 200 mL of DI water, 1 mL of H_2O_2, and 1 mL of HCl were added to above resultant mixture and stirred for 2 h. The resultant solution was washed and centrifuged three times. The resultant was washed with DI water until a neutral pH was achieved to obtain dark-yellow GO nanosheets, which were dried by vacuum freeze drying for 72 h.

2.3. Preparation of PEDOT/PSS and PEDOT/GO Nanohybrid Coatings

A PEDOT/PSS nanohybrid coating was prepared from an EDOT monomer with PSS and GO nanosheets by electrochemical polymerization on a SUS316L stainless steel substrate. The SUS316L stainless steel substrate was washed sequentially with common detergent, ethanol, and DI water, which was dried in a vacuum oven. First, 56.8 µL of EDOT and 113.6 µL of PSS were dispersed into DI water and ultrasonicated for 2 h at room temperature. Then, the solution was placed into a three-electrode electrochemical analyzer (PGSTAT12, Metrohm Autolab B.V., Utrecht, Netherlands) and polymerized at a constant potential of 1 V to fabricate a PEDOT/PSS nanohybrid coating with an average deposition density of 10–20 mC cm^{-2}. The electrochemical polymerization was performed using a three-electrode electrochemical analyzer, including the reference electrode of AgCl, the counter electrode of Pt wire, and the working electrode of a SUS316L stainless steel substrate. According to the above procedure, 56.8 µL of EDOT and 0.1 g of GO nanosheets were dispersed into DI water and ultrasonicated for 2 h. Then, the above solution was polymerized on SUS316L stainless steel substrate to obtain a PEDOT/GO nanohybrid coating by electrochemical polymerization.

2.4. Preparation of PEDOT/GO/PDDA Nanohybrid Coating

PEDOT/GO/PDDA nanohybrid coating was prepared from PEDOT/GO and PDDA by the electrostatic absorbed method, as illustrated in Scheme 1. The PEDOT/GO nanohybrid coating was immersed into PDDA solution for 30 min. Then, the coating was washed with DI water three times. The resultant PEDOT/GO/PDDA nanohybrid coating was dried at 60 °C.

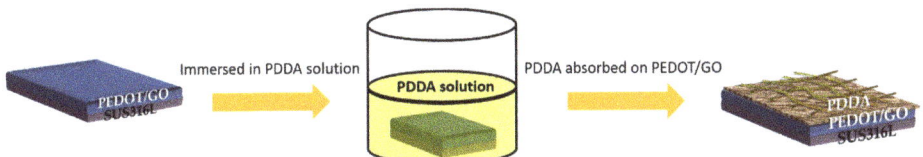

Scheme 1. Preparation of the poly(3,4-ethylenedioxythiophene)/graphene oxide/poly-diallyldimethylammonium chloride (PEDOT/GO/PDDA) nanohybrid coating.

2.5. Anti-Bacterial Capability of PEDOT Derivatives Nanohybrid Coatings

The anti-bacterial capability of PEDOT derivative nanohybrid coatings was surveyed in connection with bacteria (*Staphylococcus aureus*, *S. aureus*). For the microbiological experiment, all of the chemicals were autoclaved at 120 °C for 15 min. *S. aureus* bacteria were incubated at 37 °C for 18–24 h. The cultured

S. aureus was mixed with 1 wt % NaCl solution and stirred in a vortex mixer for 30 s. Then, 1 mL of the solution was added to 9 mL of Lysogeny broth (LB) medium and stirred in a vortex mixer for 30 s. According to the above procedure, the mixing process was repeated three times. Subsequently, 0.01 mL of suspension S. aureus (10^5 CFU/mL) was dropped and uniformly dispersed on a nutrient agar plate. The PEDOT derivative nanohybrid coatings were placed on the nutrient agar plate with S. aureus and incubated at 37 °C for 18–24 h. After 18–24 h of culturing, the inhibition zone of PEDOT derivative nanohybrid coatings was demonstrated.

2.6. Anti-Fouling Capability of PEDOT Derivatives Nanohybrid Coatings

A quantity of 10^5 CFU/mL of S. aureus was chosen as the adhering bacteria for the evaluation of anti-fouling capability. The PEDOT derivative nanohybrid coatings were immersed in the S. aureus solution and incubated at 37 °C for 18–24 h. The number of S. aureus bacteria adsorbed on the PEDOT derivative nanohybrid coatings was counted by fluorescence microscope (fluorescent nucleic acid stain-SYTO®9/propidium iodide kits).

2.7. Characterization

Raman measurement with a 532 nm He-Ne emitting laser was performed using a micro-Raman system (Renishaw Vendor, Gloucestershire, UK) in the detection range from 1000 to 3000 cm^{-1}. The chemical binding conditions of PEDOT derivative nanohybrid coatings were measured by X-ray photoelectron spectrometer (XPS, ULVAC-PHI PHI 5000 VersaProbe, Ulvac-PHI, Kanagawa, Japan). The surface charge of PEDOT derivative nanohybrid coatings was observed by zeta potential (Zetasizer 3000, Malvern Instruments, Malvern, UK). A contact angle goniometer (DSA 100, Krüss GmbH, Hamburg, Germany) was used for the static contact angle measurement of PEDOT derivative nanohybrid coatings. An atomic force microscope (AFM, Dimension Edge, Bruker, Madison, WI, USA) and field emission scanning electron microscope (FE-SEM, JEOL JSM 6701F, JEOL Co., Tokyo, Japan) were used for the morphological characterization of PEDOT derivative nanohybrid coatings. The number of bacteria was calculated from fluorescence microscope images bacteria adhered on the surface of the SUS316L substrate and PEDOT derivative nanohybrid coatings by using commercial ImageJ software (National Institutes of Health, Bethesda, MD, USA).

3. Results and Discussion

3.1. Characteristics of PEDOT Derivative Nanohybrid Coatings

The structural properties of GO nanosheets and PEDOT derivative nanohybrid coatings were measured by Raman spectroscopy, as shown in Figure 1. The GO nanosheets exhibited the characteristic peaks at 1350 cm^{-1} (D band) for the in-plane bond stretching of sp^2 carbon atoms and 1583 cm^{-1} (G band) for defects of structure and lattice distortion [30]. The Raman spectrum of the PEDOT/PSS nanohybrid coating was observed at 1433 and 1505 cm^{-1} for the asymmetric and symmetric C=C stretching vibration of PEDOT [31]. When the GO nanosheets were used in the PEDOT nanohybrid system, the Raman spectrum of the PEDOT/GO nanohybrid coating was similar to that of the PEDOT/PSS nanohybrid coating, while the shoulder peaks of the D and G bands of the GO nanosheets were located at 1350 and 1583 cm^{-1}, respectively. This confirms that GO nanosheets were doped in the chain [32]. After the electrostatically absorbed process by PDDA, the Raman spectrum of the PEDOT/GO/PDDA nanohybrid coating was similar to that of the PEDOT/GO nanohybrid coating.

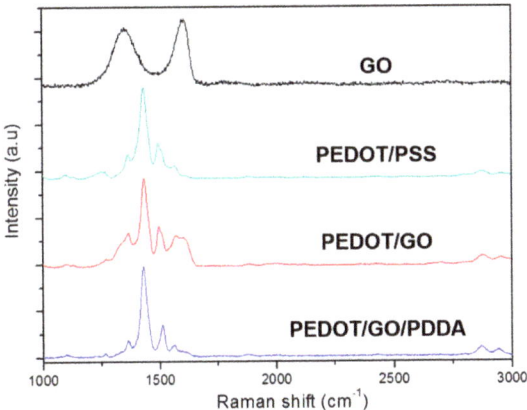

Figure 1. Raman spectra of GO nanosheets, and PEDOT/PSS (polystyrene sulfonates), PEDOT/GO, and PEDOT/GO/PDDA nanohybrid coated substrates.

The chemical binding of PEDOT derivative nanohybrid coatings was analyzed by XPS spectra, as shown in Figure 2. The XPS spectra were observed in the presence of oxygen (O-1s), carbon (C-1s), and sulfur (S-2s and S-2p) elements in the PEDOT/PSS nanohybrid coating. Compared with the PEDOT/PSS nanohybrid coating, the characteristic peaks (S-2s and S-2p) of the PEDOT/GO nanohybrid coating were almost dispersed, while the O-1s peak was increased with oxygen-containing groups of GO nanosheets incorporated into the PEDOT nanohybrid system. Moreover, a new characteristic peak (Cl-2p) was observed for PEDOT/GO/PDDA nanohybrid coatings, indicating that PDDA was absorbed on the PEDOT/GO nanohybrid coating by the electrostatic absorbed process. The curve-fitting of C-1s spectra of PEDOT/PSS and PEDOT/GO nanohybrid coatings are shown in Figure 3a,b. Compared with the PEDOT/PSS nanohybrid coating, the stronger characteristic peak at 286.8 eV for the C-O-C bond was observed for the PEDOT/GO nanohybrid coating. The results indicate that the many oxygen-containing functional groups of GO nanosheets were successfully incorporated into the PEDOT nanohybrid system [31]. In addition, the curve-fitting of Cl-2p spectra (Figure 3c) of PEDOT/GO/PDDA nanohybrid coatings demonstrates characteristic peaks at 198.8 and 197.4 eV, corresponding to Cl-$2p_{1/2}$ and Cl-$2p_{3/2}$, which further confirm the adsorption of PDDA on the PEDOT/GO nanohybrid coating.

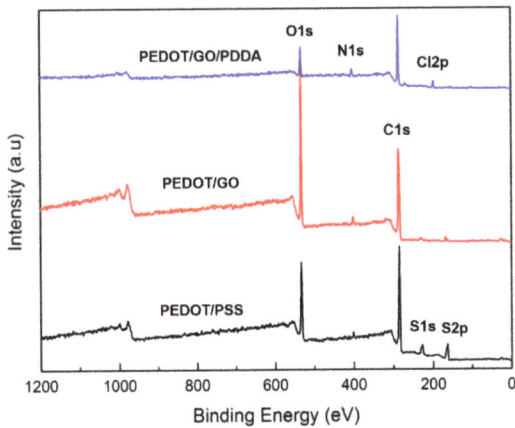

Figure 2. X-ray photoelectron spectroscopy (XPS) survey spectra of PEDOT/PSS, PEDOT/GO, and PEDOT/GO/PDDA nanohybrid coatings.

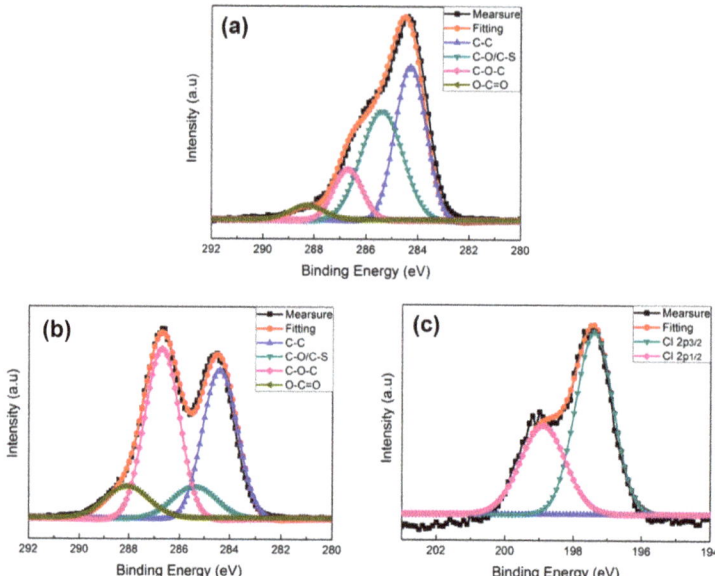

Figure 3. C-1s XPS spectra of (**a**) PEDOT/PSS and (**b**) PEDOT/GO nanohybrid coatings, and (**c**) Cl-2p spectra of PEDOT/GO/PDDA nanohybrid coated substrates.

The zeta potential results of the nanohybrid coated substrates (PEDOT/PSS, PEDOT/GO, PEDOT/GO/PDDA substrates) and bacteria (*S. aureus*) are illustrated in Figure 4. Due to the negative charge of PSS and GO nanosheets incorporated into the PEDOT nanohybrid system, the zeta potential values of PEDOT/PSS and PEDOT/GO nanohybrid coatings were −69.3 and −51.4 mV, respectively. When positively charged PDDA was absorbed on PEDOT/GO nanohybrid coating, the zeta potential value of PEDOT/GO/PDDA substrate significantly increased to +34.1 mV. In addition, the zeta potential value of *S. aureus* with the negative charge of the cell walls was −19.8 mV. Therefore, the positive charge of PEDOT/GO/PDDA substrate can physically absorb and thus kill the bacteria (*S. aureus*), thereby inducing the anti-bacterial capability.

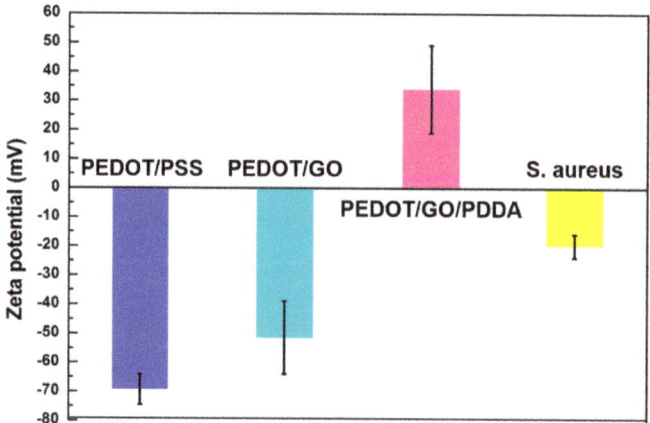

Figure 4. Zeta potential values of nanohybrid coated substrates (PEDOT/PSS, PEDOT/GO, PEDOT/GO/PDDA substrates) and bacteria (*S. aureus*).

The surface hydrophilicity of PEDOT derivative nanohybrids coated on SUS316L stainless steel substrate was investigated by water contact angle measurement, as shown in Figure 5. The surface of pristine SUS316L substrate displayed a water contact angle of 53.4°. After the PEDOT/PSS nanohybrid was coated on the SUS316L substrate, the water contact angle of the coated surface decreased to 47.2°. Moreover, the water contact angle (38.4°) of the PEDOT/GO nanohybrid coating was slightly lower than that of the PEDOT/PSS nanohybrid coating, indicating that the incorporation of GO nanosheets possessed a large number of oxygen functional groups. Furthermore, the PEDOT/GO/PDDA nanohybrid coating demonstrated the lowest water contact angle (5.7°). The results indicate the successful adsorption of hydrophilic PDDA on the PEDOT/GO nanohybrid substrate to create the hydrophilic surface.

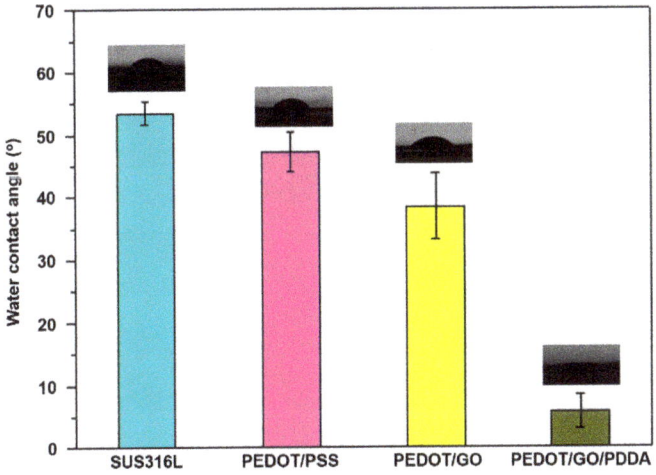

Figure 5. Water contact angle of the SUS316L substrate, and PEDOT/PSS, PEDOT/GO and PEDOT/GO/PDDA nanohybrid coatings.

The surface morphological characterization of the SUS316L substrate and PEDOT derivative nanohybrid coatings were investigated by SEM, as shown in Figure 6. Macroscopically, the pristine SUS316L stainless steel substrate showed a clear and smooth surface morphology (Figure 6a). The PEDOT/PSS nanohybrid coating was observed as a dense layer on the SUS316L substrate, as shown in Figure 6b. Moreover, the PEDOT/GO nanohybrid coating was shown as GO nanosheets with a wrinkled structure on the surface of the PEDOT nanohybrid system (Figure 6c). When the positively charged PDDA was absorbed on the PEDOT/GO nanohybrid coating (Figure 6d), the absence of GO nanosheets with wrinkled morphology on the surface of PEDOT/GO/PDDA nanohybrid coatings indicated that PDDA successfully covered the surface of the PEDOT/GO nanohybrid coated substrate. The surface roughness of coating layers was further investigated by AFM analysis.

Figure 7 shows the microscopic surface roughness and morphology of the SUS316L substrate and PEDOT derivative nanohybrid coatings by AFM. Microscopically, the pristine SUS316L substrate displayed a relatively rough morphology (Ra = 76.2 nm). By the electrochemical polymerization process, the PEDOT/PSS nanohybrid was polymerized and filled the nanostructured surface of SUS316L substrate to obtain a uniform surface morphology and lower surface roughness (Ra = 39.0 nm). Compared with the PEDOT/PSS nanohybrid coating, the surface roughness of the PEDOT/GO nanohybrid coating increased from 39.0 to 50.9 nm, which was due to the presence of GO nanosheets on the surface. When positively charged PDDA was absorbed on the PEDOT/GO nanohybrid coating, the surface roughness of the PEDOT/GO/PDDA nanohybrid coating decreased to 45.3 nm. The result shows that the PDDA covered the surface, reducing surface roughness and creating a relatively smooth morphology.

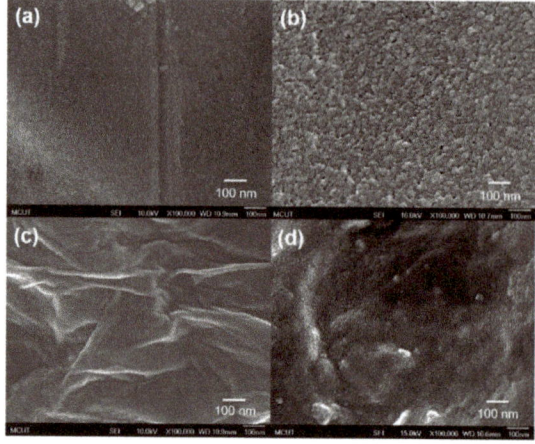

Figure 6. Top-view SEM images of (**a**) the pristine SUS316L stainless steel substrate, and (**b**) PEDOT/PSS, (**c**) PEDOT/GO, and (**d**) PEDOT/GO/PDDA modified SUS316 substrates.

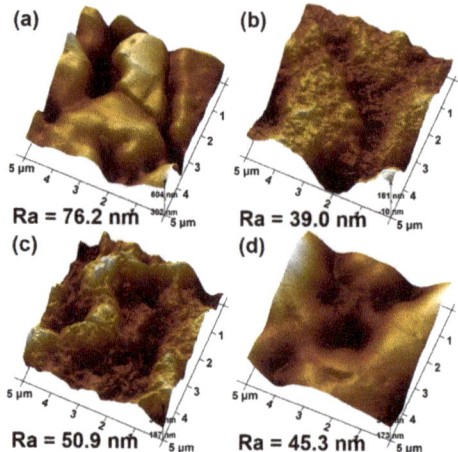

Figure 7. Atomic force microscopy (AFM) images of (**a**) the SUS316L substrate, and (**b**) PEDOT/PSS, (**c**) PEDOT/GO, and (**d**) PEDOT/GO/PDDA nanohybrid coatings.

3.2. Anti-Fouling Capability of PEDOT Derivative Nanohybrid Coatings

To evaluate the anti-fouling capability of the PEDOT derivative nanohybrid coated on the surface of SUS316L stainless steel (Figure 8), 10^5 CFU/mL of S. aureus was chosen as the model bacteria. Green dots of the fluorescence microscope images represent live *S. aureus* bacteria adhering to the surface of the SUS316L substrate and PEDOT derivative nanohybrid coatings. A dense distribution of *S. aureus* (approximately $10^8/cm^2$) adhered to the surface of the pristine SUS316L substrate, as shown in Figure 8a. Compared with the pristine SUS316 substrate (Ra = 76.2 nm), a lower adhesion density of bacteria was obtained on the PEDOT derivative nanohybrid coatings (Figure 8b–d), which was due to the lower surface roughness. Therefore, the numbers of adhered bacteria were decreased to approximately 10^6/cm^2 for PEDOT/PSS, approximately 10^6/cm^2 for PEDOT/GO, and approximately 10^7/cm^2 for PEDOT/GO/PDDA substrates. In particular, the negative charge of PEDOT/PSS (Figure 8b) and PEDOT/GO (Figure 8c) nanohybrid coatings can further inhibit and reduce the adhesion of the negative charge of S. aureus, demonstrating the anti-fouling capability. On the other hand, the numbers of S. aureus decrease from 10^5 CFU/mL (SUS316L substrate) to approximately 10^2 CFU/mL (PEDOT/PSS and PEDOT/GO nanohybrid coating), and only 0.1% of bacteria can be adhered on the surface. However, the numbers of adhering bacteria (approximately 10^3 CFU/mL) on the PEDOT/GO/PDDA nanohybrid coating were slightly higher than that of the PEDOT/PSS and PEDOT/GO nanohybrid coating. The results indicate that the bacteria seemed to slightly prefer to adhere on the positive charge of the PEDOT/GO/PDDA substrate (Figure 8d). The negative cell wall of the bacteria would be locked by the positive charge of PEDOT/GO/PDDA substrates to inhibit the growth; thus, the bacteria are going to die [6]. The substrate exhibited excellent anti-bacterial capability, as discussed in the next section.

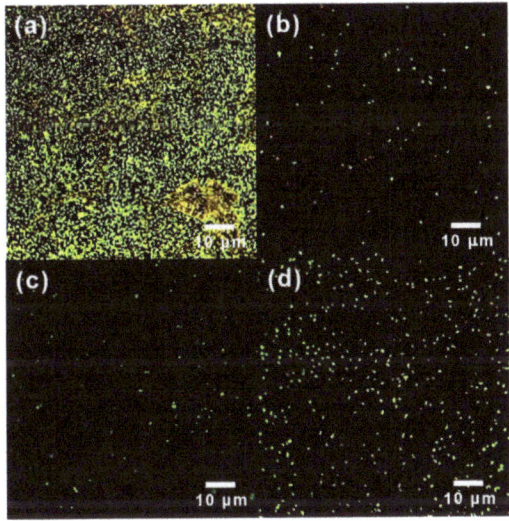

Figure 8. The number of bacteria adhering to (**a**) the SUS316L substrate, and (**b**) PEDOT/PSS, (**c**) PEDOT/GO, and (**d**) PEDOT/GO/PDDA nanohybrid coatings.

3.3. Anti-Bacterial Capability of PEDOT Derivative Nanohybrid Coatings

The anti-bacterial capability of PEDOT derivative nanohybrid coatings was evaluated using *S. aureus* as a model bacterium. As compared with other anti-bacterial tests, the zone of inhibition testing is a rapid and inexpensive test for anti-bacterial activity. Nevertheless, the zone of inhibition testing is a qualitative test to inhibit the growth of tested bacteria. In this study, the anti-bacterial capability was investigated by measuring the inhibition zone incubated on a nutrient agar plate, as shown in Figure 9. When the zone of inhibition shows on the nutrient agar plate, the results displayed that the tested bacteria were susceptible to the PEDOT derivative nanohybrid coatings. The pristine SUS316L substrate, and the PEDOT/PSS and PEDOT/GO nanohybrid coatings, showed the absence of an inhibition zone. However, a significant inhibition zone (7 mm) on the PEDOT/GO/PDDA nanohybrid coating was observed. This is due to the positive charge of PDDA absorbed on the PEDOT/GO substrate to create the positive charge surface of the PEDOT/GO/PDDA substrate, which can inhibit the growth of the bacteria. Compared with our previous studies [24,33], the negative charge of PEDOT derivatives (PEDOT/PSS, PEDOT/GO, PEDOT/Heparin, PEDOT/ chondroitin sulfate, and PEDOT/carboxymethyl-hexanoyl chitosan) can suppress the adhesion of the negative charge of proteins, platelets, and bacteria to form the anti-fouling surface. However, it would be an anti-bacterial surface after immobilizing the positive charge of PDDA on the PEDOT/GO substrate. Therefore, the inclusion of PDDA in the PEDOT/GO/PDDA substrate can effectively diffuse and then kill bacteria, as shown as the larger inhibition zone [6].

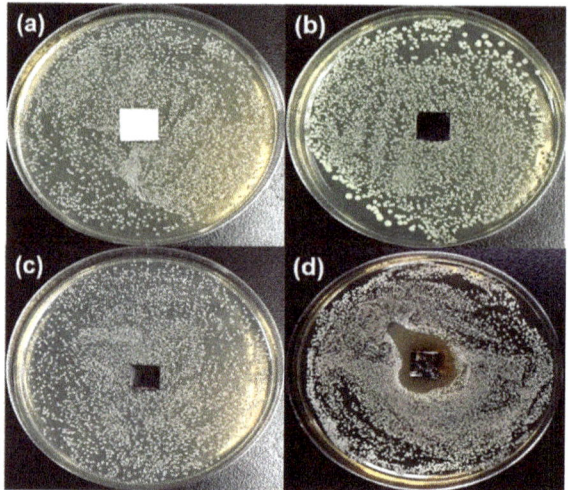

Figure 9. Anti-bacterial capability (inhibition zone) of (**a**) the SUS316L substrate, and (**b**) PEDOT/PSS, (**c**) PEDOT/GO, and (**d**) PEDOT/GO/PDDA substrates.

4. Conclusions

This study successfully fabricated PEDOT derivative nanohybrid coatings on a flexible SUS316L stainless steel substrate by electrochemical polymerization for improving the anti-bacterial and anti-fouling capabilities. Via the addition of hydrophilic derivatives, these PEDOT derivative nanohybrid coatings decrease the water contact angle and surface roughness to form a hydrophilic surface. In addition, the negatively charged surface of the PEDOT/GO nanohybrid coating can be further modified by the electrostatically absorbed process of the positively charged PDDA. PEDOT/PSS and PEDOT/GO nanohybrid coatings with lower surface roughness and negative charge led to less adsorption of bacteria (*S. aureus*). By comparison, the positively charged PEDOT/GO/PDDA nanohybrid coating inhibited the growth of bacteria, resulting in excellent anti-bacterial capability. Therefore, the electrochemically polymerized PEDOT derivative nanohybrid coatings can provide an anti-fouling and anti-bacterial surface, which offers a straightforward and rapid method to develop anti-bacterial and anti-fouling coatings for medical devices.

Author Contributions: Conceptualization, C.-C.H. and T.-Y.L.; Data curation, C.-C.H., C.-C.L. and X.-Y.P.; Funding acquisition, T.-Y.L. and M.-C.Y.; Investigation, Y.-W.C., C.-C.L. and X.-Y.P.; Methodology, Y.-W.C. and T.-Y.L.; Validation, C.-C.H. and M.-C.Y.; Formal analysis, C.-C.L. and Y.-W.C., Visualization, C.-C.H.; Project administration, T.-Y.L.; Supervision, T.-Y.L., C.-C.H. and M.-C.Y.; Resources, T.-Y.L. and M.-C.Y.; Writing—original draft, C.-C.H., Y.-W.C., C.-C.L., M.-C.Y. and T.-Y.L. Writing—revised manuscript, C.-C.H., Y.-W.C., M.-C.Y. and T.-Y.L. All authors have read and agreed to the published version of the manuscript.

Funding: This study was financially supported by the Taiwan Association of Cardiovascular Surgery Research, Research Center for Intelligent Medical Devices of Ming Chi University of Technology and Ministry of Science and Technology of Taiwan (MOST 108-2622-E-131-002-CC3; MOST 108-2218-E-002-010).

Conflicts of Interest: The authors declare no conflict of interest.

References

1. Zhang, Z.; Tang, J.; Wang, H.; Xia, Q.; Xu, S.; Han, C.C. Controlled Antibiotics Release System through Simple Blended Electrospun Fibers for Sustained Antibacterial Effects. *ACS Appl. Mater. Interfaces* **2015**, *7*, 26400–26404. [CrossRef]
2. Xue, C.-H.; Chen, J.; Yin, W.; Jia, S.-T.; Ma, J.-Z. Superhydrophobic conductive textiles with antibacterial property by coating fibers with silver nanoparticles. *Appl. Surf. Sci.* **2012**, *258*, 2468–2472. [CrossRef]
3. Wu, Q.; Li, J.; Zhang, W.; Qian, H.; She, W.; Pan, H.; Wen, J.; Zhang, X.; Liu, X.; Jiang, X. Antibacterial property, angiogenic and osteogenic activity of Cu-incorporated TiO2 coating. *J. Mater. Chem. B* **2014**, *2*, 6738–6748. [CrossRef]
4. Liu, S.; Ng, A.K.; Xu, R.; Wei, J.; Tan, C.M.; Yang, Y.; Chen, Y. Antibacterial action of dispersed single-walled carbon nanotubes on Escherichia coli and Bacillus subtilis investigated by atomic force microscopy. *Nanoscale* **2010**, *2*, 2744–2750. [CrossRef]
5. Qi, L.; Xu, Z.; Jiang, X.; Hu, C.; Zou, X. Preparation and antibacterial activity of chitosan nanoparticles. *Carbohydr. Res.* **2004**, *339*, 2693–2700. [CrossRef]
6. Cheng, Y.-W.; Wang, S.-H.; Liu, C.-M.; Chien, M.-Y.; Hsu, C.-C.; Liu, T.-Y. Amino-modified graphene oxide nanoplatelets for photo-thermal and anti-bacterial capability. *Surf. Coat. Technol.* **2020**, *385*, 125441. [CrossRef]
7. Adlhart, C.; Verran, J.; Azevedo, N.F.; Olmez, H.; Keinänen-Toivola, M.M.; Gouveia, I.; Melo, L.F.; Crijns, F. Surface modifications for antimicrobial effects in the healthcare setting: A critical overview. *J. Hosp. Infect.* **2018**, *99*, 239–249. [CrossRef] [PubMed]
8. Chen, R.; Sun, K.; Zhang, Q.; Zhou, Y.; Li, M.; Sun, Y.; Wu, Z.; Wu, Y.; Li, X.; Xi, J.; et al. Sequential Solution Polymerization of Poly(3,4-ethylenedioxythiophene) Using V2O5 as Oxidant for Flexible Touch Sensors. *iScience* **2019**, *12*, 66–75. [CrossRef] [PubMed]
9. Kim, J.; You, J.; Kim, E. Flexible Conductive Polymer Patterns from Vapor Polymerizable and Photo-Cross-Linkable EDOT. *Macromolecules* **2010**, *43*, 2322–2327. [CrossRef]
10. Cui, X.; Martin, D.C. Electrochemical deposition and characterization of poly(3,4-ethylenedioxythiophene) on neural microelectrode arrays. *Sens. Actuators B Chem.* **2003**, *89*, 92–102. [CrossRef]
11. Mantione, D.; Del Agua, I.; Sanchez-Sanchez, A.; Mecerreyes, D. Poly (3, 4-ethylenedioxythiophene)(PEDOT) derivatives: Innovative conductive polymers for bioelectronics. *Polymers* **2017**, *9*, 354. [CrossRef] [PubMed]
12. Beesley, D.J.; Price, B.K.; Hunter, S.; Shaffer, M.S.P.; De Mello, J.C. Direct dispersion of SWNTs in highly conductive solvent-enhanced PEDOT:PSS films. *Nanocomposites* **2016**, *2*, 135–140. [CrossRef]
13. Kim, Y.-S.; Chang, M.-H.; Lee, E.-J.; Ihm, D.-W.; Kim, J.-Y. Improved electrical conductivity of PEDOT-based electrode films hybridized with silver nanowires. *Synth. Metals* **2014**, *195*, 69–74. [CrossRef]
14. Horikawa, M.; Fujiki, T.; Shirosaki, T.; Ryu, N.; Sakurai, H.; Nagaoka, S.; Ihara, H. The development of a highly conductive PEDOT system by doping with partially crystalline sulfated cellulose and its electric conductivity. *J. Mater. Chem. C* **2015**, *3*, 8881–8887. [CrossRef]
15. Sun, K.; Zhang, S.; Li, P.; Xia, Y.; Zhang, X.; Du, D.; Isikgor, F.H.; Ouyang, J. Review on application of PEDOTs and PEDOT:PSS in energy conversion and storage devices. *J. Mater. Sci. Mater. Electron.* **2015**, *26*, 4438–4462. [CrossRef]
16. Lövenich, W. PEDOT-properties and applications. *Polym. Sci. Ser. C* **2014**, *56*, 135–143. [CrossRef]
17. Cho, W.; Im, S.; Kim, S.; Kim, S.; Kim, J.H. Synthesis and characterization of PEDOT: P (SS-co-VTMS) with hydrophobic properties and excellent thermal stability. *Polymers* **2016**, *8*, 189. [CrossRef]
18. Novoselov, K.S.; Geim, A.K.; Morozov, S.V.; Jiang, D.; Zhang, Y.; Dubonos, S.V.; Grigorieva, I.V.; Firsov, A.A. Electric Field Effect in Atomically Thin Carbon Films. *Science* **2004**, *306*, 666–669. [CrossRef]
19. Wei, N.; Li, Q.; Cong, S.; Ci, H.; Song, Y.; Yang, Q.; Lu, C.; Li, C.; Zou, G.; Sun, J.; et al. Direct synthesis of flexible graphene glass with macroscopic uniformity enabled by copper-foam-assisted PECVD. *J. Mater. Chem. A* **2019**, *7*, 4813–4822. [CrossRef]
20. Dikin, D.A.; Stankovich, S.; Zimney, E.J.; Piner, R.D.; Dommett, G.H.B.; Evmenenko, G.; Nguyen, S.T.; Ruoff, R.S. Preparation and characterization of graphene oxide paper. *Nature* **2007**, *448*, 457–460. [CrossRef] [PubMed]
21. Lee, J.; Kim, J.; Kim, S.; Min, D.-H. Biosensors based on graphene oxide and its biomedical application. *Adv. Drug Deliv. Rev.* **2016**, *105*, 275–287. [CrossRef] [PubMed]

22. Morimoto, N.; Kubo, T.; Nishina, Y. Tailoring the Oxygen Content of Graphite and Reduced Graphene Oxide for Specific Applications. *Sci. Rep.* **2016**, *6*, 21715. [CrossRef] [PubMed]
23. Yang, D.-Q.; Rochette, J.-F.; Sacher, E. Spectroscopic Evidence for π–π Interaction between Poly(diallyl dimethylammonium) Chloride and Multiwalled Carbon Nanotubes. *J. Phys. Chem. B* **2005**, *109*, 4481–4484. [CrossRef] [PubMed]
24. Yang, M.-C.; Tsou, H.-M.; Hsiao, Y.-S.; Cheng, Y.-W.; Liu, C.-C.; Huang, L.-Y.; Peng, X.-Y.; Liu, T.-Y.; Yung, M.-C.; Hsu, C.-C. Electrochemical Polymerization of PEDOT–Graphene Oxide–Heparin Composite Coating for Anti-Fouling and Anti-Clotting of Cardiovascular Stents. *Polymers* **2019**, *11*, 1520. [CrossRef] [PubMed]
25. Liu, Y.; Weng, B.; Razal, J.M.; Xu, Q.; Zhao, C.; Hou, Y.; Seyedin, S.; Jalili, R.; Wallace, G.G.; Chen, J. High-Performance Flexible All-Solid-State Supercapacitor from Large Free-Standing Graphene-PEDOT/PSS Films. *Sci. Rep.* **2015**, *5*, 17045. [CrossRef] [PubMed]
26. Si, W.; Lei, W.; Zhang, Y.; Xia, M.; Wang, F.; Hao, Q. Electrodeposition of graphene oxide doped poly(3,4-ethylenedioxythiophene) film and its electrochemical sensing of catechol and hydroquinone. *Electrochim. Acta* **2012**, *85*, 295–301. [CrossRef]
27. Sriprachuabwong, C.; Karuwan, C.; Wisitsorrat, A.; Phokharatkul, D.; Lomas, T.; Sritongkham, P.; Tuantranont, A. Inkjet-printed graphene-PEDOT:PSS modified screen printed carbon electrode for biochemical sensing. *J. Mater. Chem.* **2012**, *22*, 5478–5485. [CrossRef]
28. Xu, Y.; Wang, Y.; Liang, J.; Huang, Y.; Ma, Y.; Wan, X.; Chen, Y. A hybrid material of graphene and poly (3,4-ethyldioxythiophene) with high conductivity, flexibility, and transparency. *Nano Res.* **2009**, *2*, 343–348. [CrossRef]
29. Hummers, W.S.; Offeman, R.E. Preparation of Graphitic Oxide. *J. Am. Chem. Soc.* **1958**, *80*, 1339. [CrossRef]
30. Kang, Y.; Chu, Z.; Zhang, D.; Li, G.; Jiang, Z.; Cheng, H.; Li, X. Incorporate boron and nitrogen into graphene to make BCN hybrid nanosheets with enhanced microwave absorbing properties. *Carbon* **2013**, *61*, 200–208. [CrossRef]
31. Österholm, A.; Lindfors, T.; Kauppila, J.; Damlin, P.; Kvarnström, C. Electrochemical incorporation of graphene oxide into conducting polymer films. *Electrochim. Acta* **2012**, *83*, 463–470. [CrossRef]
32. Si, W.; Lei, W.; Han, Z.; Zhang, Y.; Hao, Q.; Xia, M. Electrochemical sensing of acetaminophen based on poly(3,4-ethylenedioxythiophene)/graphene oxide composites. *Sens. Actuators B Chem.* **2014**, *193*, 823–829. [CrossRef]
33. Hsu, C.-C.; Liu, T.-Y.; Peng, X.-Y.; Cheng, Y.-W.; Lin, Y.-R.; Yang, M.-C.; Huang, L.-Y.; Liu, K.-H.; Yung, M.-C. Anti-fouling and anti-coagulation capabilities of PEDOT-biopolymer coating by in-situ electrochemical copolymerization. *Surf. Coat. Technol.* **2020**, *397*, 125963. [CrossRef]

© 2020 by the authors. Licensee MDPI, Basel, Switzerland. This article is an open access article distributed under the terms and conditions of the Creative Commons Attribution (CC BY) license (http://creativecommons.org/licenses/by/4.0/).

Article

Using Methacryl-Polyhedral Oligomeric Silsesquioxane as the Thermal Stabilizer and Plasticizer in Poly(vinyl chloride) Nanocomposites

Yu-Kai Wang [1], Fang-Chang Tsai [2,*], Chao-Chen Ma [3], Min-Ling Wang [3] and Shiao-Wei Kuo [1,4,*]

1. Department of Materials and Optoelectronic Science, Center of Crystal Research, National Sun Yat-Sen University, Kaohsiung 80424, Taiwan; gba01123@gmail.com
2. Hubei Key Laboratory of Polymer Materials, Key Laboratory for the Green Preparation and Application of Functional Materials (Ministry of Education), Hubei Collaborative Innovation Center for Advanced Organic Chemical Materials, School of Materials Science and Engineering, Hubei University, Wuhan 430062, China
3. UPC Technology Corporation, No.3, Kung-Yeh 2nd Rd., Linyuan Dist., Kaohsiung 832, Taiwan; cj.maa@upc.com.tw (C.-C.M.); vini.wang@upc.com.tw (M.-L.W.)
4. Department of Medicinal and Applied Chemistry, Kaohsiung Medical University, Kaohsiung 807, Taiwan
* Correspondence: tfc0323@gmail.com (F.-C.T.); kuosw@faculty.nsysu.edu.tw (S.-W.K.); Tel.: +86-27-88661729 (F.-C.T.), +886-7-525-4099 (S.-W.K.)

Received: 2 October 2019; Accepted: 16 October 2019; Published: 18 October 2019

Abstract: In this study, we investigated the influence of methacryl-functionalized polyhedral oligomeric silsesquioxane (MA-POSS) nanoparticles as a plasticizer and thermal stabilizer for a poly(vinyl chloride) (PVC) homopolymer and for a poly(vinyl chloride)/dissononyl cyclohexane-1,2-dicarboxylate (PVC/DINCH) binary blend system. The PVC and the PVC/DINCH blend both became flexible, with decreases in their glass transition temperatures and increases in their thermal decomposition temperatures, upon an increase in MA-POSS content, the result of hydrogen bonding between the C=O groups of MA-POSS and the H–CCl units of the PVC, as determined using infrared spectroscopy. Furthermore, the first thermal decomposition temperature of the pure PVC, due to the emission of HCl, increased from 290 to 306 °C, that is, the MA-POSS nanoparticles had a retarding effect on the decomposition of the PVC matrix. In tensile tests, all the PVC/DINCH/MA-POSS ternary blends were transparent and displayed flexibility, but their modulus and tensile strength both decreased, while their elongation properties increased, upon an increase in MA-POSS concentration, both before and after thermal annealing. In contrast, the elongation decreased, but the modulus and tensile strength increased, after thermal annealing at 100 °C for 7 days.

Keywords: POSS; poly(vinyl chloride); plasticizer; nanocomposites

1. Introduction

Poly(vinyl chloride) (PVC) is one of the most commonly employed polymeric materials. It is used widely in packing materials, toys, healthcare, electric insulation, automobiles, and interior decorations, because of its high transparency, inertness, and reasonable mechanical properties [1–5]. Nonetheless, the intrinsic rigidity of PVC limits its end-user applications, because of its high glass transition temperature (T_g, ca. 89 °C). Therefore, PVC is often blended commercially with various plasticizers, e.g., di-(2-ethylhexyl)phthalate (DEHP) [1]. A serious shortcoming of DEHP, however, is its leachability from PVC materials upon contact with tissues or body fluids [6,7].

Dissononyl cyclohexane-1,2-dicarboxylate (DINCH) is a promising substitute plasticizer for DEHP in PVC; it exhibits high biodegradability and low environmental persistence compared with DEHP, the result of replacing the benzene ring in DEHP with a cyclohexane ring in DINCH [1]. Nevertheless, this substitution can result in PVC/DINCH blend systems possessing thermal stability lower than that of corresponding PVC/DEHP blends. As a result, various nanofillers have been tested as plasticizers to enhance the thermal stability of PVC [8–10]. Ideally, these nanofillers would function as plasticizers, while inhibiting the degradation of the polymer and the release of HCl.

Polyhedral oligomeric silsesquioxane (POSS) derivatives are organic/inorganic hybrid materials that have been widely dispersed in polymeric matrices by taking advantage of both chemical and physical bonding interactions [11–26]. Such POSS-related polymer nanocomposites can exhibit enhanced strength, rigidity, decomposition temperature, and modulus, and decreased viscosity, flammability, and surface free energy, depending on the degree of dispersion of POSS nanoparticles in the polymeric matrix [27–31]. In the case of physical bonding, the POSS units can be dispersed through solvent-casting or melting/mixing blending approaches (i.e., without covalent bonding). Blending with POSS as a plasticizer can increase thermal decomposition temperature, enhance impact strength, and lower the value of T_g of PVC [32–35]. Various functional groups (e.g., methacryl and ethylene glycol units) have been appended to POSS nanoparticles for incorporation into PVC matrices. Because the thermal decomposition temperature with 10 wt% loss of the ethylene glycol unit (T_{d10} = 250 °C) [35] is lower than that of the methacryl unit (T_{d10} = 420 °C) [33,34], in this study, we chose to combine methacryl-POSS (MA-POSS) nanoparticles with DINCH to use as plasticizers for a PVC matrix. The incorporation of MA-POSS in PVC reduces both the primary and secondary transition temperature. In addition, the ternary blend of PVC/MA-POSS/DOP could reduce T_g behavior near room temperature with desirable ductile behavior. To the best of our knowledge, this paper is the first to describe the incorporation of combinations of DINCH and MA-POSS within PVC matrices. We have employed differential scanning calorimetry (DSC), dynamic mechanical analysis (DMA), Fourier-transform infrared (FTIR) spectroscopy, thermogravimetric analysis (TGA), and tensile tests to characterize the glass transition and thermal decomposition temperatures, hydrogen bonding interactions, and mechanical properties of binary blends of PVC/MA-POSS and ternary blends of PVC/DINCH/MA-POSS.

2. Materials and Methods

2.1. Materials

PVC (Scheme 1a) and tetrahydrofuran (THF) were purchased from Sigma–Aldrich. DINCH (UniHydro UN899, Scheme 1b)) was supported by UPC Technology Corporation. The average molecular weight of the PVC used in this study was approximately 90,000, with a polydispersity index (PDI) of 2.25 by Sigma–Aldrich. MA-POSS (Scheme 1c) was purchased from Hybrid Plastics; this compound did not have a crystalline structure, but featured a glass transition temperature of −55 °C (Figure S1) and was a pale-yellow heavy oil [33,34].

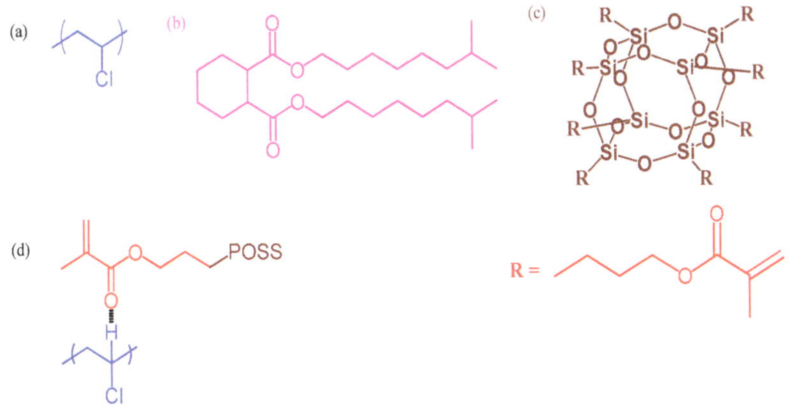

Scheme 1. (a–c) Chemical structures of (a) Poly(vinyl chloride) (PVC), (b) Dissononyl cyclohexane-1,2-dicarboxylate (DINCH), and (c) methacryl-polyhedral oligomeric silsesquioxane (MA-POSS). (d) Hydrogen bonding interaction in PVC/MA-POSS blends.

2.2. PVC/DINCH, PVC/MA-POSS, and PVC/DINCH/MA-POSS Blend Films

Binary blends of PVC/DINCH and PVC/MA-POSS and ternary blends of PVC/DINCH/MAPOSS were prepared through solvent-blending, using THF as the solvent. A solution containing 10 wt% of the polymer mixture was stirred for 24 h and then the solution was poured into a Teflon mold. The solvent (THF) was evaporated slowly over 3 days at room temperature. The resulting blend film was heated under a vacuum at 50 °C for 2 days and then at 80 °C for 2 days to ensure the total removal of any residual solvent. Samples for DSC, TGA, DMA and tensile testing were pelletized, compression-molded into disks (160 °C, 3 min), and then machined into suitable specimens for testing.

2.3. Characterization

The glass transition temperatures of all the blend systems were measured using DSC and DMA. DSC was performed using a TA-Q20 instrument, operated at a heating rate of 20 °C/min between 25 and 150 °C under a N_2 atmosphere, and the sample was quickly cooled to −90 °C after the first scan. The T_g values were determined at the midpoint of the transition temperature, with a scan rate of 20 °C/min, in the range −90–200 °C. The dynamic mechanical properties were characterized using a DuPont 2980 dynamic mechanical analyzer, with the blend sample mounted on the single cantilever clamp. A constant strain amplitude of 2.0% and a frequency of 1 Hz were applied, with a heating rate of 2 °C/min between −60 and +120 °C. The thermal stabilities of the blend samples were determined using a TA Q-50 thermogravimetric analyzer, operating within the temperature range 30–800 °C at a heating rate of 20 °C/min, under N_2. FTIR spectra samples of the blend were recorded using a Bruker Tensor 27 spectrophotometer. A total of 32 scans were collected at a spectral resolution of 4 cm^{-1}. The FTIR samples were measured using conventional KBr disk method, similarly to the preparation of the bulk sample. The film used in this work was sufficiently thin to obey the Beer–Lambert law. Tensile tests were performed using an Instron machine. Five different blend samples of each composition were examined at a tensile rate of 1 cm/s.

3. Results

3.1. Thermal Analysis of PVC/DINCH Blends

Because DINCH is among the most promising plasticizers used as a replacement for DEHP in PVC, we first investigated the thermal properties of PVC/DINCH blends through TGA analyses. Figure 1 displays the TGA traces of pure PVC, pure DINCH, and the PVC/DINCH = 100/60 blend, measured from 25 to 800 °C at a heating rate of 20 °C/min. Pure DINCH displayed its thermal decomposition at a temperature of T_{d10} = 213 °C, without providing a char yield, as displayed in Figure 1c. The pure PVC displayed two main thermal decomposition temperatures during the degradation process. The pyrolysis of PVC is complex when compared with those of other thermoplastic polymeric materials; it depends on the type and amount of plasticizer used [1,35]. The first thermal decomposition of PVC was initiated by the elimination of HCl at a relatively low temperature (300 °C and T_{d10} = 290 °C)—the so-called "dehydrochlorination" stage, during which polyene radicals and unsubstituted aromatics are released, as displayed in Figure S2a. Aromatic products and various degrees of char were formed during the second thermal decomposition stage at approximately 460 °C from DTG curves (Figure S2a) [35]. The solid residue of the char yield was approximately 2.7 wt% after completion of the PVC pyrolysis, corresponding to ash formation. After blending with DINCH, the value of T_{d10} was 265 °C, because of the lower thermal decomposition temperature of pure DINCH. The char yield increased to 6 wt%, presumably because of ash formation from the reaction of DINCH with PVC. The lower thermal stability of the PVC/DINCH blend system, compared with that of pure PVC, encouraged us to add MA-POSS to potentially improve the thermal stability. Because of the possibility of intermolecular hydrogen bonding between PVC and the MA-POSS nanoparticles, we used DSC, TGA, DMA, and FTIR spectroscopy to perform initial investigations into the miscibility, thermal properties, and hydrogen bonding interactions of binary blends of PVC/MA-POSS.

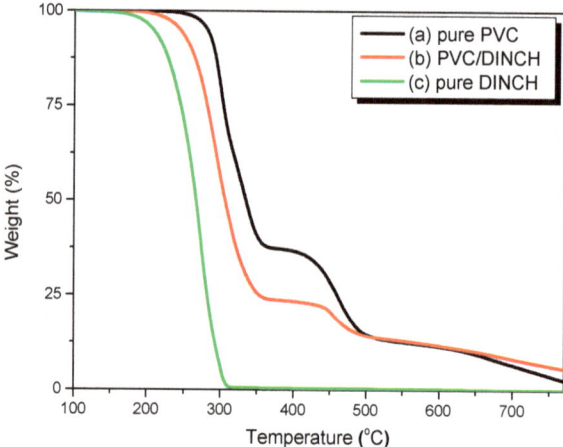

Figure 1. Thermogravimetric analyses (TGA) of (a) the pure PVC, (b) the PVC/DINCH = 100/60 blend, and (c) pure DINCH.

3.2. Thermal Properties and FTIR Spectra of PVC/MA-POSS Blends

Figure 2 presents the results of DSC and DMA analyses of PVC/MA-POSS blends in the range 30–110 °C. The pure PVC exhibited a value of T_g of 89 °C, based on DSC analysis (Figure 2a), and 94 °C, based on DMA analysis (Figure 2b). It is not unusual for the value of T_g determined using DMA to be higher than that from DSC analysis, typically by 5–10 °C. In addition, pure MA-POSS has a reported glass transition temperature of −55 °C (Figure S1), and is a pale-yellow heavy oil [33,34].

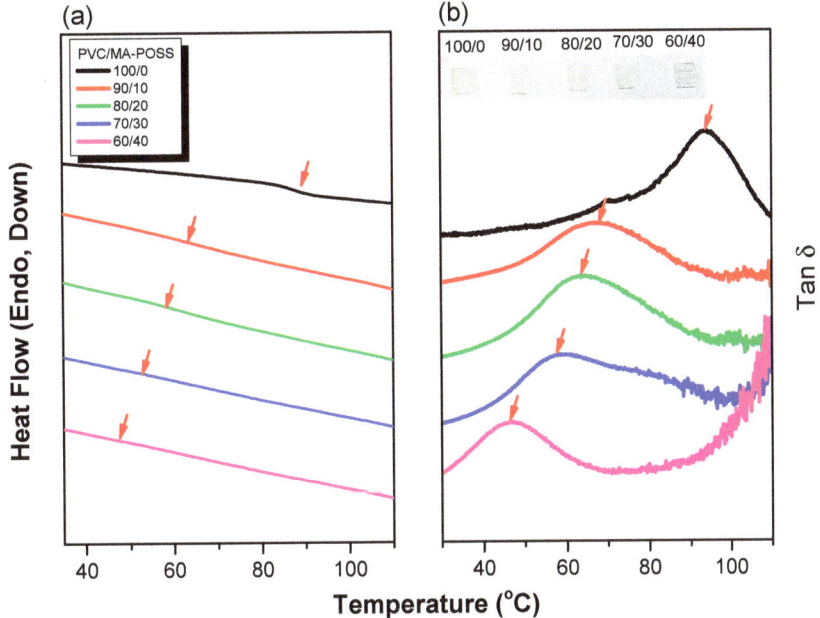

Figure 2. (a) Differential scanning calorimetry (DSC) and (b) dynamic mechanical analysis (DMA) of various PVC/MA-POSS blends; the inset to (b) displays the transparent films obtained after DMA analyses.

The values of T_g, determined using both DSC and DMA, decreased significantly upon increasing the MA-POSS concentration. When 10 wt% of MA-POSS was incorporated into the PVC matrix, the value of T_g decreased to 63 (DSC) and 68 (DMA) °C, approximately 26 °C lower than that of pure PVC, indicating that the plasticizing effect of MA-POSS towards PVC was significant. Further increasing the MA-POSS content to 20, 30, and 40 wt% caused the values of T_g to decrease to 64, 58, and 47 °C, respectively, based on DMA analysis, confirming that MA-POSS is a promising plasticizer for PVC, consistent with previous results [33,34]. In addition, all the PVC/MA-POSS blends possessed only a single glass transition temperature, implying that these blends were fully miscible. Furthermore, the values of T_g were between those of the two individual compounds, as expected. More importantly, all the PVC/MA-POSS blends were transparent and flexible, even after thermal treatment (DMA), as displayed in the inset to Figure 2b. Figure 3 presents the values of T_g of the PVC/MA-POSS blends, as determined using DMA. Although the glass transition temperatures of the PVC/MA-POSS blends were higher than the values predicted by the Fox rule and the linear rule, they were predicted using the Kwei equation [36]:

$$T_g = \frac{W_1 T_{g1} + k W_2 T_{g2}}{W_1 + k W_2} + q W_1 W_2 \qquad (1)$$

where W_1 and W_2 are the weight fractions and T_{g1} and T_{g2} are the glass transition temperatures of the MA-POSS and the PVC, respectively, while k and q are fitting constants. We obtain values of k and q of 1 and 50, respectively, for the PVC/MA-POSS blends, indicating that specific intermolecular interactions were occurring between the PVC and MA-POSS nanoparticles.

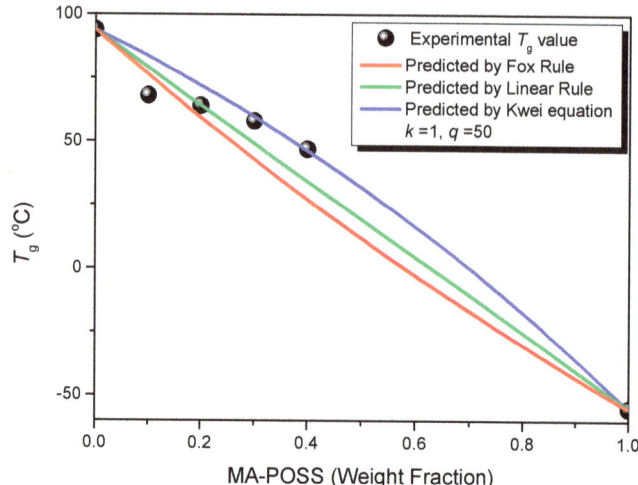

Figure 3. Experimental values of T_g of PVC/MA-POSS blends and those predicted by the Fox and linear rules and the Kwei equation.

FTIR spectroscopy can provide information about the noncovalent interactions of both polar and nonpolar groups through hydrogen bonding or dipole–dipole interactions [37–40]. Figure 4 presents the characteristic absorption signals for the (a) C=O units of MA-POSS, (b) C–H deformations of the CH_2 units of PVC, and (c) C–Cl stretching vibrations of PVC in various PVC/MA-POSS blends. The free C=O groups of MA-POSS provided an absorption at 1717 cm^{-1}; this signal shifted to 1714 cm^{-1} at 90 wt% of PVC concentration, as displayed in Figure 4a. In addition, the signal for the CH_2 units of PVC shifted from 1332 to 1321 cm^{-1} (Figure 4b), while its C–Cl units shifted significantly from 614 to 606 cm^{-1} (Figure 4c), after blending with the MA-POSS nanoparticles. These results indicate that intermolecular hydrogen bonding existed between the H–CCl units of the PVC and the C=O groups of the MA-POSS, as displayed in Scheme 1d. Such weak hydrogen bonding interactions have also been observed in many other binary blend systems [41–43].

Figure 5 displays the TGA curves of various PVC/MA-POSS blends, recorded from 25 to 800 °C at a heating rate of 20 °C/min, and the corresponding DTG curves are displayed in Figure S3. The thermal stability of pure MA-POSS was greater than that of pure PVC, which possessed a value of T_{d10} of 421 °C and a char yield of 44.5 wt%. Upon increasing the content of MA-POSS, the first thermal decomposition temperature and the char yield both increased. When the MA-POSS concentration reached 50 wt%, the value of T_{d10} was close to 306 °C, approximately 16 °C higher than that of pure PVC, while the char yield also increased to 33.7 wt%, higher than the value [23.6 wt% = (44.5 + 2.7)/2] suggested by the simplified relationship of pure MA-POSS (44.5 wt%) + pure PVC (2.7 wt%)—again, probably because of ash formation from the reaction of MA-POSS with PVC. This result suggests that the rate of HCl emission during the first thermal decomposition procedure had decreased as a result of the presence of MA-POSS nanoparticles. The main thermal decomposition of MA-POSS overlapped with the second thermal decomposition process of PVC, making it difficult to identify, as displayed in Figure S3. In addition, the decomposition process of MA-POSS becomes obvious after increasing the MA-POSS concentration, becoming comparable with that of pure MA-POSS. The final step in the TGA curves corresponds to the pyrolysis of conjugated double bonds (e.g., of the polyenes formed in the first step of PVC decomposition), with the decomposition temperature and char yield both increasing upon increasing the MA-POSS concentration, because of the retarding effect of MA-POSS nanoparticles toward the possible residual ash [33,34]. Thus, our DSC, DMA, FTIR spectroscopic, and TGA analyses confirmed that the incorporation of MA-POSS nanoparticles could decrease the value of T_g of PVC, and increase both its thermal decomposition temperature and the char yield, while providing full miscibility with PVC, because of weak hydrogen bonding in the PVC/MA-POSS blends. As a result, MA-POSS has the potential to be a suitable replacement for traditional plasticizers, e.g., DEHP, and could be used as a blood bag, to avoid the DEHP coming into contact with tissues or body fluids.

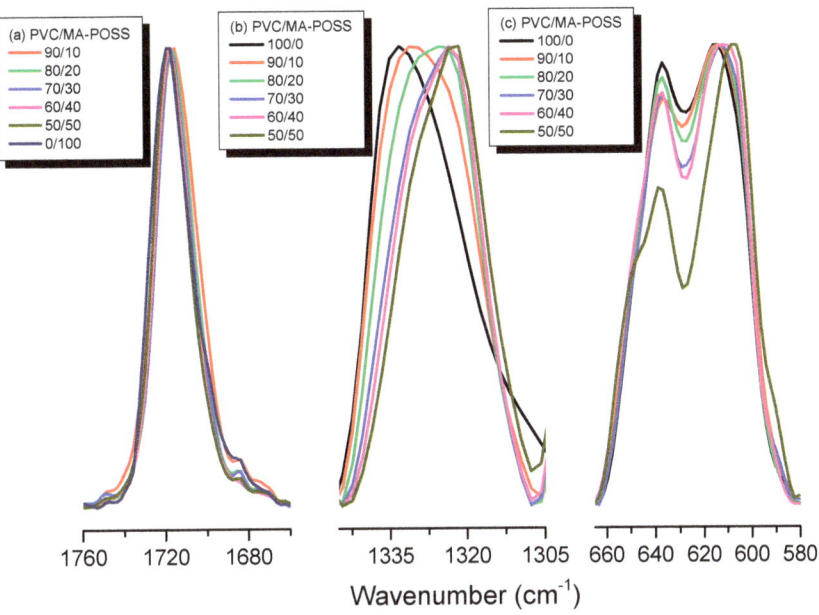

Figure 4. FTIR spectral analyses of the (**a**) C=O groups of MA-POSS, (**b**) C–H deformation of the CH_2 units of the PVC, and (**c**) C–Cl stretching vibration of the PVC, for various PVC/MA-POSS blends recorded at room temperature.

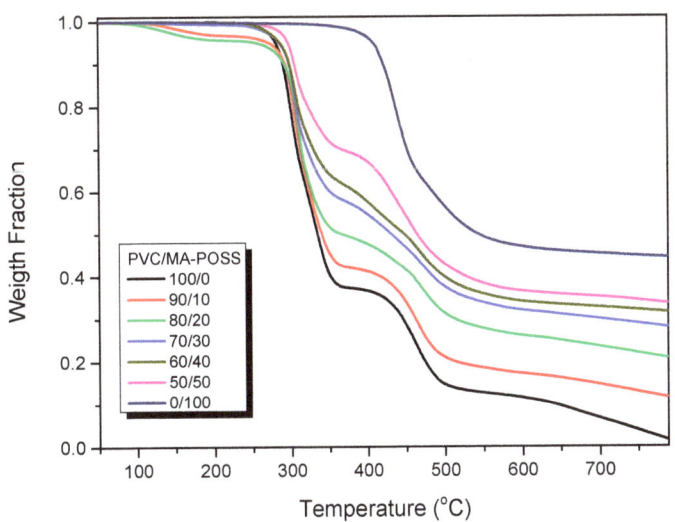

Figure 5. TGA analyses of PVC/MA-POSS blends of various compositions.

3.3. Thermal and Mechanical Properties of PVC/DINCH/MA-POSS Ternary Blends

To solve the problem of the lower thermal stability of PVC/DINCH blend systems compared to pure PVC, we tested the effect of adding MA-POSS nanoparticles. Figure 6 presents DMA analyses of PVC/DINCH/MA-POSS ternary blends at various MA-POSS contents. The value of T_g of the binary PVC/DINCH blend was 13.4 °C. Increasing the MA-POSS content to 5, 10, and 15 phr caused the value of T_g to decrease to 6.6, 4.7, and 2.6 °C, respectively, based on DMA analysis, suggesting that MA-POSS remains a promising plasticizer for PVC/DINCH blends. In addition, we observed transparent behavior for all the PVC/DINCH/MA-POSS ternary blends, with flexibility, even after thermal treatment through DMA analysis (see inset to Figure 6).

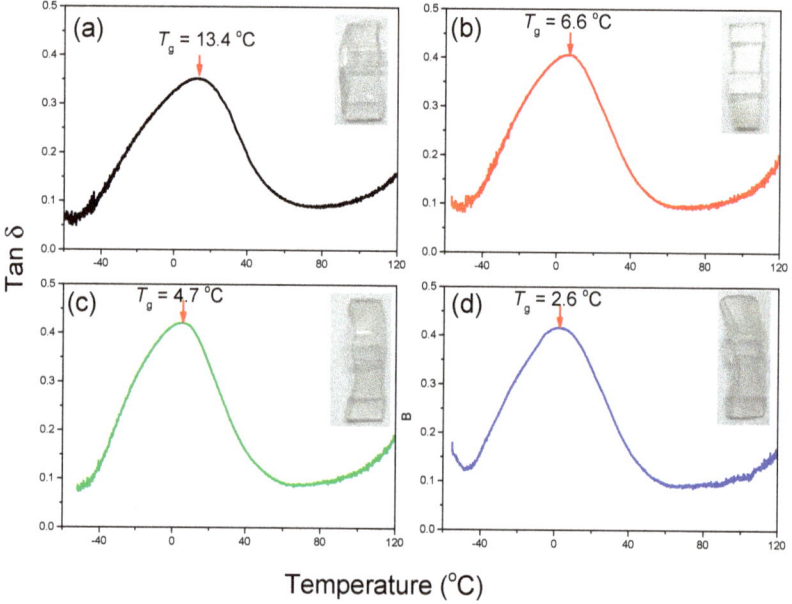

Figure 6. DMA analyses of various PVC/DINCH/MA-POSS blends: (**a**) 100/60/0, (**b**) 100/60/5, (**c**) 100/60/10, and (**d**) 100/60/15. The inset displays the transparent films obtained after DMA analysis of each composition.

Figure 7 presents the TGA curves of PVC/DINCH/MA-POSS ternary blends at various MA-POSS contents, measured from 25 to 800 °C at a heating rate of 20 °C/min. The corresponding DTG curves are also displayed in Figure S4. Increasing the MA-POSS content in the PVC/DINCH binary blend caused the first thermal decomposition temperature and the char yield to increase, as displayed in Figure S4. When the MA-POSS content reached 15 phr, the value of T_d was close to 271 °C, approximately 6 °C higher than that of the PVC/DINCH binary blend. Furthermore, the char yield also increased to 17.3 wt%, 11.3 wt% higher than that of the PVC/DINCH binary blend.

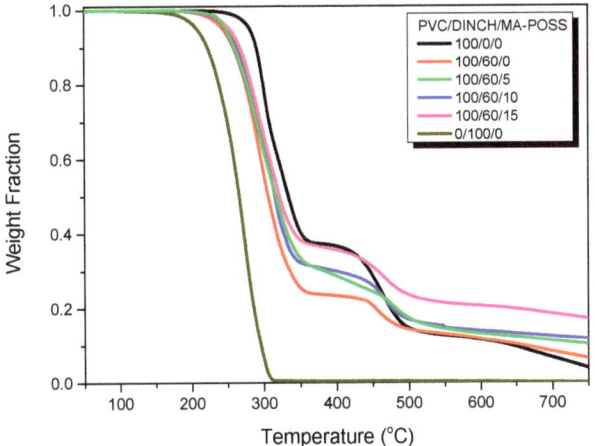

Figure 7. TGA analyses of PVC/DINCH/MA-POSS blends of various compositions.

Figure 8 displays the results of tensile tests of PVC/DINCH/MA-POSS ternary blends at various MA-POSS contents (a) before and (b) after thermal annealing at 100 °C for 7 days. All the PVC/DINCH/MA-POSS blends exhibited high elongation behavior, even after thermal annealing, and we could briefly observe that the modulus and tensile strength decreased, and elongation increased, upon increasing the content of MA-POSS, both before and after thermal annealing.

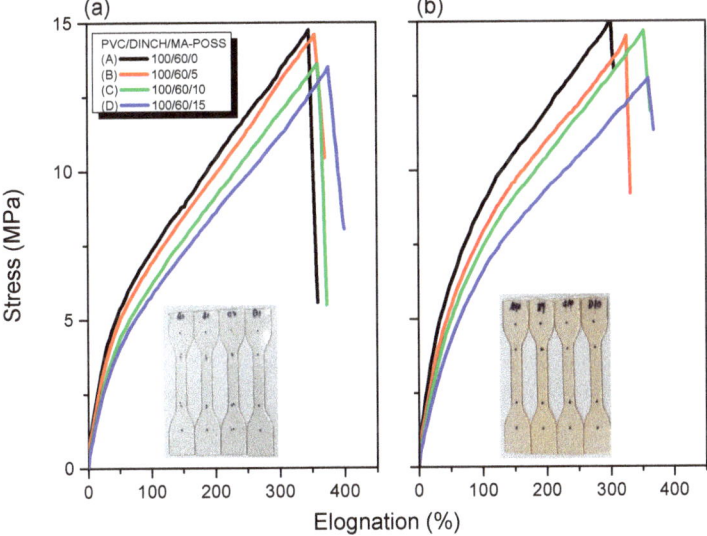

Figure 8. Stress–strain curves of PVC/DINCH/MA-POSS blends at various MA-POSS contents (a) before and (b) after thermal annealing at 100 °C for 7 days. The inset displays the tensile test samples used in this study (a) before and (b) after thermal annealing.

Figure 9 summarizes the mechanical properties. Elongation decreased, but the modulus and tensile strength increased, after thermal annealing. We also observed transparency for all the

PVC/DINCH/MA-POSS ternary blends, as well as flexibility. They became pale yellow after thermal annealing at 100 °C for 7 days, as displayed in Figure 8.

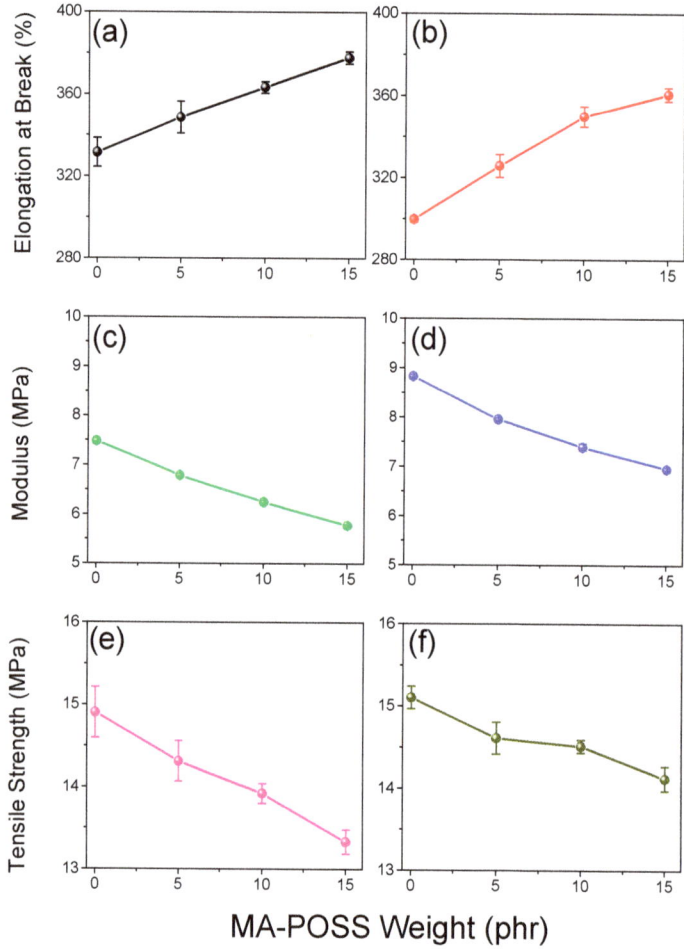

Figure 9. Mechanical properties of PVC/DINCH/MA-POSS blends at various MA-POSS contents (**a,c,e**) before and (**b,d,f**) after thermal annealing at 100 °C for 7 days: (**a,b**) elongation properties, (**b,d**) moduli, and (**c,f**) tensile strengths.

4. Conclusions

We have examined the thermal properties, miscibility, hydrogen bonding interactions, and mechanical properties of PVC/MA-POSS binary blends and PVC/DINCH/MA-POSS ternary blends. Employing DSC, DMA, TGA, and tensile tests, we confirmed that the incorporation of MA-POSS nanoparticles into the blends occurred with full miscibility; they decreased the value of T_g, the modulus, and tensile strength, but increased the thermal decomposition temperature, char yield, and elongation properties, because weak hydrogen bonding existed between the C=O groups of the MA-POSS and the H–CCl units of the PVC, as determined through FTIR spectral analyses. As a result, MA-POSS appears to be a suitable candidate for replacing traditional plasticizers, e.g., DEHP, in PVC matrices.

Supplementary Materials: The following are available online at http://www.mdpi.com/2073-4360/11/10/1711/s1.

Author Contributions: Y.-K.W. did the experiment of Figures 1–7, C.-C.M. and M.-L.W. did the experimental of Figures 8 and 9, F.-C.T., and S.-W.K. contributed to the literature review and to the writing of this paper.

Funding: This research was funded by the Ministry of Science and Technology, Taiwan, under contracts MOST 106-2221-E-110-067-MY3, MOST 105-2221-E-110-092-MY3, industry-academia-research cooperation project between UPC and NSYSU, and also supported by the National Natural Science Foundation of China (Grants U1805253 and 51773214).

Conflicts of Interest: The authors declare no conflict of interest.

References

1. Chiellni, F.; Ferri, M.; Morelli, A.; Dipaola, L.; Latini, G. Perspectives on alternatives to phthalate plasticized poly(vinyl chloride) in medical devices applications. *Prog. Polym. Sci.* **2013**, *38*, 1067–1088. [CrossRef]
2. Huynh, N.M.N.; Boeva, Z.A.; Smatt, J.H.; Pesonen, M.P.; Lindfors, T. Reduced graphene oxide as a water, carbon dioxide and oxygen barrier in plasticized poly(vinyl chloride) films. *RSC Adv.* **2018**, *8*, 17645–17655. [CrossRef]
3. Jia, P.Y.; Xia, H.Y.; Tang, K.H.; Zhou, Y.H. Plasticizers derived from biomass resources: A short review. *Polymers* **2018**, *10*, 1303. [CrossRef] [PubMed]
4. Suresh, S.S.; Mohanty, S.; Nayak, S.K. Bio-based epoxidised oil for compatibilization and value addition of poly (vinyl chloride) (PVC) and poly(methyl methacrylate) (PMMA) in recycled blend. *J. Polym. Res.* **2017**, *24*, 120. [CrossRef]
5. Li, D.Y.; Panchal, K.; Mafi, R.; Xi, L. An Atomistic Evaluation of the Compatibility and Plasticization Efficacy of Phthalates in Poly(vinyl chloride). *Macromolecules* **2018**, *51*, 6997–7012. [CrossRef]
6. Jaeger, R.J.; Rubin, R.J. Migration of a Phthalate Ester Plasticizer from Polyvinyl Chloride Blood Bags into Stored Human Blood and Its Localization in Human Tissues. *N. Engl. J. Med.* **1972**, *287*, 1114–1118. [CrossRef]
7. Ozeren, H.O.; Balcik, M.; Ahunbay, M.G.; Elliott, J.R. In Silico Screening of Green Plasticizers for Poly(vinyl chloride). *Macromolecules* **2019**, *52*, 2421–2430. [CrossRef]
8. Deshmukh, K.; Khatake, S.M.; Joshi, G.M. Surface properties of graphene oxide reinforced polyvinyl chloride nanocomposites. *J. Polym. Res.* **2013**, *20*, 286. [CrossRef]
9. Mansour, S.A.; Elsad, R.A. Dielectric spectroscopic analysis of polyvinyl chloride nanocomposites loaded with Fe_2O_3 nanocrystals. *Polym. Adv. Tech.* **2018**, *29*, 2477–2485. [CrossRef]
10. Hajibeygi, M.; Maleki, M.; Shabanian, M.; Ducos, F.; Vahab, H. New polyvinyl chloride (PVC) nanocomposite consisting of aromatic polyamide and chitosan modified ZnO nanoparticles with enhanced thermal stability, low heat release rate and improved mechanical properties. *Appl. Sur. Sci.* **2018**, *439*, 1163–1179. [CrossRef]
11. Kuo, S.W.; Chang, F.C. POSS related polymer nanocomposites. *Prog. Polym. Sci.* **2011**, *36*, 1649–1696. [CrossRef]
12. Cordes, D.B.; Lickiss, P.D.; Rataboul, F. Recent developments in the chemistry of cubic polyhedral oligosilsesquioxanes. *Chem. Rev.* **2010**, *110*, 2081–2173. [CrossRef] [PubMed]
13. Wu, J.; Mather, P.T. POSS polymers: Physical properties and biomaterials applications. *Polym. Rev.* **2009**, *49*, 25–63. [CrossRef]
14. Mohamed, M.G.; Kuo, S.W. Functional Polyimide/Polyhedral Oligomeric Silsesquioxane Nanocomposites. *Polymers* **2019**, *11*, 26. [CrossRef]
15. Mohamed, G.M.; Kuo, S.W. Functional Silica and Carbon Nanocomposites Based on Polybenzoxazines. *Macromol. Chem. Phys.* **2019**, *220*, 1800306. [CrossRef]
16. Kuo, S.W. Building Blocks Precisely from Polyhedral Oligomeric Silsesquioxane Nanoparticles. *ACS Cent. Sci.* **2016**, *2*, 62–64. [CrossRef]
17. Mohamed, M.G.; Jheng, Y.R.; Yeh, S.L.; Chen, T.; Kuo, S.W. Unusual Emission of Polystyrene-Based Alternating Copolymers Incorporating Aminobutyl Maleimide Fluorophore-Containing Polyhedral Oligomeric Silsesquioxane NPs. *Polymers* **2017**, *9*, 103. [CrossRef]
18. Li, Z.; Kong, J.; Wang, F.; He, C. Polyhedral oligomeric silsesquioxanes (POSSs): An important building block for organic optoelectronic materials. *J. Mater. Chem. C* **2017**, *5*, 5283–5298. [CrossRef]
19. Liu, Y.; Wu, X.; Sun, Y.; Xie, W. POSS Dental Nanocomposite Resin: Synthesis, Shrinkage, Double Bond Conversion, Hardness, and Resistance Properties. *Polymers* **2018**, *10*, 369. [CrossRef]
20. Blanco, I. The Rediscovery of POSS: A Molecule Rather than a Filler. *Polymers* **2018**, *10*, 904. [CrossRef]

21. Yang, X.M.; Li, Y.Y.; Wang, Y.L.; Yang, Y.G.; Hao, J.W. Nitrocellulose-based hybrid materials with T7-POSS as a modifier: Effective reinforcement for thermal stability, combustion safety, and mechanical properties. *J. Polym. Res.* **2018**, *24*, 50. [CrossRef]
22. Liao, Y.T.; Lin, Y.C.; Kuo, S.W. Highly Thermally Stable, Transparent, and Flexible Polybenzoxazine Nanocomposites by Combination of Double-Decker-Shaped Polyhedral Silsesquioxanes and Polydimethylsiloxane. *Macromolecules* **2017**, *50*, 5739–5747. [CrossRef]
23. Yu, C.Y.; Kuo, S.W. Phenolic Functionality of Polyhedral Oligomeric Silsesquioxane Nanoparticles Affects Self-Assembly Supramolecular Structures of Block Copolymer Hybrid Complexes. *Ind. Eng. Chem. Res.* **2018**, *57*, 2546–2559. [CrossRef]
24. Zhou, H.; Ye, Q.; Xu, J. Polyhedral oligomeric silsesquioxane-based hybrid materials and their applications. *Mater. Chem. Front.* **2017**, *1*, 212–230. [CrossRef]
25. Li, Y.; Dong, X.H.; Zou, Y.; Wang, Z.; Yue, K.; Huang, M.; Liu, H.; Feng, X.; Lin, Z.; Zhang, W.; et al. Polyhedral oligomeric silsesquioxane meets "click" chemistry: Rational design and facile preparation of functional hybrid materials. *Polymer* **2017**, *125*, 303–329. [CrossRef]
26. Huang, K.W.; Tsai, L.W.; Kuo, S.W. Influence of Octakis-Functionalized Polyhedral Oligomeric Silsesquioxanes on the Physical Properties of Their Polymer Nanocomposites. *Polymer* **2009**, *50*, 4876–4887. [CrossRef]
27. Mohamed, M.G.; Kuo, S.W. Polybenzoxazine/Polyhedral Oligomeric Silsesquioxane (POSS) Nanocomposites. *Polymers* **2016**, *8*, 225. [CrossRef]
28. Ayandele, E.; Sarkar, B.; Alexandridis, P. Polyhedral Oligomeric Silsesquioxane (POSS)-Containing Polymer Nanocomposites. *Nanomaterials* **2012**, *2*, 445–475. [CrossRef]
29. Chen, W.C.; Lin, R.C.; Tseng, S.M.; Kuo, S.W. Minimizing the Strong Screening Effect of Polyhedral Oligomeric Silsesquioxane Nanoparticles in Hydrogen-Bonded Random Copolymers. *Polymers* **2018**, *10*, 303. [CrossRef]
30. Yeh, S.L.; Zhu, C.Y.; Kuo, S.W. Strong Hydrogen Bonding with Inorganic Pendant Polyhedral Oligomeric Silsesquioxane Nanoparticles Provides High Glass Transition Temperature Poly (methyl methacrylate) Copolymers. *J. Nanosci. Nanotechnol.* **2018**, *18*, 188–194. [CrossRef]
31. Chen, W.C.; Kuo, S.W. *Othro*-Imide and Allyl Groups Effect on Highly Thermal Stable Polybenzoxazine/Double-Decker–Shaped Polyhedral Silsesquioxanes (DDSQ) Nanocomposites. *Macromolecules* **2018**, *51*, 9602–9612. [CrossRef]
32. Sterzynski, T.; Tomaszewska, J.; Andrzejewski, J.; Skorczewska, K. Evaluation of glass transition temperature of PVC/POSS nanocomposites. *Compos. Sci. Tech.* **2015**, *117*, 398–403. [CrossRef]
33. Soong, S.Y.; Cohen, R.E.; Boyce, M.C. Polyhedral oligomeric silsesquioxane as a novel plasticizer for oly(vinyl chloride). *Polymer* **2007**, *48*, 1410–1418. [CrossRef]
34. Soong, S.Y.; Cohen, R.E.; Boyce, M.C.; Mulliken, A.D. Rate-dependent deformation behavior of POSS-filled and plasticized poly(vinyl chloride). *Macromolecules* **2006**, *39*, 2900–2908. [CrossRef]
35. Yang, S.; Wu, W.; Jiao, Y.; Fan, H.; Cai, Z. Poly(ethylene glycol)–polyhedral oligomeric sesquioxane as a novel plasticizer and thermal stabilizer for poly(vinyl chloride) nanocomposites. *Polym. Inter.* **2016**, *65*, 1172–1178. [CrossRef]
36. Kwei, T. The effect of hydrogen bonding on the glass transition temperatures of polymer mixtures. *J. Polym. Sci. Polym. Lett. Ed.* **1984**, *22*, 307–313. [CrossRef]
37. Kuo, S.W. *Hydrogen Bonding in Polymeric Materials*; John Wiley & Sons: Hoboken, NJ, USA, 2018.
38. Pan, C.T.; Yen, C.K.; Wang, S.Y.; Fan, S.K.; Chiou, F.Y.; Lin, L.; Huang, J.C.; Kuo, S.W. Energy Harvester and Cell Proliferation from Biocompatible PMLG Nanofibers Prepared Using Near-Field Electrospinning and Electrospray Technology. *J. Nanosci. Nanotechnol.* **2018**, *18*, 156–164. [CrossRef]
39. Jheng, Y.R.; Mohamed, M.G.; Kuo, S.W. Supramolecular Interactions Induce Unexpectedly Strong Emissions from Triphenylamine-Functionalized Polytyrosine Blended with Poly (4-vinylpyridine). *Polymers* **2017**, *9*, 503. [CrossRef]
40. Lin, R.C.; Mohamed, M.G.; Kuo, S.W. Coumarin-and Carboxyl-Functionalized Supramolecular Polybenzoxazines Form Miscible Blends with Polyvinylpyrrolidone. *Polymers* **2017**, *9*, 146. [CrossRef]
41. Kuo, S.W.; Huang, W.J.; Huang, C.F.; Chan, S.C.; Chang, F.C. Miscibility, Specific Interactions, and Spherulite Growth Rates of Binary Poly(acetoxystyrene)/Poly(ethylene oxide) Blends. *Macromolecules* **2004**, *37*, 4164–4173. [CrossRef]

42. Kuo, S.W.; Huang, C.F.; Tung, P.H.; Huang, W.J.; Huang, J.M.; Chang, F.C. Synthesis, Thermal Properties and Specific Interactions of High T_g Increase in Poly(2,6-dimethyl-1,4-phenylene oxide)-block-Polystyrene Copolymers. *Polymer* **2005**, *46*, 9348–9361. [CrossRef]
43. Mao, B.H.; El-Mahdy, A.F.M.; Kuo, S.W. Miscibility Enhancement of Conjugated Polymers Blending with Polystyrene Derivatives through Bio-Inspired Multiple Complementary Hydrogen Bonds. *J. Polym. Res.* **2019**, *26*, 208. [CrossRef]

© 2019 by the authors. Licensee MDPI, Basel, Switzerland. This article is an open access article distributed under the terms and conditions of the Creative Commons Attribution (CC BY) license (http://creativecommons.org/licenses/by/4.0/).

Article

Transparent Polyimide/Organoclay Nanocomposite Films Containing Different Diamine Monomers

Hyeon Il Shin and Jin-Hae Chang *

Department of Polymer Science and Engineering, Kumoh National Institute of Technology, Gumi 39177, Korea; poweril55@naver.com
* Correspondence: changjinhae@hanmail.net

Received: 29 November 2019; Accepted: 1 January 2020; Published: 6 January 2020

Abstract: Poly (amic acid) s (PAAs) were synthesized using 4,4'-(hexafluoroisopropyl-idene) diphthalic anhydride (6FDA) and two types of diamines—bis(3-aminophenyl) sulfone (BAS) and bis(3-amino-4-hydroxyphenyl) sulfone (BAS-OH). Two series of transparent polyimide (PI) hybrid films were synthesized by solution intercalation polymerization and thermal imidization using various concentrations (from 0 to 1 wt%) of organically modified clay Cloisite 30B in PAA solution. The thermo-mechanical properties, morphology, and optical transparency of the hybrid films were observed. The transmission electronic microscopy (TEM) results showed that some of the clays were agglomerated, but most of them showed dispersed nanoscale clay. The effects of -OH groups on the properties of the two PI hybrids synthesized using BAS and BAS-OH monomers were compared. The BAS PI hybrids were superior to the BAS-OH PI hybrids in terms of thermal stability and optical transparency, but the BAS-OH PI hybrids exhibited higher glass transition temperatures (T_g) and mechanical properties. Analysis of the thermal properties and tensile strength showed that the highest critical concentration of organoclay was 0.50 wt%.

Keywords: transparent polyimide; nanocomposite; film; organoclay; thermo-mechanical properties; optical transparency

1. Introduction

Polyimide (PI), which has been used for a long time, was mainly studied for military purposes when it was first developed [1,2]. It has also been studied in various applications in the aerospace industry and as electric materials. Moreover, PI thin films are easy to synthesize and do not require crosslinking agents for curing [3–5].

Recently, PI has been applied as an integrated material for semiconductor materials such as liquid crystal display (LCD) and plasma display panel (PDP) because of its low weight and its ability to improve electronic products. In addition, PI, which is lightweight and flexible, has been extensively studied for use in flexible plastic display substrates to overcome disadvantages such as heavy weight and tendency of shattering of glass substrates used in the field [6–8].

Although PI is a high-performance polymer material with excellent properties, including high thermal stability and good mechanical properties, chemical resistance, and electrical properties, it does not meet the basic requirements for display applications. Many synthesized PIs are insoluble and infusible and thus have poor processability. Therefore, efforts are now being made to improve these optical properties and the processability of PI. For example, PIs containing trifluoromethyl groups have been synthesized that show a high modulus, low thermal expansion coefficient, and good solubility in conventional organic solvents [9–11]. Another method is to use a copolyimide (Co-PI) using a specific monomer. A Co-PI typically possesses much lower molecular regularity than the corresponding homopolyimide [12,13]. This decreased regularity leads to fewer intermolecular interactions that, in turn, results in new characteristics, such as modified thermo-optical and gas permeation properties, and solubilities. Furthermore, the properties of Co-PIs can be adjusted by varying the ratio of the dianhydride and diamine comonomers.

PI is dark brown because electrons forming the double bonds in the benzene in the imide structure are transferred to the charge carriers of the π electrons generated by intermolecular bonding between the chain transfer complex (CT-complex) [14,15]. To reduce the CT-complex, a strong electron withdrawing group such as fluorine (F) or sulfone ($-SO_2-$) is required that can effectively introduce a bending structure which can interfere with the interaction between the PI main chains, thus aiding in the fabrication of a PI film with high transparency. For example, a $-CF_3$ group, which is a strong electron-withdrawing group, is often used as a substituent, or bent monomer structures are used to prevent CT complexes from being formed in linear structures [16,17].

It is also possible to improve the processability of PI films by introducing couplers with small polarity and free rotation in the main chain or by introducing bulky substituents to reduce the crystallinity and molecular packing density [18–20].

Among various layered inorganic materials used as nanofillers, clay has a layered structure with ionic bonds and Van der Waals forces acting between the layers. Clay has platelets with a thickness of 1 nm and a width of 100–1000 nm [21,22]. In general, layered inorganic materials are hydrophilic because the anions on the surface exist in a form that is stabilized with an alkali metal cation. Therefore, polymer materials including -OH groups are known to exhibit very good dispersibility and compatibility with clay [23,24]. Several methods are used to increase the physical properties of the blend, including hydrogen bonding using the hydroxyl (-OH) group between the polymer chain and the filler. For example, it is known that PVA can form a hydrogen bond between a polymer chain and a filler to ensure they are more tightly bonded, resulting in better physical properties than pure polymers [25,26].

In this study, the diamine monomers—bis(3-aminophenyl) sulfone (BAS) and bis(3-amino-4-hydroxyphenyl) (BAS-OH)—were allowed to react with 4,4'-(hexafluoroisopropyl (6FDA)) as a dianhydride to synthesize two PIs. PI hybrids were prepared from the synthesized PIs and Cloisite 30B as an organoclay. Cloisite 30B is hydrophilic because it has -OH groups. Furthermore, it is expected that the affinity of this clay will increase if it is blended with PI containing -OH groups. BAS-OH containing a hydroxyl group can enhance the reactivity, dispersibility, and compatibility between the synthesized PI and clay.

In this study, the thermo-mechanical properties, morphology, and optical transparency of the PI hybrids with various concentrations of organoclay (from 0 wt% to 1.00 wt%) were investigated, and their results were compared. We also described the effect of -OH groups in PI hybrids and compared the properties of the two PI hybrids synthesized using BAS and BAS-OH diamine monomers.

2. Experimental Details

2.1. Materials

Cloisite 30B was purchased from Southern Clay Product, Co. Most solvents used in this study were purchased from Daehan Chemical (Daegu, Korea). 6FDA, BAS, and BAS-OH were purchased from TCI (Tokyo, Japan) and Aldrich Chemical Co. (Yongin, Korea). DMAc was purified with 4Å molecular sieve and stored for water absorption treatment.

2.2. Preparation of PI Hybrid Films

Because the synthesis procedures of the two PAAs using BAS and BAS-OH diamine monomers are the same, only the method of fabricating BAS-OH PAA will be described here. BAS-OH (4.65 g, 1.85×10^{-2} mol) and DMAc (30 mL) were placed in a flask and stirred at 0 °C for 1 h while supplying nitrogen. A solution of 6FDA (7.38 g; 1.85×10^{-2} mol) in DMAc (40 mL) was added to the BAS-OH/DMAc solution. This solution was stirred at a moderate speed at 25 °C for 14 h to synthesize a PAA solution having a solid content of 15.5 wt%. The inherent viscosities of the synthesized PAAs of

BAS and BAS-OH were 1.02 and 0.94, respectively. The total 15.5 wt% PAA solution produced by this method was 77.62 g. When 0.39 g of Cloisite 30 B was added to the solution, the total solution weight was 78.01 g.

Because the methods of synthesizing PI hybrids using different concentrations of organoclay were the same, the synthetic method of synthesizing PI/Cloisite 30 B (0.50 wt%) as a representative example will be described here. We first added 0.064 g of Cloisite 30 B, 13.0 g of PAA solution, and DMAc (20 mL) to a flask and vigorously stirred the mixture at room temperature for 3 h. The solution was also washed in an ultrasonic cleaner for 3 h, and the solvent was removed in a vacuum oven at 50 °C for 2 h. The obtained PAA film was dried in a vacuum oven at 80 °C for 1 h to completely remove the solvent.

The PAA film was further imidized on the glass plate by sequential heating at 110 °C, 140 °C, and 170 °C for 30 min at each temperature, followed by 50 min at 195 °C and 220 °C each. Finally, the reaction was completed by heat treatment at 235 °C for 2 h.

The thickness of the heat-treated film was approximately 67–70 μm. Table 1 summarizes the detailed heat treatment conditions for obtaining PAA, PAA hybrid, and PI hybrid film. The structural formula for the synthetic route is shown in Figure 1.

Table 1. Heat treatment conditions of PI hybrid films.

Samples	Temp. (°C)/Time (h)/Pressure (Torr)
PAA	0/1/760 → 25/14/760
PAA hybrid	25/6/760
PI hybrid	50/2/1 → 80/1/1 → 110/0.5/1 → 140/0.5/1 → 170/0.5/1 → 195/0.8/1 → 220/0.8/1 → 235/2/1

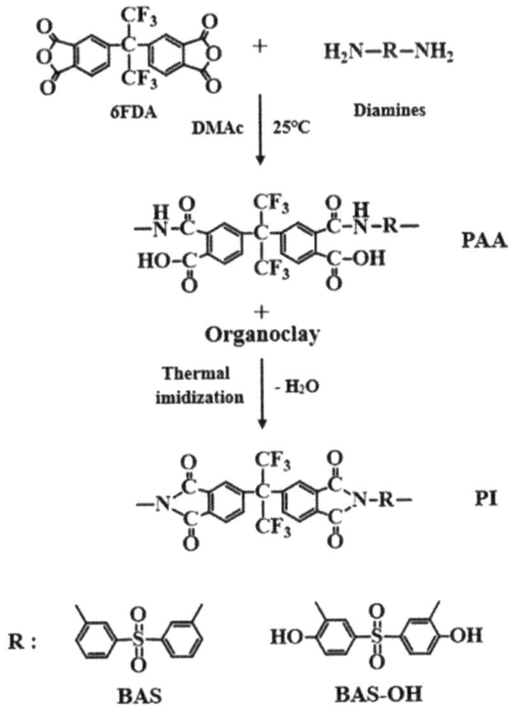

Figure 1. Synthetic route of PI hybrids with various diamines.

2.3. Characterization

The ^{13}C chemical shifts for the BAS and BAS-OH were obtained using ^{13}C cross-polarization/magic angle spinning nuclear magnetic resonance (CP/MAS NMR) at Larmor frequencies of $\omega_0/2\pi$ = 100.61 MHz using Bruker AVANCE II+ 400 MHz NMR spectrometers (Berlin, Germany) at the Korea Basic Science Institute, Western Seoul Center. The chemical shifts were referenced to tetramethylsilane (TMS). Powdered samples were placed in a 4-mm magic angle spinning (MAS) probe and the MAS rate was set to 10 kHz to minimize the spinning sideband overlap.

Differential scanning calorimeter (DSC) was measured using a NETZSCH DSC 200F3 (Berlin, Germany) instrument, and thermogravimetric analysis (Auto TGA 1000) was performed on a TA instrument (New Castle, DE, USA) at a heating rate of 20 °C/min. Both DSC and TGA were operated under nitrogen conditions. The thermal expansion coefficient (CTE) of the film was obtained using a macroexpansion probe (TMA-2940) with a 0.1-N expansion force at a heating rate of 5 °C/min and a temperature range from 50 °C to 150 °C.

Wide angle X-ray diffraction (XRD) values were obtained on a Rigaku (Tokyo, Japan) (D/Max-IIIB) X-ray diffractometer using Ni-filtered Cu-Kα radiation at room temperature. The scanning speed was 2°/min in the 2θ range = 2–12°. Transmission electron microscopy (TEM) photographs of ultra-thin sections of PI hybrid films containing various concentrations of Cloisite 30B were obtained using a Leo 912 OMEGA TEM (Tokyo, Japan) at an accelerating voltage of 120 kV. The yellow index (YI) value of the polymer film was measured using a Minolta spectrophotometer (Model CM-3500d, Tokyo, Japan), and UV-Vis spectral results were obtained using Shimadzu UV-3600 (Tokyo, Japan).

The tensile properties of the films were measured using an Instron (Seoul, Korea) mechanical tester (Model 5564) with a crosshead speed of 20 mm/min at room temperature. Experimental errors of the obtained tensile strength and elastic modulus were within ± 1 MPa and ± 0.05 GPa, respectively. Average values of at least 20 samples were finally used for analysis.

3. Results and Discussion

3.1. FT-IR and NMR Spectra

Figure 2 shows the IR results of monomers and polymers. In the BAS polymer, C=O peaks were observed at 1714 and 1762 cm^{-1}. C-N-C peaks indicating the imidization of the polymers were observed at 1377 cm^{-1}. The spectrum of the BAS-OH monomer, the primary amine -NH2 was observed at 3374 cm^{-1}, and the -OH peak was also observed at 3225 cm^{-1}. In addition, C=C aromatic stretch. Peaks appeared at 1746, 1667, and, 1536 cm^{-1}, respectively.

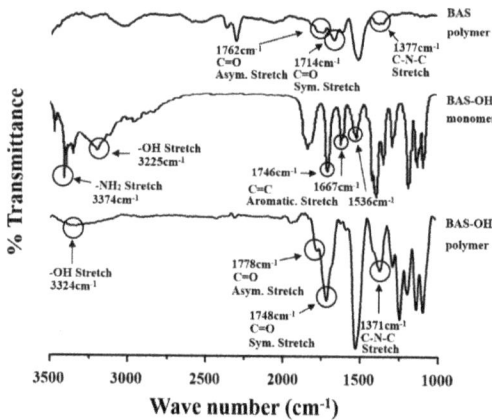

Figure 2. FT-IR spectra of monomer and PIs.

In the BAS-OH polymer, the peak of O-H stretching was observed at 3324 cm^{-1}. The -OH peaks in BAS-OH were smaller than typical hydroxyl peaks. The O-H stretching absorption of the hydroxyl group is sensitive to hydrogen bonding. Molecules with hydrogen donors and acceptors capable of intramolecular hydrogen bonding in the PI main chain show a broad O-H stretching absorption in the range from 3000 to 3500 cm^{-1}. The spectrum of the BAS-OH polymer in Figure 2 shows the hydrogen-bonded peak between -OH in the phenols and the nitrogen in the adjacent imides [27]. In addition, similar to the BAS polymer, the C-N-C peak was observed at 1371 cm^{-1} [28]. These results show that both PIs exhibited a completed imidization reaction.

Structural analyses of the BAS and BAS-OH polymers were carried out by solid state ^{13}C CP/MAS NMR [26]. The ^{13}C chemical shifts of the BAS polymer were obtained at room temperature. The chemical shift for carbon in 4, 4′-hexafluoroisopropylidene (HFP) was present at 65.38 ppm, as shown in Figure 3a. Here, the peaks of 126.38, 131.73, and 138.51 ppm were attributable to the carbon in aromatic ring and CF$_3$, and the chemical shift of C=O was 165.23 ppm. The resonance peak for the carbon in 4,4′-hexafluoroisopropylidene (HFP) had a smaller intensity. The spinning sidebands are marked with an asterisk. The chemical shifts for all carbons were consistent with the structure shown in Figure 3a.

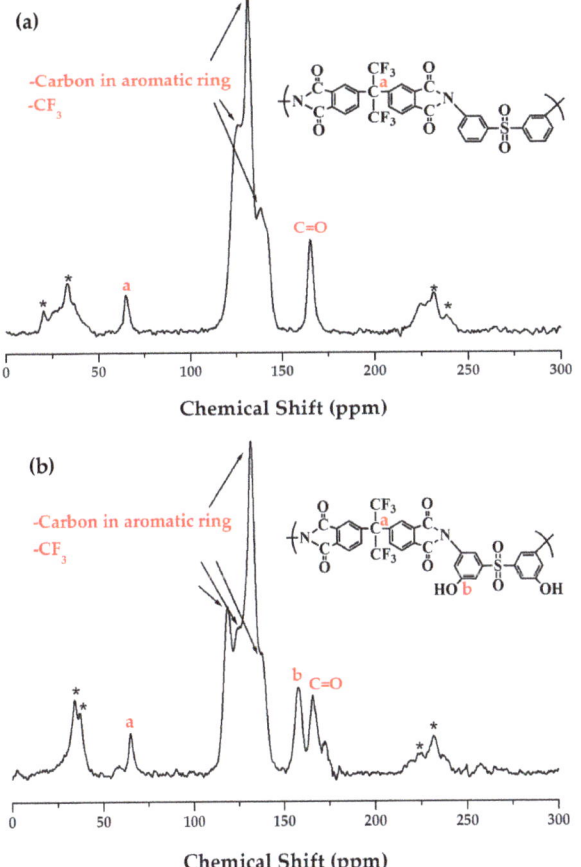

Figure 3. ^{13}C-NMR chemical shifts of (**a**) BAS and (**b**) BAS-OH polymers at room temperature. The spinning sidebands are marked with asterisks.

On the other hand, the chemical shift for carbon in 4,4′-hexafluoroisopropylidene (HFP) in the BAS-OH polymer was present at 64.99 ppm, as shown in Figure 3b. The signals at 119.21, 124.95, 132.44, and 137.41 ppm were attributable to the aromatic ring and -CF$_3$. In addition, the ^{13}C chemical shift at 157.74 and 165.94 ppm corresponding to C-OH and C = O, respectively, was consistent with the structure shown in Figure 3b.

3.2. XRD

Figure 4 shows the XRD patterns of pure organoclay, PI, and PI hybrid films with organoclay concentrations ranging from 0.25 wt% to 1.00 wt%. In the case of the BAS/PI hybrid (Figure 4), the XRD peak of Cloisite 30 B was observed at 2θ = 4.76°, which represents an interlayer distance of 18.56 Å. For all PI hybrids containing 0.25 wt% to 1.00 wt% of Cloisite 30 B, no clay peaks were observed in the XRD curves. Similarly, in the case of BAS-OH/PI (Figure 4), clay peaks were not observed at all in the XRD curves for all PI hybrids containing 0.25–0.75 wt% of Cloisite 30 B for PI. However, when 1.00 wt% Cloisite 30 B was used, a very weak peak was observed at 2θ = 6.62° (d = 13.35 Å). This peak is due to the agglomeration of clay because of the use of excess organoclay.

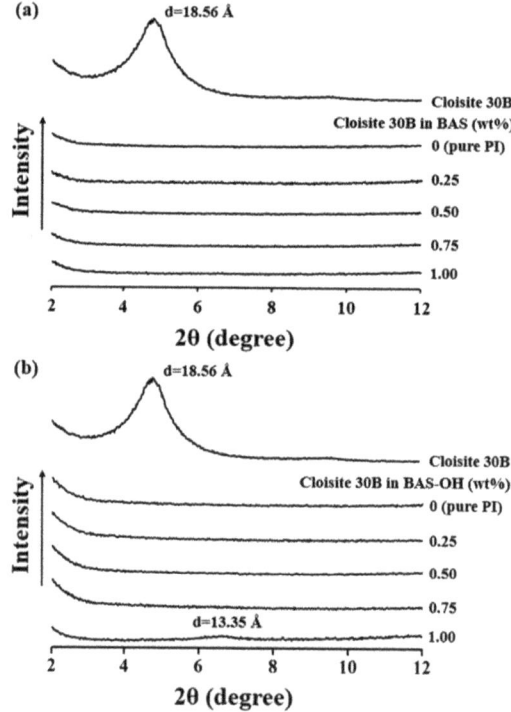

Figure 4. Wide angle X-ray patterns of PI hybrid films with various organoclay. contents: (a) BAS and (b) BAS-OH.

In both the hybrids, we found that clay was very uniformly distributed in the PI matrix regardless of the clay concentration. This denotes that the clay layers in the hybrid material are homogeneously dispersed in the PI matrix [29,30].

XRD is a useful technique for measuring d-spacing in dispersed clay layers of hybrids. In particular, studies using small-angle X-ray scattering can provide detailed information on composites. Schneider et al. [31] reported the influence of different interactions between silica surface and rubber chains depending on the fraction of silica.

3.3. Morphological Analysis by TEM

Direct evidence regarding the formation of nano-sized composites can be confirmed via TEM of an ultra-microtomed section. TEM was used to more accurately measure the dispersion of clay layers in PI hybrids. Figure 5 shows TEM images of the BAS hybrids with 0.5 wt% and 1.00 wt% Cloisite 30 B. In the case of the 0.50 wt% hybrid, some clay was exfoliated and some was intercalated, as shown in Figure 5, but, for the 1.00 wt% hybrid, clay was more agglomerated rather than dispersed. The result is that clay is not evenly dispersed and some of the clay is agglomerated (see Figure 5). As the content of Cloisite 30 B in the PI increased, clay aggregation also increased. The agglomerated clay exhibited poor cohesion and compatibility with PI, thus weakening the interfacial adhesion between the polymer matrix and the filler [22,32,33]. These factors eventually reduce the thermal properties of the hybrid, as described in the next section.

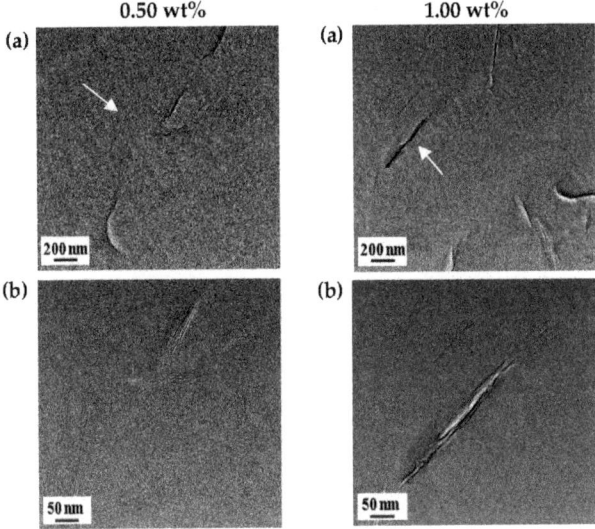

Figure 5. TEM micrographs of BAS hybrid films containing 0.5 wt% and 1.00 wt%. Cloisite 30 B with increasing magnification levels from (**a**) to (**b**).

Hydrophilic clay has good affinity with PI synthesized with BAS-OH monomers, which include an -OH group. Thus, good affinity of clay and PI can result in excellent dispersion. The TEM results of the BAS-OH hybrid are shown in Figure 6. Unlike the BAS hybrid, BAS-OH showed good dispersion in both 0.50 wt% and 1.00 wt% Cloisite 30 B. At 0.50 wt% concentration (see Figure 6), it was confirmed that the clay was uniformly distributed in the PI matrix with a size of less than 10 nm, and this result was similar at 1.00 wt%. Compared with the results of 0.50 wt% hybrid, some clay was agglomerated to a size of less than 20 nm, but most clay was distributed evenly below 20 nm (see Figure 6). When BAS and BAS-OH hybrids with the same clay concentration were compared, the degree of the BAS-OH clay dispersion was better than that of the BAS hybrid. These results suggest that the good hydrophilicity of BAS-OH can be explained by its high affinity for clay.

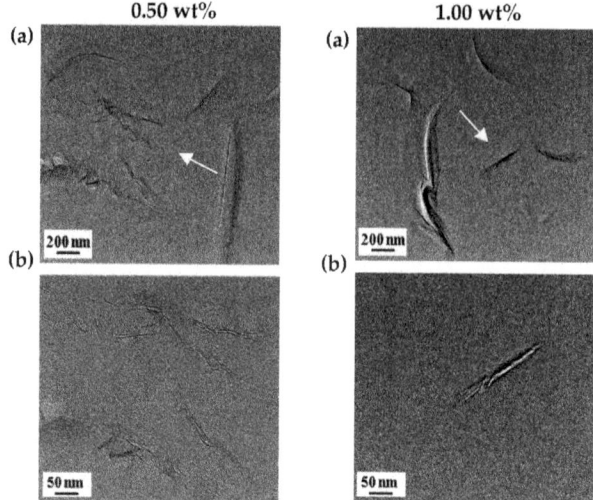

Figure 6. TEM micrographs of BAS-OH hybrid films containing 0.5 wt% and 1.00 wt%. Cloisite 30 B with increasing magnification levels from (a) to (b).

3.4. Thermal Behavior

Table 2 shows the thermal properties of PI hybrids with various clay contents. The T_gs of PI hybrid films containing BAS monomer increased gradually from 227 °C to 245 °C as the clay content increased from 0 wt% to 0.50 wt%. This increase in T_g is caused by the difficulty of movement of the inserted polymer chains in the clay layer, which results in difficulty in segmental motion [34,35]. However, the T_g of the PI hybrid increased when organic clay content was increased to a certain critical concentration but decreased when the critical concentration was crossed. For example, when Cloisite 30 B was increased to 0.75 wt%, the T_g of the PI hybrid decreased to 240 °C, and, when the organoclay content of PI reached 1.00 wt%, the T_g further decreased to 236 °C.

Table 2. Thermal properties of PI hybrid films with various organoclay contents.

Cloisite 30B (wt%)	BAS				BAS-OH			
	T_g (°C)	T_D^i [a] (°C)	wt_R^{600} [b] (%)	CTE [c] (ppm/°C)	T_g (°C)	T_D^i (°C)	wt_R^{600} (%)	CTE (ppm/°C)
0 (pure PI)	227	456	55	47.21	259	313	60	53.17
0.25	231	526	79	41.64	263	321	61	49.95
0.50	245	533	84	38.48	270	330	61	48.61
0.75	240	530	82	42.37	261	324	60	54.33
1.00	236	521	83	45.92	257	316	62	61.17

[a] At a 2% initial weight-loss temperature. [b] Weight percent of residue at 600 °C. [c] Coefficient of thermal expansion for 2nd heating is 50–150 °C.

The trends exhibited by PI hybrids containing BAS-OH monomers were similar to those exhibited by PI hybrids containing BAS. When Cloisite 30B content was increased from 0 wt% to 0.50 wt%, T_g increased from 259 °C to 270 °C. However, when the organoclay content was increased to 1.00 wt%, T_g decreased to 257 °C. This reduction in T_g was because of the agglomeration of excess clay above the critical concentration in the PI matrix. Figure 7 shows the DSC thermogram of PI hybrids according to the concentration of Cloisite 30 B. Clay aggregation above the critical concentration has already been demonstrated using TEM (see Figures 5 and 6).

In general, T_g is affected by various factors in addition to the critical concentration of the filler. That is, the T_g of a polymer depends on the structural differences within the chain, chemical interactions such as hydrogen bonding and curing reaction, the chain flow according to the free volume, and the presence of additives [36]. When the T_g values of the PI hybrids—BAS and the BAS-OH— and those of the two monomers were compared, the T_g value of the BAS-OH PI hybrids were found to be higher than those of the BAS PI hybrid, regardless of the concentration of organic clay. In addition, when the T_g values of the hybrids with the same Cloisite 30 B concentrations were compared, the T_g value of BAS-OH was found to be higher than that of BAS. These results show that the -OH groups of the BAS-OH monomers increase the dispersibility and compatibility of the monomers through hydrogen bonding with the -OH groups present in the clay, thereby increasing the T_g values of the PI hybrids. Similar results have been obtained in previous studies [37,38].

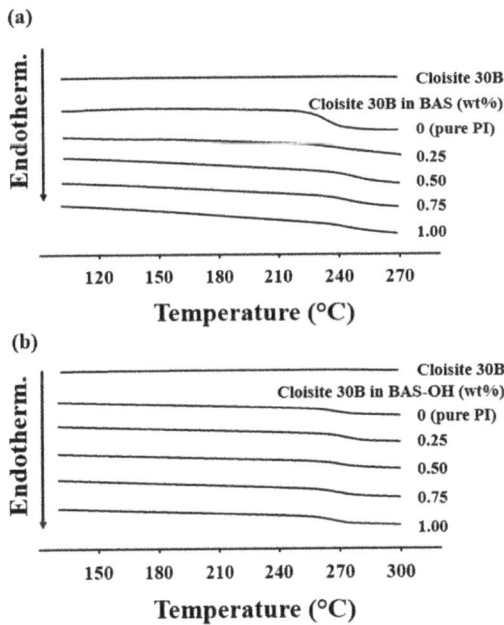

Figure 7. DSC thermograms of PI hybrid films with various organoclay contents: (a) BAS and (b) BAS-OH.

Table 2 shows the initial decomposition temperatures ($T_D{}^i$) of the PI hybrid film containing BAS monomer according to the concentration of the organoclay. As Cloisite 30 B content increased from 0 wt% to 0.50 wt%, the $T_D{}^i$ value of the PI hybrids increased gradually from 456 °C to 533 °C (see Table 2). This suggests that the clay in the polymer chains acts as an insulator and a barrier against volatile products generated during heating, thereby increasing the initial decomposition temperature. This increase in thermal stability is caused by the high thermal stability of the clay itself and the interaction between clay particles and the polymer matrix [24,39]. However, when 1.00 wt% organoclay was added to the PI hybrid (521 °C), the $T_D{}^i$ value was lower than that of the 0.50 wt% hybrid by 12 °C. As has already been explained, the decrease in the $T_D{}^i$ value was because of the aggregation of the excess clay used. Figure 8 shows the TGA thermogram of the PI hybrids having various concentrations of organoclay. The same tendency was also observed for PI hybrids containing BAS-OH monomers. That is, $T_D{}^i$ showed a maximum value of 330 °C for the 0.50 wt% hybrid B, but the $T_D{}^i$ value for the 1.00 wt% hybrid was reduced to 316 °C (see Table 2). The values of the weight residue at 600 °C ($wt_R{}^{600}$) were almost the same regardless of the clay loading, as shown in Table 2 and Figure 8.

In the case of PI containing BAS without organic clay, the weight residue was 55%. When Cloisite 30 B content increased from 0.25 wt% to 1.00 wt%, $wt_R{}^{600}$ was 79–84%. However, when the content of Cloisite was 0–1.00 wt%, the value of $wt_R{}^{600}$ was almost constant at 60–62% for PI containing BAS-OH.

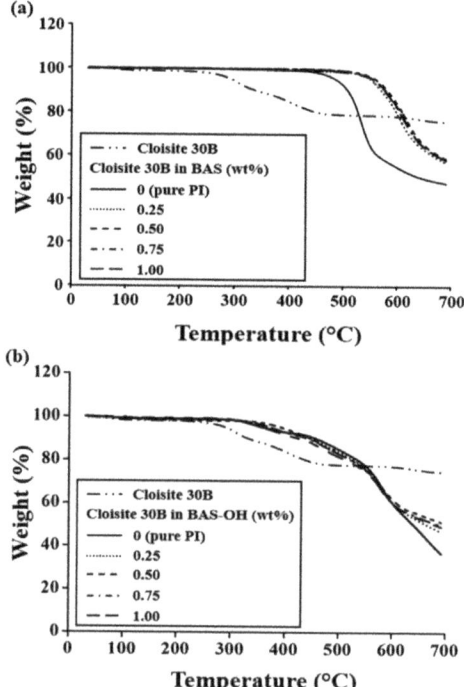

Figure 8. TGA thermograms of PI hybrid films with various organoclay contents: (a) BAS and (b) BAS-OH.

We compared the $T_D{}^i$ and $wt_R{}^{600}$ values of the two different PIs containing BAS and BAS-OH monomers. The BAS hybrid showed an overall better thermal stability than the BAS-OH PI hybrid regardless of the concentration of Cloisite 30 B. These values conflicted with the T_g results. This result can be explained by the weak thermal stability of the -OH group present in the main PI chain containing BAS-OH. In the curve of Figure 8, the first weight loss at around 300 °C was thought to be due to the thermal decomposition of the -OH group of the BAS-OH monomer [25].

The CTE of the BAS PI hybrids decreased to the minimum at the critical content of 0.50% clay but then increased when the concentration of organoclay increased to 1.00% by weight. For example, the CTE of both PI hybrids decreased from 47.21 ppm/°C to 38.48 ppm/°C when the clay concentration increased from 0 wt% to 0.50 wt% but increased to 45.92 ppm/°C when the clay concentration increased to 1.00 wt%. The CTE values of the PI hybrids according to different organoclay concentrations are summarized in Table 2. The same tendency was also observed in the BAS-OH/PI hybrids. The CTE value of the was decreased from 53.17 ppm/°C to 49.95 ppm/°C when up to 0.50 wt% of clay was added but increased to 61.17 ppm/°C when clay concentration increased to 1.00 wt%.

These observations regarding CTE values depend on the orientation of the straight plate-like clay itself, the shape of the PI polymer embedded in the clay layer, and the cohesive strength of clay and PI [40]. When heated, the PI molecules oriented in plane tended to expand in a direction perpendicular to the original direction and thus expanded mainly in the out-of-plane direction [41]. However, because the clay layer in the hybrid is much straighter and stronger than the PI molecule, it is not easily deformed or expanded like a PI molecule. As a result, the existing clay layer in the hybrid can very effectively inhibit the thermal expansion of the PI molecules in the out-of-plane direction [42,43]. However, in our system, the CTE values increased when the concentration of organoclay exceeded 0.50 wt%. This result may be because of clay aggregation, as has been already explained. These results are supported by the T_g and $T_D{}^i$ results previously described (Table 2). The CTE results of the BAS hybrids were better than those of the BAS-OH hybrids at the same organoclay concentration, as seen from the thermal stability analysis. These results are also due to the -OH groups with low thermal stability.

3.5. Mechanical Tensile Properties

The ultimate tensile strength, initial tensile modulus, and elongation percentage at break of the PI hybrid films with various organoclay contents were measured using a universal tensile machine (UTM), and the obtained results are summarized in Table 3. The mechanical properties of the PI hybrid film were improved when the clay concentration was below the critical concentration but deteriorated when the critical clay content was exceeded, as already confirmed from the thermal properties according to the critical concentration of clay. For example, when the Cloisite 30 B content increased from 0 wt% to 0.5 wt%, the tensile strength increased from 53 MPa to 103 MPa for BAS hybrids and from 59 MPa to 112 MPa for BAS-OH hybrids. These results show a remarkable increase of 194% and 190%, respectively, using a small amount of organoclay (0.50 wt%). However, when the concentration of organoclay became 1.00 wt%, the tensile strength decreased to 69 MPa for the BAS PI hybrid and 82 MPa for the BAS-OH PI hybrid, as shown in Table 3. Similar results have been reported for other polymer hybrids. Yano et al. [44,45] reported that the mechanical properties of cellulose/silica composites were improved up to a certain concentration of silica, but the tensile strengths of the hybrids reduced at higher filler concentrations [46]. These tendencies occur mainly because the filler particles existing above the critical concentration are not dispersed evenly and are aggregated with each other [20]. This phenomenon has already been confirmed from the TEM photograph in Figures 5 and 6.

However, the initial modulus increased steadily with increasing organoclay concentration. As the concentration of Cloisite 30 B increased from 0 wt% to 1.00 wt%, the tensile modulus increased steadily from 2.24 GPa to 3.54 GPa for the BAS PI hybrid and from 2.80 GPa to 6.16 GPa for the BAS-OH PI hybrid. In contrast to the results of ultimate tensile strength, this improvement in the initial tensile modulus of the two hybrids was due to the high aspect ratio and orientation of the clay layer as well as the resistance of the clay itself to external forces [47,48].

Table 3. Tensile properties of PI hybrid films with various organoclay contents.

Cloisite 30B (wt%)	BAS			BAS-OH		
	Ult. Str. (MPa)	Ini. Mod. (GPa)	E.B. [a] (%)	Ult. Str. (MPa)	Ini. Mod. (GPa)	E.B. (%)
0 (pure PI)	53	2.24	2	59	2.80	2
0.25	60	2.85	2	81	4.09	3
0.50	103	3.16	4	112	4.44	3
0.75	77	3.30	3	87	5.43	2
1.00	69	3.54	2	82	6.16	2

[a] Elongation percentage at break.

We compared the mechanical properties of the two PI hybrids and those of the BAS and BAS-OH monomers. The mechanical properties of BAS-OH hybrids were better than those of the BAS hybrids regardless of the concentration of organoclay. This tendency was in contrast to the tendency of thermal stability and CTE. This result is attributed to the formation of a hybrid film with stronger bonds that can withstand external tensile forces because the hydroxyl groups between PI and organoclay form hydrogen bonds with each other. Figure 9 shows the change in the mechanical properties of the two PI hybrid films according to the various organoclay contents.

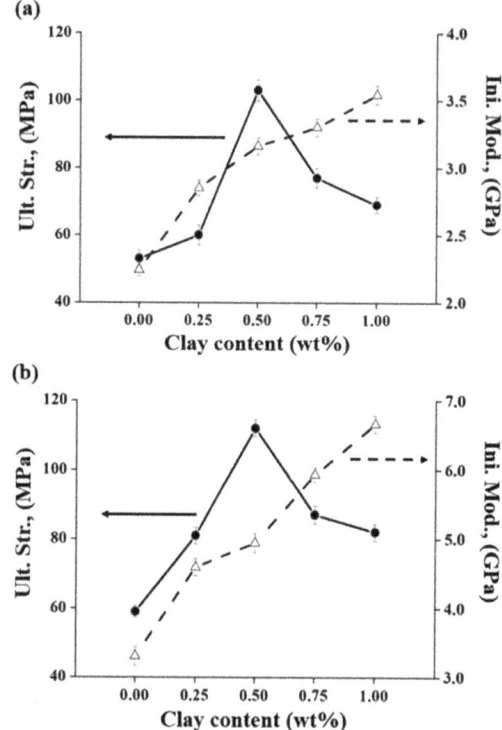

Figure 9. Tensile properties of PI hybrid films with various organoclay contents: (a) BAS and (b) BAS-OH.

The elongation at break of the two hybrid films was almost constant regardless of the organoclay concentration. That is, when the organoclay content increased up to 1.00 wt%, the elongation at break varied from 2% to 4% (see Table 3). This result is characteristic of hybrid materials reinforced with a rigid and hard inorganic material.

3.6. Optical Transparency

The color intensity can be determined by measuring the cut-off wavelength (λ_0) using UV-Vis. absorption spectra, as shown in Figure 10. The results of the absorption spectrum of the BAS hybrid are listed in Table 4. As the Cloisite 30 B concentration increased from 0 wt% to 1.00 wt% in the BAS and BAS-OH hybrids, the value of λ_0 increased steadily from 330 nm to 346 nm and from 344 nm to 350 nm, respectively. PI hybrids showed a small λ_0 value because they have a -CF_3 substituent and a kinked monomer structure that can reduce the interactions between the CT-complexes and the intermolecular interaction in the polymer main chain [15,16,49].

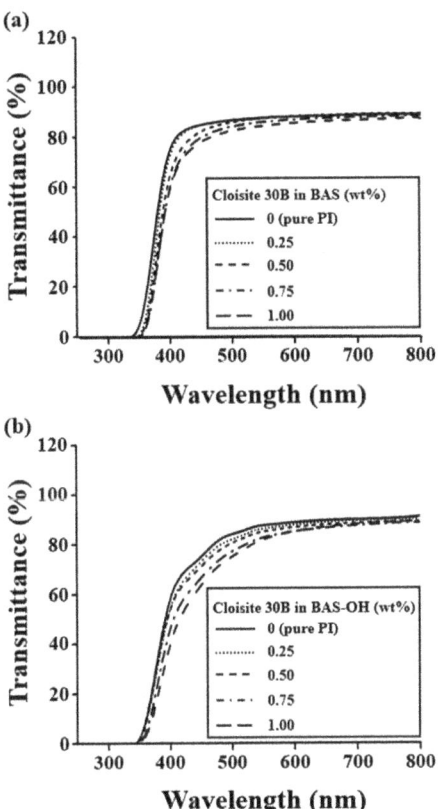

Figure 10. UV-Vis transmittance of the PI hybrid films with various organoclay contents: (a) BAS and (b) BAS-OH.

Table 4. Optical transparencies of PI hybrid films with various organoclay contents.

Cloisite 30B (wt%)	BAS				BAS-OH			
	T [a] (μm)	λ_0 (nm)	500 nm^{trans} (%)	YI [b]	T (μm)	λ_0 (nm)	500 nm^{trans} (%)	YI
0 (pure PI)	67	330	87	2	70	344	84	14
0.25	68	334	87	2	69	344	82	14
0.50	68	336	86	3	68	345	81	14
0.75	67	340	84	3	69	349	77	16
1.00	69	346	82	4	69	350	75	20

[a] Film thickness. [b] Yellow index.

The transmittance at 500 nm gradually decreased as the organoclay concentration increased to 1.00 wt%. For example, for BAS and BAS-OH hybrids, the transmittance at 500 nm decreased from 87% to 82% and from 84% to 75%, respectively. The color intensities of pure PI and PI hybrid films containing various concentrations of Cloisite 30 B are summarized in Table 4. Color intensity can be described as an index indicating the value of yellow index (YI). The YI of the pure PI film was 2, but those of the BAS hybrid films slightly increased as the organoclay content increased because of the agglomeration of the clay particles [43]. However, the difference was not large, so it showed a value of 2–4. In the case of BAS-OH hybrids, the YI of the hybrids increased as Cloisite 30 B content from 0 to 0.50 wt% and then remained constant at 14. However, when the organoclay content increased to 1.00 wt%, the YI increased significantly to 20.

The reason that the optical transparency is gradually lowered as the clay concentration increases is that clay agglomeration affects optical properties. Evidence for clay agglomeration has already been demonstrated using XRD and TEM (see Figures 4–6). The optical transparency of BAS-OH hybrids is lower than that of the BAS hybrids because of the strong intermolecular bonding due to the hydrogen bonding of -OH present in BAS-OH and the consequent increase in the CT-complex, as described above.

As shown in Figure 11, the BAS hybrids with a 0 wt% to 1.00 wt% concentration of Cloisite 30 B were almost colorless and transparent regardless of the organoclay concentration. This transparency and colorlessness suggest that, even when 1.00 wt% of organoclay is added, the phase region of the hybrid film is significantly smaller than the visible light wavelength (i.e., 400 nm to 800 nm). In contrast, in the case of BAS-OH hybrids, a yellowish film was obtained as a whole, and this color became darker as the concentration of Cloisite 30 B increased to 1.00 wt% (see Figure 12). However, both hybrids exhibited excellent transparency, so that there was no problem in reading the text through the film for all organoclay concentrations.

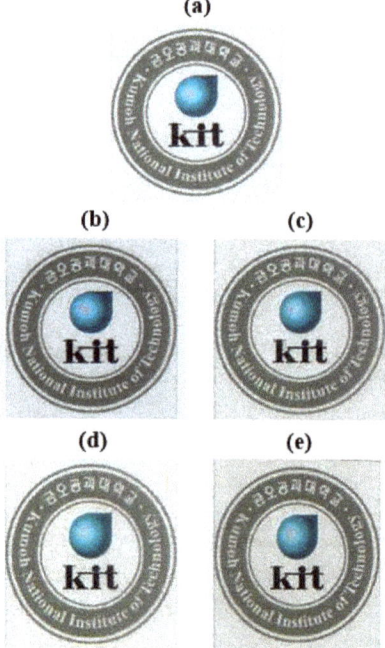

Figure 11. Photographs of PI hybrid films with BAS monomer containing. (**a**) 0 (pure PI), (**b**) 0.25, (**c**) 0.50, (**d**) 0.75, and (**e**) 1.00 wt% of Cloisite 30 B.

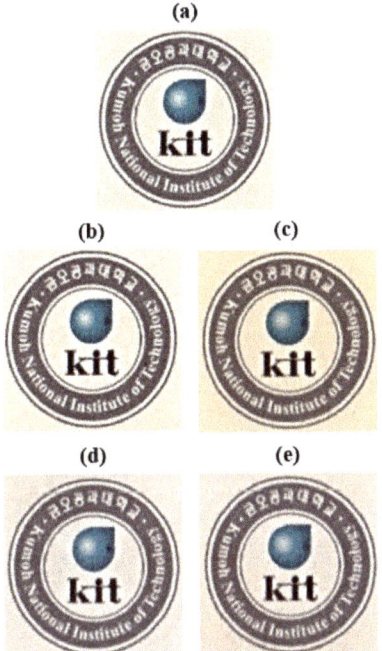

Figure 12. Photographs of PI hybrid films with BAS-OH monomer containing. (**a**) 0 (pure PI), (**b**) 0.25, (**c**) 0.50, (**d**) 0.75, and (**e**) 1.00 wt% of Cloisite 30 B.

4. Conclusions

Two PI hybrid films were fabricated from dianhydrides (6FDA), diamines (BAS and BAS-OH), and various concentrations of organoclay by solution intercalation. The PI hybrid films containing -CF_3 substituents and bent monomer structures showed higher optical transparency than conventional linear non-fluorinated PI films. In particular, the -CF_3 groups present in the main chain were effective in strongly attracting electrons to reduce the CT-complex between polymer chains through steric hindrance and the induction effect.

In this study, the PI hybrids synthesized using BAS and BAS-OH as diamines were compared with each other according to various Cloisite 30 B concentrations. In terms of T_g and mechanical properties, BAS-OH hybrids showed better performance than BAS hybrids. However, BAS hybrids were superior to BAS-OH hybrids in terms of thermal stability, CTE, and optical transparency over the range of organoclay concentrations. The thermal properties and ultimate strengths of the two PI hybrid films were highest when the organic clay was 0.5 wt%.

In addition, these PI hybrid films are expected to be useful as high-performance polymer materials because of their excellent thermo-mechanical properties and optical transparencies as compared with general engineering plastics.

Author Contributions: J.-H.C. designed the project and wrote the manuscript. H.I.S. prepared the samples and participated in the data analysis. All authors have read and agreed to the published version of the manuscript.

Funding: This research was supported by the Basic Science Research Program of the National Research Foundation of Korea (NRF), funded by the Ministry of Education (2018R1D1A1B07045502).

Conflicts of Interest: The authors declare no conflict of interest.

References

1. Yokota, R. Recent trends and space applications of polyimides. *J. Photopolym. Sci. Technol.* **1999**, *12*, 209–216. [CrossRef]
2. Ho, J.S.; Greenbaum, S.G. Polymer capacitor dielectrics for high temperature applications. *ACS Appl. Mater. Interfaces* **2018**, *10*, 29189–29218. [CrossRef]
3. Guan, Y.; Wang, C.; Wang, D.; Dang, G.; Chen, C.; Zhou, H.; Zhao, X. High transparent polyimides containing pyridine and biphenyl units: Synthesis, thermal, mechanical, crystal and optical properties. *Polymer* **2015**, *62*, 1–10. [CrossRef]
4. Wozniak, A.I.; Yegorov, A.S.; Ivanov, V.S.; Igumnov, S.M.; Tcarkova, K.V. Recent progress in synthesis of fluorine containing monomers for polyimides. *J. Fluor. Chem.* **2015**, *180*, 45–54. [CrossRef]
5. Liaw, D.-J.; Wang, K.-L.; Huang, Y.-C.; Lee, K.-R.; Lai, J.-Y.; Ha, C.-S. Advanced polyimide materials: Syntheses, physical properties and applications. *Prog. Polym. Sci.* **2012**, *37*, 907–974. [CrossRef]
6. Liu, Z.; Xu, J.; Chen, D.; Shen, G. Flexible electronics based on inorganic nanowires. *Chem. Soc. Rev.* **2015**, *44*, 161–192. [CrossRef] [PubMed]
7. Liu, J.-M.; Lee, T.M.; Wen, C.-H.; Leu, C.-M. High-performance organic-inorganic hybrid plastic substrate for flexible displays and electronics. *J. Soc. Inf. Display* **2011**, *19*, 63–69. [CrossRef]
8. Mativenga, M.; Choi, M.H.; Choi, J.W.; Jang, J. Transparent flexible circuits based on amorphous-indium–gallium–zinc–oxide thin-film transistors. *IEEE Electron Device Lett.* **2011**, *32*, 170–172. [CrossRef]
9. Yu, X.; Zhao, X.; Liu, C.; Bai, Z.; Wang, D.; Dang, G.; Zhou, H.; Chen, C. Synthesis and properties of thermoplastic polyimides with ether and ketone moieties. *J. Polym. Sci. Part A Polym. Chem.* **2010**, *48*, 2878–2884. [CrossRef]
10. Garg, P.; Singh, R.P.; Pandey, L.K.; Choudhary, V. Pervaporative studies using polyimide-filled PDMS membrane. *J. Appl. Polym. Sci.* **2010**, *115*, 1967–1974. [CrossRef]
11. Lu, Y.H.; Kang, W.J.; Hu, Z.Z.; Wang, Y.F.; Fang, Q.X. Synthesis and properties of fluorinated polyimide films based on 1,2,3,4-cyclobutanetetracarboxylic dianhydride. *Adv. Mater. Res.* **2011**, *150*, 1758–1763. [CrossRef]
12. Liou, H.C.; Willecke, R.; Ho, P.S. Study of out-of-plane elastic properties of PMDA-ODA and BPDA-PDA polyimide thin films. *Thin Solid Films* **1998**, *323*, 203–208. [CrossRef]
13. Hwang, H.Y.; Kim, S.J.; Oh, D.Y.; Hong, Y.T.; Nam, S.Y. Proton conduction and methanol transport through sulfonated poly (styrene-b-ethylene/butylene-b-styrene)/clay nanocomposite. *Macromol. Res.* **2011**, *19*, 84–89. [CrossRef]
14. Chen, C.-J.; Yen, H.-J.; Hu, Y.-C.; Liou, G.-S. Novel programmable functional polyimides: Preparation, mechanism of CT induced memory, and ambipolar electrochromic behavior. *J. Mater. Chem. C* **2013**, *1*, 7623–7634. [CrossRef]
15. Nishihara, M.; Christiani, L.; Staykov, A.; Sasaki, K. Experimental and theoretical study of charge-transfer complex hybrid polyimide membranes. *J. Polym. Sci. Part B Polym. Phys.* **2014**, *52*, 293–298. [CrossRef]
16. Choi, I.H.; Chang, J.-H. Colorless polyimide nanocomposite films containing hexafluoroisopropylidene group. *Polym. Adv. Technol.* **2011**, *22*, 682–689. [CrossRef]
17. Chen, G.; Pei, X.; Liu, J.; Fang, X. Synthesis and properties of transparent polyimides derived from trans- and cis-1,4-bis(3,4-dicarboxyphenoxy)cyclohexane dianhydrides. *J. Polym. Res.* **2013**, *20*, 159. [CrossRef]
18. Wiegand, J.R.; Smith, Z.P.; Liu, Q.; Patterson, C.T.; Freeman, B.D.; Guo, R. Synthesis and characterization of triptycene-based polyimides with tunable high fractional free volume for gas separation membranes. *J. Mater. Chem. A* **2014**, *2*, 13309–13320. [CrossRef]
19. Yeo, H.; Goh, M.; Ku, B.-C.; You, N.-H. Synthesis and characterization of highly-fluorinated colorless polyimides derived from 4,4′-((perfluoro-[1,1′-biphenyl]-4,4′-diyl)bis(oxy))bis(2,6-dimethylaniline) and aromatic dianhydrides. *Polymer* **2015**, *76*, 280–286. [CrossRef]
20. Damaceanu, M.-D.; Constantin, C.-P.; Nicolescu, A.; Bruma, M.; Belomoina, N.; Begunov, R.S. Highly transparent and hydrophobic fluorinated polyimide films with ortho-kink structure. *Eur. Polym. J.* **2014**, *50*, 200–213. [CrossRef]
21. Min, U.; Chang, J.-H. Colorless and transparent polyimide nanocomposite films containing organoclay. *J. Nanosci. Nanotechnol.* **2011**, *11*, 6404–6409. [CrossRef] [PubMed]

22. Kim, Y.; Chang, J.-H. Colorless and transparent polyimide nanocomposites: Thermo-optical properties, morphology, and gas permeation. *Macromol. Res.* **2013**, *21*, 228–233. [CrossRef]
23. Tan, B.; Thomas, N.L. A review of the water barrier properties of polymer/clay and polymer/graphene nanocomposites. *J. Membr. Sci.* **2016**, *514*, 595–612. [CrossRef]
24. Sapalidis, A.A.; Katsaros, F.K.; Steriotis, T.A.; Kanellopoulos, N.K. Properties of poly(vinyl alcohol)—Bentonite clay nanocomposite films in relation to polymer–clay interactions. *J. Appl. Polym. Sci.* **2012**, *123*, 1812–1821. [CrossRef]
25. Shin, J.-E.; Ham, M.-R.; Kim, J.-C.; Chang, J.-H. Characterizations of flexible clay-PVA hybrid films: Thermo-optical properties, morphology, and gas permeability. *Polym. Korea* **2011**, *35*, 402–408. [CrossRef]
26. Yeun, J.-H.; Bang, G.-S.; Park, B.J.; Ham, S.K.; Chang, J.-H. Poly(vinyl alcohol) nanocomposite films: Thermooptical properties, morphology, and gas permeability. *J. Appl. Polym. Sci.* **2006**, *101*, 591–596. [CrossRef]
27. Joshi, K.R.; Rojivadiya, A.J.; Pandya, J.H. Synthesis and spectroscopic and antimicrobial studies of schiff base metal complexes derived from 2-hydroxy-3-methoxy-5-nitrobenzaldehyde. *Int. J. Inorg. Chem.* **2014**, *2014*, 817412. [CrossRef]
28. Pavia, D.L.; Lampman, G.M.; Kriz, G.S.; Vyvyan, J.A. *Introduction to Spectroscopy*; Cengage Learning: Boston, MA, USA, 2008.
29. Hsiao, S.-H.; Liou, G.-S.; Chang, L.-M. Synthesis and properties of organosoluble polyimide/clay hybrids. *J. Appl. Polym. Sci.* **2001**, *80*, 2067–2072. [CrossRef]
30. Ke, Y.; Lü, J.; Yi, X.; Zhao, J.; Qi, Z. The effects of promoter and curing process on exfoliation behavior of epoxy/clay nanocomposites. *J. Appl. Polym. Sci.* **2000**, *78*, 808–815. [CrossRef]
31. Schneider, G.J.; Hengl, K.B.; Brandt, S.V.; Roth, R.S.; Goritz, D. Influence of the matrix on the fractal properties of precipitated silica in composites. *J. Appl. Cryst.* **2012**, *45*, 430–438. [CrossRef]
32. Zare, Y. Modeling the yield strength of polymer nanocomposites based upon nanoparticle agglomeration and polymer–filler interphase. *J. Colloid Interface Sci.* **2016**, *467*, 165–169. [CrossRef] [PubMed]
33. Zare, Y. Effects of imperfect interfacial adhesion between polymer and nanoparticles on the tensile modulus of clay/polymer nanocomposites. *Appl. Clay Sci.* **2016**, *129*, 65–70. [CrossRef]
34. Kim, Y.; Chang, J.-H.; Kim, J.-C. Optically transparent and colorless polyimide hybrid films with various clay contents. *Macromol. Res.* **2012**, *20*, 1257–1263. [CrossRef]
35. Chang, J.-H.; Seo, B.-S.; Hwang, D.-H. An exfoliation of organoclay in thermotropic liquid crystalline polyester nanocomposites. *Polymer* **2002**, *43*, 2969–2974. [CrossRef]
36. Agag, T.; Takeichi, T. Polybenzoxazine–montmorillonite hybrid nanocomposites: Synthesis and characterization. *Polymer* **2000**, *41*, 7083–7090. [CrossRef]
37. Giannakas, A.; Grigoriadi, K.; Leontiou, A.; Barkoula, N.-M.; Ladavos, A. Preparation, characterization, mechanical and barrier properties investigation of chitosan–clay nanocomposites. *Carbohydr. Polym.* **2014**, *108*, 103–111. [CrossRef]
38. Holder, K.M.; Priolo, M.A.; Secrist, K.E.; Greenlee, S.M.; Nolte, A.J.; Grunlan, J.C. Humidity-Responsive gas barrier of hydrogen-bonded polymer–clay multilayer thin films. *J. Phys. Chem. C* **2012**, *116*, 19851–19856. [CrossRef]
39. Ray, S.S. Recent trends and future outlooks in the field of clay-containing polymer nanocomposites. *Macromol. Chem. Phys.* **2014**, *215*, 1162–1179. [CrossRef]
40. Chiu, C.-W.; Lin, J.-J. Self-assembly behavior of polymer-assisted clays. *Prog. Polym. Sci.* **2012**, *37*, 406–444. [CrossRef]
41. Tyan, H.-L.; Liu, Y.-C.; Wei, K.-H. Thermally and mechanically enhanced clay/polyimide nanocomposite via reactive organoclay. *Chem. Mater.* **1999**, *11*, 1942–1947. [CrossRef]
42. Hsu, S.L.-C.; Wang, U.; King, J.-S.; Jeng, J.-L. Photosensitive poly(amic acid)/organoclay nanocomposites. *Polymer* **2003**, *44*, 5533–5540. [CrossRef]
43. Min, U.; Kim, J.-C.; Chang, J.-H. Transparent polyimide nanocomposite films: Thermo-optical properties, morphology, and gas permeability. *Polym. Eng. Sci.* **2011**, *51*, 2143–2150. [CrossRef]
44. Yano, K.; Usuki, A.; Okada, A. Synthesis and properties of polyimide-clay hybrid films. *J. Polym. Sci. Part A Polym. Chem.* **1997**, *35*, 2289–2294. [CrossRef]
45. Yano, K.; Usuki, A.; Okada, A.; Kurauchi, T.; Kamigaito, O. Synthesis and properties of polyimide–clay hybrid. *J. Polym. Sci. Part A Polym. Chem.* **1993**, *31*, 2493–2498. [CrossRef]

46. Sequeira, S.; Evtuguin, D.V.; Portugal, I. Preparation and properties of cellulose/silica hybrid composites. *Polym. Compos.* **2009**, *30*, 1275–1282. [CrossRef]
47. Galgali, G.; Agarwal, S.; Lele, A. Effect of clay orientation on the tensile modulus of polypropylene–nanoclay composites. *Polymer* **2004**, *45*, 6059–6069. [CrossRef]
48. Luo, J.-J.; Daniel, I.M. Characterization and modeling of mechanical behavior of polymer/clay nanocomposites. *Compos. Sci. Technol.* **2003**, *63*, 1607–1616. [CrossRef]
49. Do, K.; Saleem, Q.; Ravva, M.K.; Cruciani, F.; Kan, Z.; Wolf, J.; Hansen, M.R.; Beaujuge, P.M.; Brédas, J.-L. Impact of fluorine substituents on π-conjugated polymer main-chain conformations, packing, and electronic couplings. *Adv. Mater.* **2016**, *28*, 8197–8205. [CrossRef]

© 2020 by the authors. Licensee MDPI, Basel, Switzerland. This article is an open access article distributed under the terms and conditions of the Creative Commons Attribution (CC BY) license (http://creativecommons.org/licenses/by/4.0/).

Article

Effect of CNTs Additives on the Energy Balance of Carbon/Epoxy Nanocomposites during Dynamic Compression Test

Manel Chihi [1,2,*], Mostapha Tarfaoui [1,*], Chokri Bouraoui [2] and Ahmed El Moumen [1]

1. ENSTA Bretagne, IRDL (UMR CNRS 6027) / PTR-1, F-29200 Brest, France; ahmed.el_moumen@ensta-bretagne.fr
2. LMS, ENISo, University of Sousse, Sousse 4023, Tunisia; chokri.bouraoui@enim.rnu.tn
* Correspondence: manel.chihi@ensta-bretagne.org (M.C.); mostapha.tarfaoui@ensta-bretagne.fr (M.T.)

Received: 14 November 2019; Accepted: 9 January 2020; Published: 11 January 2020

Abstract: Previous research has shown that nanocomposites show not only enhancements in mechanical properties (stiffness, fracture toughness) but also possess remarkable energy absorption characteristics. However, the potential of carbon nanotubes (CNTs) as nanofiller in reinforced epoxy composites like glass fiber-reinforced polymers (GFRP) or carbon fiber-reinforced polymers (CFRP) under dynamic testing is still underdeveloped. The goal of this study is to investigate the effect of integrating nanofillers such as CNTs into the epoxy matrix of carbon fiber reinforced polymer composites (CFRP) on their dynamic energy absorption potential under impact. An out-of-plane compressive test at high strain rates was performed using a Split Hopkinson Pressure Bar (SHPB), and the results were analyzed to study the effect of changing the concentration of CNTs on the energy absorption properties of the nanocomposites. A strong correlation between strain rates and CNT mass fractions was found out, showing that an increase in percentage of CNTs could enhance the dynamic properties and energy absorption capabilities of fiber-reinforced composites.

Keywords: CFRP; Carbon nanotubes; Nanocomposites; Split Hopkinson Pressure Bar; Energy absorption

---

1. Introduction

Energy absorption is considered to be one of the most important functions of structural materials, especially when subjected to an accidental collision or sudden shock. It is a crucial condition for structural crashworthiness and damage assessment for example, in designing rail cars, aircraft, automobiles and rotorcraft. During the design phase, the crashworthy structure is manufactured in such a manner that it can halt the transfer of the energy to the passenger compartment by absorbing all the impact energy in a controlled manner during crash. Moreover, in civil construction, many structures lack energy-absorbing capabilities, resulting in catastrophic failure during events like explosions. This can cause massive human causalities and property loss. However, this can be avoided by improving the blast resistance of buildings, using sacrificial cladding structures with cores made of highly shock-absorbing material.

Traditionally, structural components used in crashworthy applications and armors were commonly manufactured using metals, because metals are able to absorb the impact energy in a controlled way because of their high toughness [1]. Then, researchers started using cellular forms such as honeycomb structures, foams and sandwich structures, which have demonstrated excellent resistance to dynamic loading due to their bulking and collapse mechanisms [2–6]. Furthermore, composite structures have shown excellent resistance against vibrations during impact thanks to their unique internal structure, which displays good internal damping behaviors [7,8]. These composites have been further enhanced by the introduction of nanotechnology and nanofillers. The introduction of nanoparticles into polymer matrices improved their thermal and electrical properties as well as enhancing their mechanical characteristics, such as strength and toughness. In addition, recent research showed that reinforcing the polymeric materials with nano-fillers such as CNTs resulted in enhancement of structural damping of the composites because of the large surface-to volume ratio of nanofillers which can result in exceptional performance of interfacial bond between the nanofillers and matrix resulting in an increase of the energy dissipation capability of the material [9–13]. Furthermore, members of the fullerenes family exhibit extraordinary energy absorption behavior because of their high strength, stiffness, and large surface area [14–20]. This is why nanocomposites with distinct matrices and filler materials show improvement not only in stiffness and fracture toughness during experimental studies, but also in impact energy absorption and vibration damping. For this reason, these nanocomposites have significant importance in civil and military applications like automobile, airplane structures and biomedical [11].

Among these nano-sized inclusions, carbon nanotubes (CNTs), being members of the carbon nanomaterials family, have shown unique energy absorption performance when used in 3D sponge-array and foams architectures because of their unique mechanical properties [20–23]. Additionally, CNTs have an ultra-high stiffness, strength and an extremely large surface area. They are also extremely lightweight in comparison to traditional materials. In fact, both experimental and computational results have shown that they had about tensile strength of 200 GPa, Young's modulus of 1 TPa, shear modulus of 1 GPa, bulk modulus about 462–546 GPa and bending strength approximately 14.2 GPa [15,16,24]. Recent studies also proved that nanocomposite laminates based on nanocharge materials have good overall energy absorption characteristics. Numerous previous experimental works have shown that nano-fillers are able to enhance not only stiffness but also the energy absorption behavior of polymers and/or conventional composites [25].

Drdlova and Prachař [26] studied the mechanical performance of lightweight porous foams reinforced with carbon nanotubes with 1–5 vol.% for structural applications under high strain rate loading using SHPB and results had shown that energy absorption capability of the material was greatly enhanced up to 4 vol.%, and then there was a significant decrease.

Chen et al. [27] developed a numerical model using dynamic simulation to study the energy absorption ability of CNT bucky paper under high-velocity impacts. Their study revealed that this bucky paper showed extremely high kinetic energy dissipation efficiency within its elastic limit, and that this depended directly on the impact velocity. In addition, Chen et al. [28] also studied the energy dissipation behavior of CNTs with nested bucky balls during impact using a dynamic simulation model. The simulated results showed that dissipated energy was mostly converted into the thermal energy at low velocity impact while bucky balls showed permanent strain deformation at high velocity impact; thus, dissipation energy was dominated by the strain energy of the energy absorption system.

Weidt et al. [29] performed a study using 2D and 3D computational modelling on aligned CNT/epoxy nanocomposites under compressive strain rates. The results revealed that, by increasing the wt.% of CNTs, the nanocomposites showed a noticeable increase in their mechanical performances including energy absorption behavior.

Although considerable research has been devoted to the energy absorption behavior of CNT-based nanocomposites, rather less attention has been paid to the dynamic/impact properties. As far as the compressive response of the composite is concerned, the Split Hopkinson Pressure Bar (SHPB) technique is one of the examples and has been extensively used to evaluate the impact behavior of different materials at high strain rates [30–40]. For instance, Gardea et al. [41] showed in their investigation that strain energy dissipation of CNT reinforced polymers under low strain was not dependent on the alignment of CNTs; however, damping factor increased monotonically with the wt.%. of CNTs, which showed the occurrence of friction dissipation mechanisms within the CNT–CNT interface; however, interfacial slip contributed to energy dissipation at higher strain rates. Moreover, they showed that the tearing and plasticity of the matrix caused by the misaligned CNTs within the loading direction played a vital role in energy dissipation. Gardea et al. [42] in another study showed that carbon nanotube (CNT)-reinforced acrylonitrile-butadiene-styrene (ABS) composites fabricated by additive manufacturing exhibited of strain energy dissipation ability with reduced damage because of the CNTs. Their results showed that CNTs altered the energy dissipation mechanism and controlled the structural damping behavior under dynamic loading. El Moumen et al. [43] evaluated the effect of integrating the CNTs in epoxy on its shock wave absorption under dynamic compressive loading using SHPB. The results showed that as the wt.% of CNTs increased, the nanocomposites were able to absorb more mechanical shock waves. Thus, this highlights the importance of CNTs in enhancing the impact resistance behavior of the composite structures. It was also found in a study that the aspect ratio and mass fraction of CNTs played a vital role in defining the energy absorption characteristics of a nanocomposite under high strain rate impacts [29].

However, the potential of CNTs, as nanofiller in reinforced epoxy composites like glass fiber-reinforced polymers (GFRP) or carbon fiber-reinforced polymers (CFRP) under dynamic testing, is still underdeveloped. There is very little, if any, information available in the literature. Therefore, in this context, the object of this paper is to study the effect of using various weight percentages of CNTs in CFRP composites on their shock wave or energy absorption performances. An experimental study was performed to investigate the energy absorption behavior of CNTs-based CFRP nanocomposites under out-of-plane dynamic loading, using the SHPB device. Samples were fabricated with 1 and 2% mass fractions and specimen with 0% was considered as a reference. Moreover, these dynamic compression tests were executed at three different impact pressures, i.e., 2, 3 and 4 bar, to further analyze the energy absorption ability of these nanocomposites.

2. Materials and Manufacturing Process

The polymer used in this study was a low-viscosity liquid epoxy resin, Epon 862 (Diglycidyl Ether of Bisphenol F), acquired from Momentive Specialty Chemicals Inc. (Cleveland, OH, USA). The carbon fiber was provided by Hexcel Company and multi-walled carbon nanotubes (MWNTCs) were produced by Nanocyl Belgium Company (Sambreville, Belgium), they were synthesized with no surface functionalization; they had an average diameter of 10 nm and length of 1.5 μm. Mechanical properties of each constituent are listed in Table 1.

Table 1. Material properties.

Carbon fiber		Epoxy matrix		CNT	
E_{11} (GPa)	230	E (GPa)	2.72	E (GPa)	500
E_{22} (GPa)	15	ν	0.3	ν	0.261
E_{33} (GPa)	15				
ν_{12}	0.28				
ν_{13}	0.28				
ν_{23}	0.28				
G_{12} (GPa)	15				
G_{13} (GPa)	15				
G_{23} (GPa)	15				

Figure 1 shows the SEM and TEM (University of Dayton, United States) characterization of CNTs in epoxy resin at micro and nano scales. The multiwall nanotubes were tube-shaped materials and considered as long curved cylindrical fibers (snake-like shapes). The CNTs are randomly distributed into matrix, Figure 1a. Transmission electron microscopy (TEM) of CNTs shows the fiber shape, see Figure 1b.

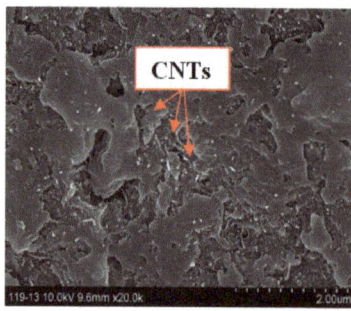
(a) SEM morphology of CNTs into Epoxy

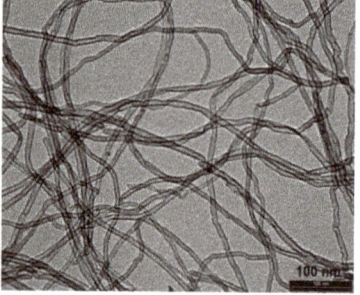
(b) TEM image of CNTs

Figure 1. The morphology of multiwall CNTs by (**a**) SEM and (**b**) TEM images.

The fabrication of the nanocomposites consisted of, first, dispersing CNTs in the polymer matrix, varying the weight fraction of MWNTCs between 0 and 2%, and then mixing this material using an T25 digital ULTRA-TURRAX increased shear laboratory mixer for a total of 30 min at 2000 rpm. Afterwards, an ultrasonic bath was also used, and the mixed material was further processed in a Lehmann Mills three-roll mixer (University of Dayton, United States) to guarantee a homogeneous dispersion of CNTs (Figure 2), the film with 120 μm in thickness containing CNTs was manufactured using film line, Figure 3a. The reinforced epoxy was introduced with the 5 HS (satin) T300 6k carbon fiber fabric, using the infusion process; Figure 3b,c. The reinforced epoxy resin flowed between the fiber plies, and the press curing condition was set to 200 MPa. All panels manufactured consisted of 24 carbon fiber fabric layers interleaved with 25 layers of CNTs/epoxy film to accomplish an overall fiber volume fraction of 50%. The panels were then cooled. SEM characterization was performed to demonstrate the CNT distribution with 500 nm resolution. The SEM image confirms the random distribution of CNTs with variable length; Figure 3d,e.

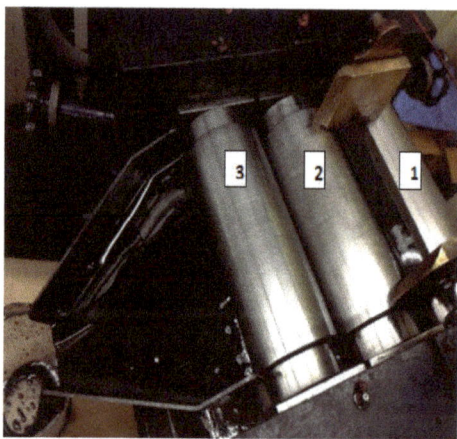

Figure 2. Lehmann Mills three-roll mixer.

(a) Resin film on white release ply

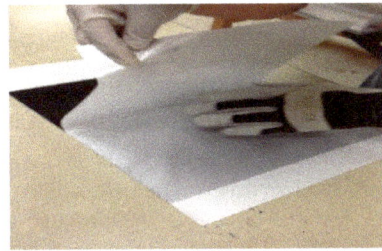
(b) Resin film with nano-additives between two release plies

(c) Sample Preparation

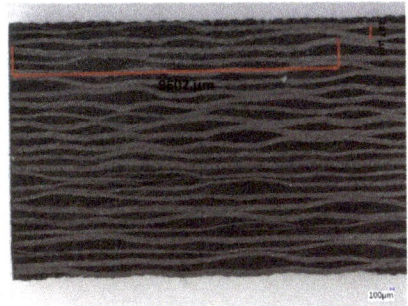
(e) SEM of cross section of the sample

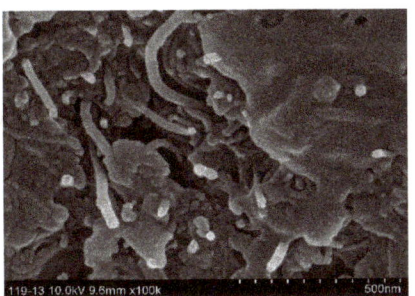
(f) CNTs distribution

Figure 3. Manufacturing steps.

Samples with dimensions of 13 mm × 13 mm × 8 mm were then cut from the prepared specimen plates for out-of-plane compression test on SHPB, Figure 4.

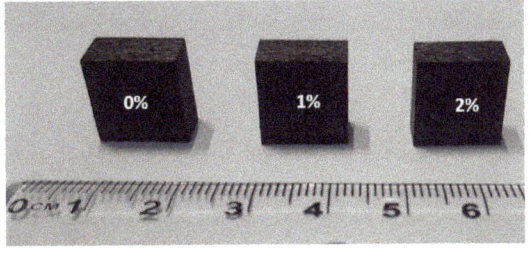

Figure 4. Specimens with different percentages dedicated to dynamic compression tests.

3. Test Procedure

3.1. SHPB Test Method

Figure 5 is a schematic of The Split Hopkinson Pressure Bar (SHPB) apparatus that was used in this study to assess the shock wave absorption characteristics of the specimens. The experimental setup was composed of striker, incident (input) and transmitted (output) steel bars. A compressive longitudinal wave was induced in the incident pressure bar by impacting it with the striker bar at a specified velocity (Impact energy). A compressive incident wave $\varepsilon_I(t)$ was generated when a striker bar impacted the free end of the input bar and travelled across the input bar until it got to the bar-specimen interface. Once the specimen was hit by the incident wave, the wave was split into two parts. One part was transferred to the output bar as a compressive transmitted wave $\varepsilon_T(t)$ and the other part was reflected to the input bar as a tensile reflected wave $\varepsilon_R(t)$. These three pulses $\varepsilon_I(t)$, $\varepsilon_R(t)$ and $\varepsilon_T(t)$ were measured using strain gauges mounted at the middle of each pressure bar, and a digital oscilloscope was used for data acquisition; Figure 6. Recorded data were then treated using by means of the Maple Software algorithm to acquire all dynamic parameters like, for example, forces and velocities, as functions of time at the two faces of the specimen, which had already been sandwiched between the two pressure bars.

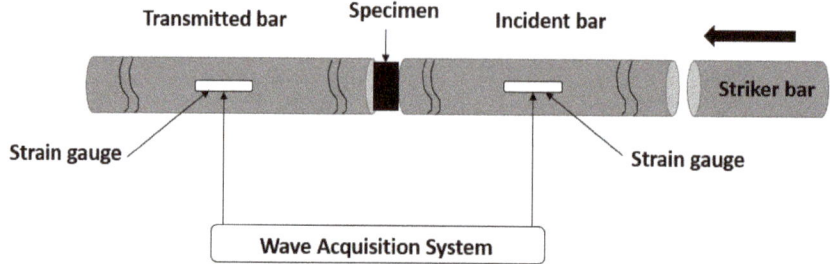

Figure 5. Split Hopkinson Pressure Bar.

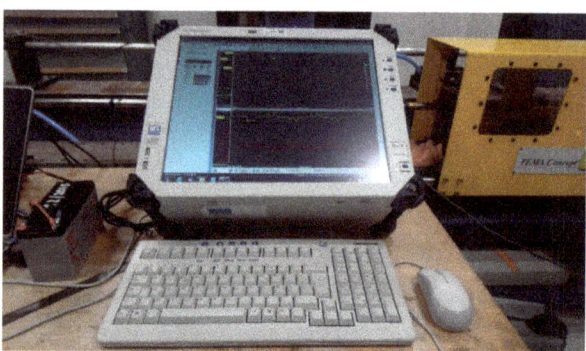

Figure 6. Digital oscilloscope.

3.2. Dynamic Compression Testing

During the experimental investigation, the velocity of the striker bar (impact pressure) was controlled to obtain a wide range of impact energy magnitudes. Variation of the incident velocity as a function of time was plotted to assess the dynamic response of each specimen at impact pressure P = 4 bar, and the results confirmed the test reproducibility, which was common for all CNTs weight percentages; Figure 7. We performed tests at different pressures ranging from 2 to 4 bar but 4 bar was the pressure at which damage was exhibited in the samples. We wanted to see the effect of CNTs on the improvement of the damage mechanism of fiber-reinforced composites, so 4 bar was chosen to present the diversity of the behavior of our material with respect to the addition of CNTs.

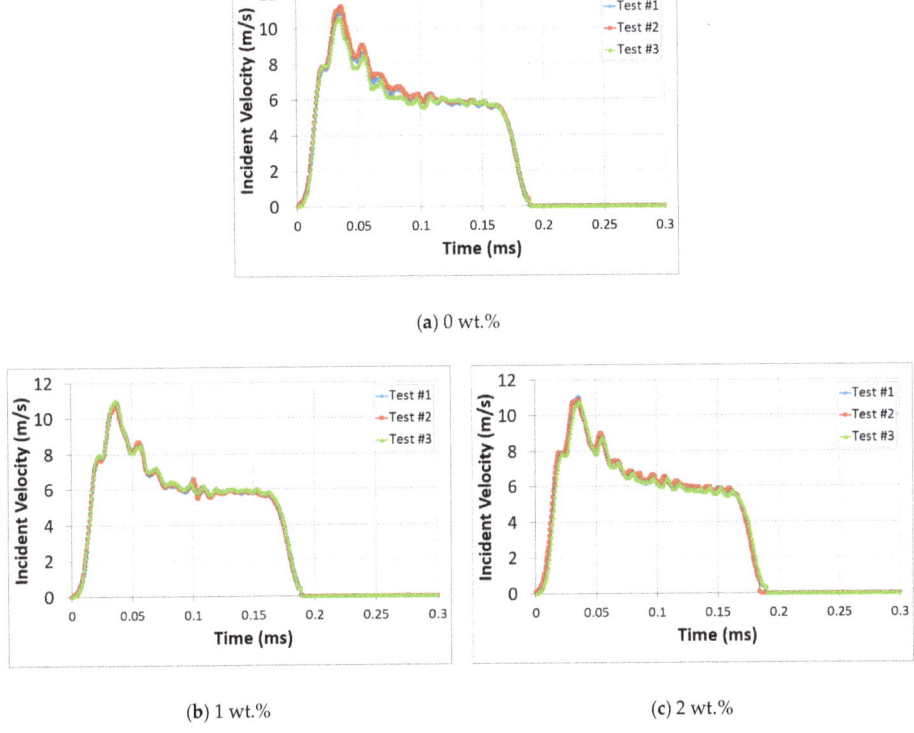

Figure 7. Test reproducibility of the nanocomposites with different CNTs mass fractions, P = 4 bar.

3.3. Theorical Characterization of Absorbed Energy

During the dynamic compressive tests, the input energy (impact energy) corresponding to the transferred kinetic energy of the striker bar to incident bar was obtained by varying its initial velocity. Once the input energy reached the bar specimen interface, it was split into two parts. One part was absorbed by the sample and could have caused plastic deformation or damage in different forms, which in turn could have led to heat generation if macroscopic damage occurred. The other part was transferred to the output bar as the transmitted energy [44]. Absorbed energy was the difference between the work transferred to the specimen from the incident bar and the mechanical work done by the specimen and transmitted to output bar [45]. The mechanical powers at the interfaces of two pressure bars were obtained by multiplying the corresponding velocity by the contact force.

Velocity (V) was determined using the incident and transmitted strains (ε_I and ε_T) as stated by Park et al. [45]:

$$V(x,t) = c\left[-\varepsilon_i\left(t - \frac{x}{c}\right) + \varepsilon_r\left(t + \frac{x}{c}\right)\right] \tag{1}$$

where $c = (E/\rho)^{\frac{1}{2}}$ is the longitudinal wave velocity of the bar, E is the Young's modulus, and ρ is the mass density of the pressure bar.

The normal force (F) on any cross-section x of the incident and transmitted bars is:

$$F(x,t) = AE\varepsilon(x,t) \quad (2)$$

where A represents the cross-sectional area of the bar.

It was obvious that the physical properties could be obtained only if the values of ε_i and ε_r were known. Hence, the main purpose of this experimentation was to find out these functions using strain gauges, mounted at the middle of each pressure bar. A digital oscilloscope was also used for data acquisition.

The overall mechanical work was calculated by integrating the mechanical power with respect to time. Thus, both the work transferred from the input bar to the CNTs-based nanocomposites sample (W_{inc}) and the work done by the sample and transferred to the output bar (W_{trans}) are given by [45]:

$$W_{inc} = -\int_{t_i}^{t_f} F_{inc}(t) V_{inc}(t) dt \quad (3)$$

$$W_{trans} = \int_{t_i}^{t_f} F_{trans}(t) V_{trans}(t) dt \quad (4)$$

where the incident and transmitted physical parameters are presented by the subscript "inc" and "trans".

The absorbed energy of the sample (E_{abs}) was calculated using the equation below, given in [45]:

$$E_{abs} = W_{inc}(t) - W_{trans}(t) \quad (5)$$

Figure 8 shows an example of a typical absorbed energy curve for the test conducted on 0% CNTs sample. Results showed an increase in the absorbed energy during the impact and this energy absorbed by the material was the combination of two different energies. The first part was the elastic part, which was released until it reached a constant value (gradual unloading cycle). The second part was the inelastic unrecoverable energy, represented by the constant value, that was dissipated permanently through damage (the end of the cycle). Thus, the absorbed energy can be given as:

$$E_{abs} = E_{elas} + E_{diss} \quad (6)$$

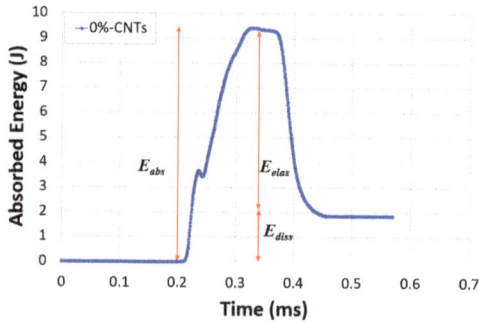

Figure 8. Typical profile of the energy absorption of the tested specimen.

4. Results and Discussion

A wide variety of input energies (or incident energy) was used to impact the carbon/epoxy nanocomposite specimens in order to understand the effect of CNTs on their energy absorption capability. Therefore, the energy absorption behavior of each specimen was studied at impact energy of 15 J, 34 J and 53 J. The absorbed energy results showed that all the samples had similar characteristics, Figure 9. This response indicated that no macroscopic damage was occurred as shown by Tarfaoui et al. [46]. Moreover, the fluctuation in the curves was caused by the storage and release of strain energy during the out-of-plane compression tests. Another interesting phenomenon observed is that the absorbed energy was very much influenced by increasing the impact energy, and the maximum peak also increased as the input energy was increasing. This behavior was common for all mass fractions of CNTs.

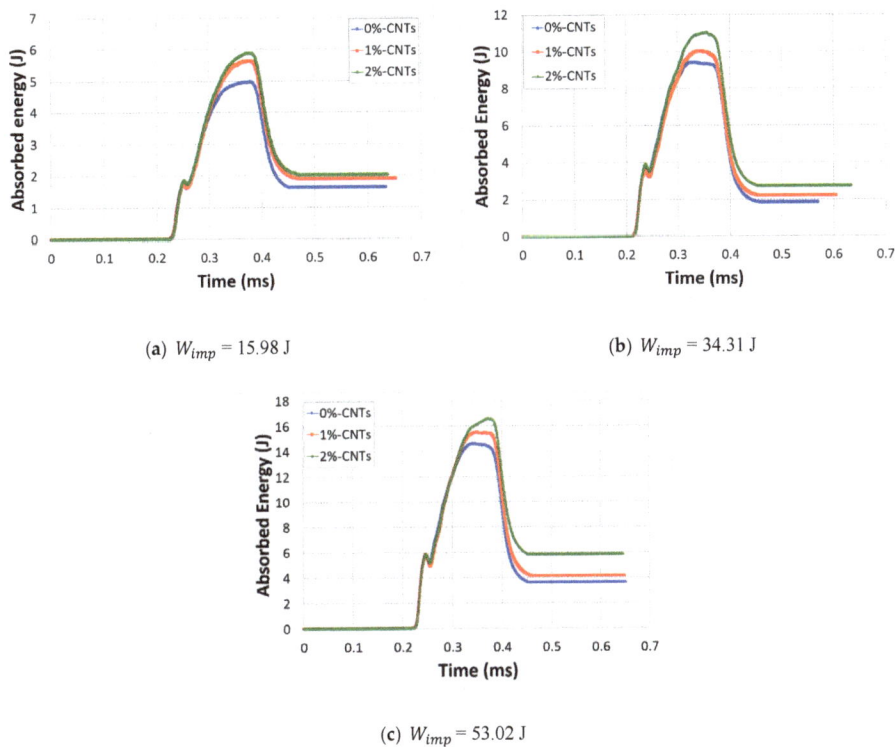

Figure 9. Absorbed energy vs. time for different CNTs mass fractions.

It can be seen that there was a noticeable effect due to the introduction of CNTs. An important portion of the impact energy W_{imp} was absorbed during the impact. Actually, this absorbed energy was the energy stored in the specimen during the elastic deformation and it was released in the form of recoverable elastic strain energy and dissipated energy. Moreover, the experimental behavior of the CNT-based nanocomposites also confirmed that an increase in absorbed energy was observed because of the increase in both elastic and dissipated energies. However, the greatest portion of the absorbed energy was stored and released as elastic strain energy (E_{elas}) and only a small portion was dissipated energy (E_{diss}). This response indicated that neat epoxy showed more plastic deformation instead of elastic behavior, compared to samples with a different mass fraction of CNTs. Addition of CNTs improved the elastic behavior of composites and reduced their plastic deformation as it became more resilient. In fiber reinforced composites the matrix is responsible for the plasticity because of their ductile nature. However, when CNTs were added as nanofillers in the matrix epoxy the material became more rigid with an increase in its elastic properties and reduction in its plasticity. In addition, CNT-reinforced nanocomposites had higher energy dissipation performance because of the increase in micro cracks. The CNTs behave as barrier for any crack propagation. Thus, they stop the propagation of any crack initiation, which can result in significant micro cracks instead of fatal macro damage and could delay the final fracture. This showed that, even with small weight percentage such as 1%, CNTs could improve the energy absorption of the CFRP laminate composites and delay the final fracture. Figure 10 gives a summary of the obtained results.

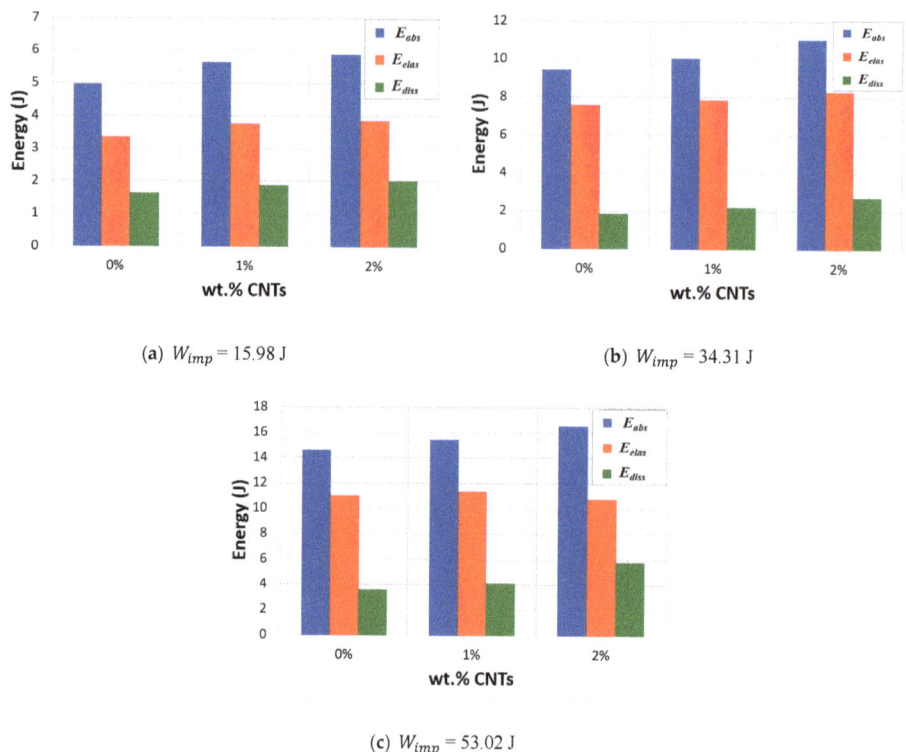

Figure 10. Energy balance vs. CNT mass fraction at different impact energies.

The energy dissipation caused in a nanocomposite during out-of-plane compressive loading was calculated by taking the area under the curve of the absorbed energy profiles for each specimen. Results revealed that there was an increase in dissipated energy as the wt.% of CNTs increased. The plausible scenario for the augmentation of the energy dissipation was the sliding phenomenon at the interface between CNTs and polymer matrix. Low mass density of CNTs and exceptionally large contact area at the interface between CNTs and matrix caused the frictional sliding at the CNT-CNT and CNT-matrix interfaces which could be the main cause of the increase in energy dissipation with minimal weight penalty [47]. Moreover, this sliding between the matrix and the CNTs could enhance the structural damping of the material. Recent studies of polymer material reinforced with nanofillers have also demonstrated that integrating nanofillers in the polymer matrix increased the damping of composite structures more efficiently [48].

Another method for augmenting the frictional energy dissipation is by boosting the weight fraction of CNTs in the composite; Figure 11.

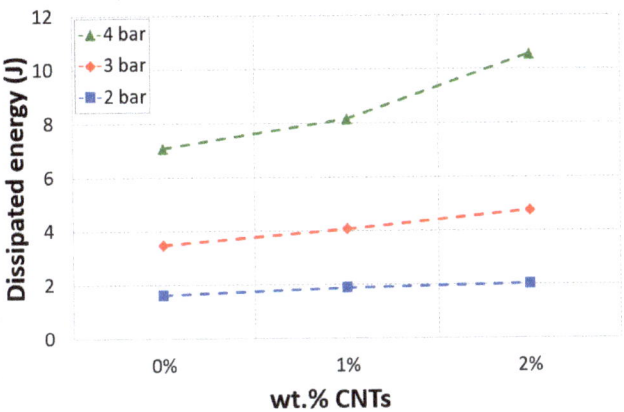

Figure 11. Dissipated energy vs CNTs mass fractions.

5. Failure Mode

The experimental investigation was studied in detail to improve the understanding of the dynamic behavior of these nanocomposites during out-of-plane compression tests. The results showed that the dynamic properties of CNTs reinforced nanocomposite were significantly affected by increasing CNTs mass fraction. Strain rate and stress as functions of time were superposed for each mass fraction of CNTs at impact pressure of 4 bar, Figure 12. According to this figure, we are able to differentiate a variety of zones for each specimen and each zone is described individually as follow:

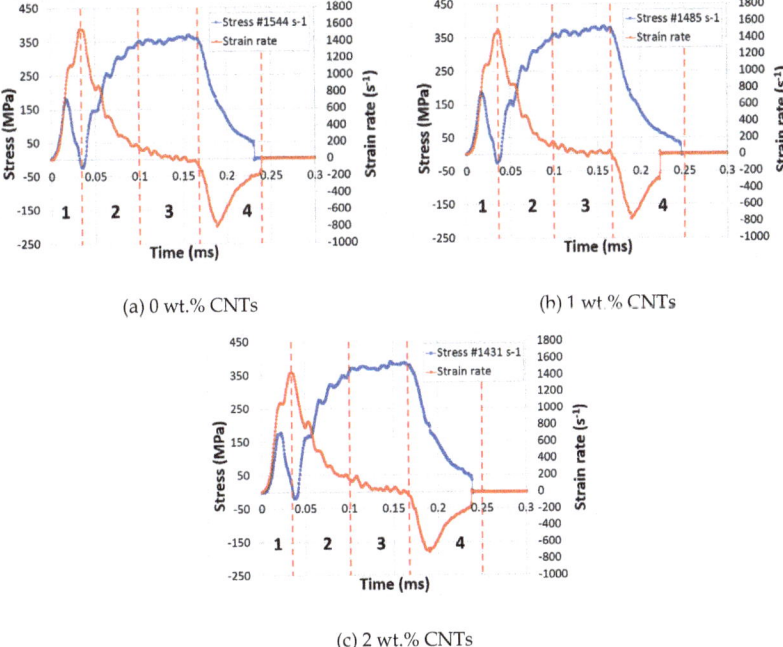

Figure 12. Evolution of the stress and strain rate vs time curves for different mass fraction, P = 4 bar.

Zone 1: strain rate increases quickly before attaining the highest peak. This maximum value was reduced through the increase of CNTs mass fraction and this can be explained by the increase in the rigidity of the matrix material because of the presence of CNTs.

Zone 2: once perfect contact was guaranteed, there was a drop in strain rate and an increase in strength.

Zone 3: an increase in strength became stable and reached the saturation level while the strain rate gradually decreased to zero value. The sample reached maximum compression stress in this zone.

Zone 4: in this zone, the specimen rebounded and started to relax. The strain rate started to decrease below zero value and there was a drop in the stress of the specimen. This situation could be justified by the spring-back behavior of the specimen. At the end of this zone, both signals were negated at approximately the same time.

The damage tolerance is a significant criterion for the composites to be used in the civil and military applications like naval or aerospace. This characteristic goes through the damage behavior of structure from the initial state to final fracture; and numerous methods have been utilized to verify the magnitude of the damage. A high-speed camera was used to monitor and record the behavior of the specimen during the dynamic compression tests performed at 4 bar. The images, which were taken in real-time, show the progression of damage, Figure 13. However, it should be kept in mind that no macroscopic damage in the nanocomposite specimens was noticed at the 4 bar; and the absence of the second peak in the strain rate vs. time curve validated this phenomenon. However, damage at microscopic and nano scale, such as plastic deformation, micro-buckling, kink-bands, and crack could have happened. A damaged zone was observed at 0.12 ms of impact time in the case of 0% CNTs sample, but no damage was obtained in the case of reinforced specimens with CNTs. The incorporation of CNTs not only increase the strength of the material but also played a vital role in delaying the crack propagation phenomena, thus increasing the resistance of material to final fracture.

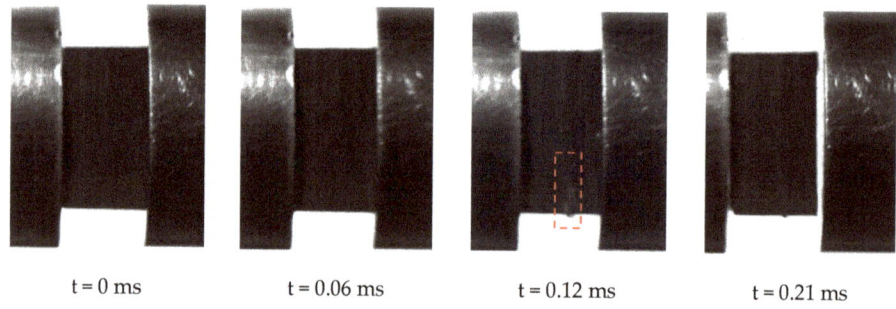

t = 0 ms   t = 0.06 ms   t = 0.12 ms   t = 0.21 ms

(a) 0 wt.% CNTs

Figure 13. *Cont.*

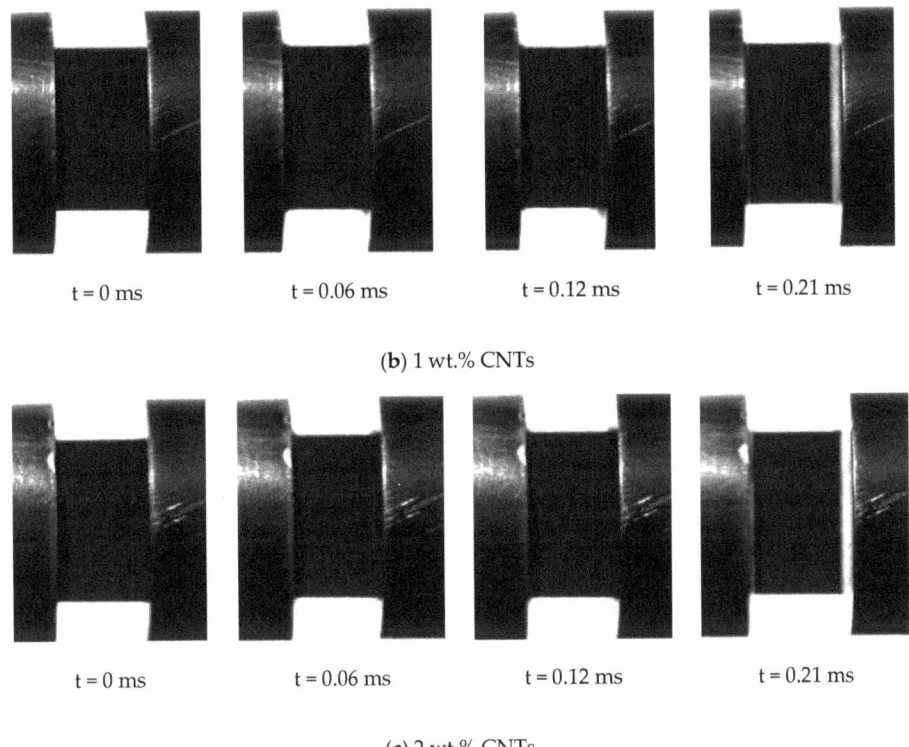

Figure 13. High-speed photograph of real-time dynamic compression test of nanocomposites with different CNTs mass fraction, P = 4 bar.

6. Conclusion

An experimental investigation was carried out to study the effect of different wt.% of CNTs on the mechanical energy balance of CFRP nanocomposites using split Hopkinson pressure bars. Samples with different CNT weight percentages (0% as reference, 1% and 2%) were subjected to different incident waves; and results showed that the ability of the material to absorb a mechanical shock wave was improved by increasing the CNTs mass fractions. This increase was due to the enhancement of elastic strain behavior of the composite and decrease in its plasticity with the addition of CNTs in the matrix of the composite. Moreover, damage modes were evaluated, and the results indicated that no macroscopic damage was observed in the specimen under the impact pressures, because CNTs act as a barrier to crack propagation; however, micro cracks and permanent plastic deformation could be present within the nano composites. Thus, the presence of CNTs resulted in greater energy dissipation and increase in energy absorption behavior of these nanocomposites. Results confirmed that for out-of-plane tests, CNT-based nanocomposites exhibited better stiffness and resistance to damage compared to neat material; and dynamic response revealed that the composite final failure was delayed by increasing the CNT % in addition to improvement in energy absorption and dissipation system. Therefore, CNTs might be good nanofillers, which improves the dynamic properties of the composites and enhance the resistance to damage and the energy absorption capability of composite materials for high velocity impact loadings.

Author Contributions: Design and conduction of Experimentation, data curation, data analysis, investigation and data interpretation, writing—original draft preparation, M.C.; Supervision, writing—review and editing, project administration, M.T.; Writing—review & editing, C.B. and A.E.M. All authors have read and agreed to the published version of the manuscript.

Funding: This research received no external funding.

Conflicts of Interest: The authors declare no conflict of interest.

References

1. Ramakrishna, S. Microstructural design of composite materials for crashworthy structural applications. *Mater. Des.* **1997**, *18*, 167–173. [CrossRef]
2. Zhang, L.; Becton, M.; Wang, X. Mechanical analysis of graphene-based woven nano-fabric. *Mater. Sci. Eng.* **2015**, *620*, 367–374. [CrossRef]
3. Banhart, J. Manufacture characterisation and application of cellular metals and metal foams. *Prog. Mater. Sci.* **2001**, *46*, 559–632. [CrossRef]
4. Zhao, H.; Gary, G. Crushing behaviour of aluminium honeycombs under impact loading. *Int. J. Impact Eng.* **1998**, *21*, 827–836. [CrossRef]
5. Nemat-Nasser, S.; Kang, W.J.; McGee, J.; Guo, W.G.; Isaacs, J.B. Experimental investigation of energy-absorption characteristics of components of sandwich structures. *Int. J. Impact Eng.* **2007**, *34*, 1119–1146. [CrossRef]
6. Hazizan, M.A.; Cantwell, W.J. The low velocity impact response of an aluminium honeycomb sandwich structure. *Compos. Part B Eng.* **2003**, *34*, 679–687. [CrossRef]
7. Chandra, R.; Singh, S.P.; Gupta, K. Damping studies in fiber-reinforced composites e a review. *Compos. Struct.* **1999**, *46*, 41–51. [CrossRef]
8. Sun, C.T.; Chaturvedi, S.K.; Gibson, R.F. Special issue: Advances and trends in structures and dynamics internal damping of short-fiber reinforced polymer matrix composites. *Compos. Struct.* **1985**, *20*, 391–400. [CrossRef]
9. Eitan, A.; Fisher, F.T.; Andrews, R.; Brinson, L.C.; Schadler, L.S. Reinforcement mechanisms in MWCNT-filled polycarbonate. *Compos. Sci. Technol.* **2005**, *66*, 1159–1170. [CrossRef]
10. Ding, W.; Eitan, A.; Fisher, F.T.; Chen, X.; Dikin, D.A.; Andrews, R.; Brinson, L.C.; Schadler, L.S.; Ruoff, R.S. Direct observation of polymer sheathing in carbon nanotube–polycarbonate composites. *Nano Lett.* **2003**, *3*, 1593–1597. [CrossRef]
11. Schadler, L.S.; Brinson, L.C.; Sawyer, W.G. Polymer nanocomposites: A small part of the story. *JOM* **2007**, *59*, 53–60. [CrossRef]
12. Watcharotone, S.; Wood, C.D.; Friedrich, R.; Chen, X.; Qiao, R.; Putz, K.; Brinson, L.C. Interfacial and substrate effects on local elastic properties of polymers using coupled experiments and modeling of nanoindentation. *Adv. Eng. Mater.* **2011**, *13*, 400–404. [CrossRef]
13. Liu, Y.; Kumar, S. Polymer/Carbon nanotube nano composite fibers—A review. *ACS Appl. Mater. Interfaces* **2014**, *6*, 6069–6087. [CrossRef]
14. Chen, X.; Huang, Y. Nanomechanics modeling and simulation of carbon nanotubes. *J. Eng. Mech.* **2008**, *134*, 211–216. [CrossRef]
15. Delfani, M.R.; Shodja, H.M.; Ojaghnezhad, F. Mechanics and morphology of single-walled carbon nanotubes: From graphene to the elastic. *Philos. Mag.* **2013**, *93*, 2057–2088. [CrossRef]
16. Wang, C.M.; Zhang, Y.Y.; Xiang, Y.; Reddy, J.N. Recent studies on buckling of carbon nanotubes. *Appl. Mech. Rev.* **2010**, *63*. [CrossRef]
17. Becton, M.; Zhang, L.; Wang, X. Molecular dynamics study of programmable nanoporous grapheme. *J. Nanomech. Micromech.* **2014**, *4*. [CrossRef]
18. Zhang, Z.; Wang, X.; Li, J. Simulation of collisions between buckyballs and graphene sheets. *Int. J. Smart Nano Mater.* **2012**, *3*, 14–22. [CrossRef]
19. Zhang, L.; Wang, X. Atomistic insights into the nanohelix of hydrogenated graphene: Formation, characterization and application. *Phys. Chem. Chem. Phys.* **2014**, *16*, 2981–2988. [CrossRef]

20. Zhang, L.; Wang, X. Tailoring pull-out properties of single-walled carbon nanotube bundles by varying binding structures through molecular dynamics simulation. *J. Chem. Theory Comput.* **2014**, *10*, 3200–3206. [CrossRef]
21. Lattanzi, L.; De Nardo, L.; Raney, J.R.; Daraio, C. Mechanical properties of carbon nanotube foams. *Adv. Eng. Mater.* **2014**, *16*, 1026–1031. [CrossRef]
22. Gui, X.; Zeng, Z.; Zhu, Y.; Li, H.; Lin, Z.; Gan, Q.; Xiang, R.; Cao, A.; Tang, Z. Three-dimensional carbon nanotube sponge-array architectures with high-energy dissipation. *Adv. Mater.* **2014**, *26*, 1248–1253. [CrossRef] [PubMed]
23. Liu, L.Q.; Ma, W.; Zhang, Z. Macroscopic carbon nanotube assemblies: Preparation, properties, and potential applications. *Small* **2011**, *7*, 1504–1520. [CrossRef] [PubMed]
24. Thostenson, E.T.; Ren, Z.F.; Chou, T.W. Advances in the science and technology of carbon nanotubes and their composites: A review. *Compos. Sci. Technol.* **2001**, *61*, 1899–1912. [CrossRef]
25. Sun, L.F.; Gibson, R.; Gordaninejad, F.; Suhr, J. Energy absorption capability of nanocomposites: A review. *Compos. Sci. Technol.* **2009**, *69*, 2392–2409. [CrossRef]
26. Drdlová, M.; Prachař, V. High strain rate characteristics of nanoparticle modified blast energy absorbing materials. *Procedia Eng.* **2016**, *151*, 214–221. [CrossRef]
27. Chen, H.; Zhang, L.; Chen, J. Energy dissipation capability and impact response of carbon nanotube buckypaper: A coarse-grained molecular dynamics study. *Carbon* **2016**, *103*, 242–252. [CrossRef]
28. Chen, H.; Zhang, L.; Becton, M. Molecular dynamics study of a CNT-buckyball-enabled energy absorption system. *Phys. Chem. Chem. Phys.* **2015**, *17*, 17311–17321. [CrossRef]
29. Weidt, D.; Buggy, M. Prediction of energy absorption characteristics of aligned carbon nanotube/epoxy nanocomposites. *IOP Conf. Ser. Mater. Sci. Eng.* **2012**, *40*, 2028. [CrossRef]
30. Arbaoui, J.; Tarfaoui, M.; Alaoui, A.E.M. Mechanical behavior and damage kinetics of woven E-glass/Vinylester laminate composites under high strain rate dynamic compressive loading: Experimental and numerical investigation. *Int. J. Impact Eng.* **2016**, *87*, 44–54. [CrossRef]
31. Arbaoui, J.; Tarfaoui, M.; Alaoui, A.E.M. Dynamical characterization and damage mechanisms of E-glass/Vinylester woven composites at high strain rates compression. *J. Compos. Mater.* **2016**, *50*, 3313–3323. [CrossRef]
32. Arbaoui, J.; Tarfaoui, M.; Bouery, C.; Alaoui, A.E.M. Comparison study of mechanical properties and damage kinetics of 2D and 3D woven composites under high-strain rate dynamic compressive loading. *Int. J. Damage Mech.* **2016**, *25*, 1–22. [CrossRef]
33. Al-Lafi, W.; Jie, J.; Xu, S.; Song, M. Performance of MWCNT/HDPE nanocomposites at high strain rates. *Macromol. Mater. Eng.* **2010**, *295*, 519–522. [CrossRef]
34. Tarfaoui, M.; Neme, A.; Choukri, S. Damage kinetics of glass/epoxy composite materials under dynamic compression. *J. Compos. Mater.* **2009**, *43*, 1137–1154. [CrossRef]
35. Sassi, S.; Tarfaoui, M.; Nachtane, M.; Benyahia, H. Strain rate effects on the dynamic compressive response and the failure behavior of polyester matrix. *Compos. Part B Eng.* **2019**, *174*, 107040. [CrossRef]
36. Tarfaoui, M.; Choukri, S.; Neme, A. Dynamic response of symmetric and asymmetric e-glass/epoxy laminates at high strain rates. *Key Eng. Mater.* **2010**, *116*, 73–82. [CrossRef]
37. Sassi, S.; Tarfaoui, M.; Benyahia, H. Experimental study of the out-of-plane dynamic behaviour of adhesively bonded composite joints using split Hopkinson pressure bars. *J. Compos. Mater.* **2018**, *52*, 2875–2885. [CrossRef]
38. Sassi, S.; Tarfaoui, M.; Benyahia, H. An investigation of in-plane dynamic behavior of adhesively-bonded composite joints under dynamic compression at high strain rate. *Compos. Struct.* **2018**, *191*, 168–179. [CrossRef]
39. Feng, B.; Fang, X.; Wang, H.-X.; Dong, W.; Li, Y.-C. The effect of crystallinity on compressive properties of Al-PTFE. *Polymers* **2016**, *8*, 356. [CrossRef]
40. Bai, Y.; Liu, C.; Huang, G.; Li, W.; Feng, S. A hyper-viscoelastic constitutive model for polyurea under uniaxial compressive loading. *Polymers* **2016**, *8*, 133. [CrossRef]
41. Gardea, F.; Glaz, B.; Riddick, J.; Lagoudas, D.C.; Naraghi, M. Identification of energy dissipation mechanisms in CNT-reinforced nanocomposites. *Nanotechnology* **2016**, *27*, 10. [CrossRef]
42. Gardea, F.; Cole, D.; Glaz, B.; Riddick, J. Strain energy dissipation mechanisms in carbon nanotube composites fabricated by additive manufacturing. *Mech. Addit. Adv. Manuf.* **2018**, *9*, 29–36.
43. El Moumen, A.; Tarfaoui, M.; Nachtane, M.; Lafdi, K. Carbon nanotubes as a player to improve mechanical shock wave absorption. *Compos. Part B Eng.* **2019**, *164*, 67–71. [CrossRef]

44. Tarfaoui, M.; Nachtane, M.; El Moumen, A. Energy dissipation of stitched and unstitched woven composite materials during dynamic compression test. *Compos. Part B Eng.* **2019**, *167*, 487–496. [CrossRef]
45. Park, S.W.; Zhou, M.; Veazie, D.R. Time-resolved impact response and damage of fiber reinforced Composite laminates. *J. Compos. Mater.* **2000**, *34*, 879–904. [CrossRef]
46. Tarfaoui, M.; Choukri, S.; Neme, A. Effect of fibre orientation on mechanical properties of the laminated polymer composites subjected to out-of-plane high strain rate compressive loadings. *Compos. Sci. Technol.* **2008**, *68*, 477–485. [CrossRef]
47. Suhr, J.; Koratkar, N.A. Energy dissipation in carbon nanotube composites: A review. *J. Mater. Sci.* **2008**, *43*, 4370–4382. [CrossRef]
48. Gardea, F.; Glaz, B.; Riddick, J.; Lagoudas, D.C.; Naraghi, M. Energy dissipation due to interfacial slip in nanocomposites reinforced with aligned carbon nanotubes. *ACS Appl. Mater. Interfaces* **2015**, *7*, 9725–9735. [CrossRef]

© 2020 by the authors. Licensee MDPI, Basel, Switzerland. This article is an open access article distributed under the terms and conditions of the Creative Commons Attribution (CC BY) license (http://creativecommons.org/licenses/by/4.0/).

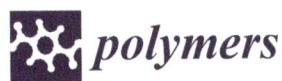

Article

Improvement of the Heat-Dissipating Performance of Powder Coating with Graphene

Fei Kung and Ming-Chien Yang *

Department of Materials Science and Engineering, National Taiwan University of Science and Technology, Taipei 10607, Taiwan; philip@allightec.com
* Correspondence: myang@mail.ntust.edu.tw; Tel.: +886-2-2737-6528; Fax: +886-2-2737-6544

Received: 18 May 2020; Accepted: 8 June 2020; Published: 10 June 2020

Abstract: In this study, the epoxy powder was blended with graphene to improve its thermal conductivity and heat dissipation efficiency. The thermal conductivity of the graphene-loaded coating was increased by 167 folds. In addition, the emissivity of the graphene-loaded coating was 0.88. The epoxy powder was further coated on aluminum plate through powder coating process in order to study the effect on the performance of heat dissipation. In the case of natural convective heat transfer, the surface temperature of the graphene-loaded coated aluminum plate was 96.7 °C, which was 27.4 °C lower than that of bare aluminum plate (124.1 °C) at a heat flux of 16 W. In the case of forced convective heat transfer, the surface temperature decreased from 77.8 and 68.3 °C for a heat flux of 16 W. The decrease in temperature can be attributed to the thermal radiation. These results show that the addition of graphene nanoparticles in the coating can increase the emissivity of the aluminum plate and thus improving the heat dissipation.

Keywords: graphene; powder coating; thermal conductivity; heat dissipation; thermal radiation

1. Introduction

Heat-dissipating coating is important for the stabilization and miniaturization of electronic components. As the aggregate density and power intensity of electronic components continue to increase, large amount of heat generated from these devices must be dissipated in a timely manner. However, the heat dissipation performance of today's electronic components cannot meet the requirements, thereby limiting the efficiency and service life of certain electronic components. To resolve this problem, heat-dissipating coating enhances the heat dissipation efficiency of the surface of a component [1]. It lowers the temperature of the heat-generating component in time and hence extends the service time and stability of components.

Literatures and patents on graphene heat-dissipating powder coating have been sparse; most of them confuse "heat dissipation" with "heat conduction" [2]. In general, the most important functions of a heat dissipation module in an electronic product include not only a rapid transfer of heat from the thermal source to the surface of the heat sink but also the ability to quickly disperse heat into the atmosphere through convection and radiation. A high thermal conductivity can only solve the problem of quick heat conduction. On the other hand, heat dissipation depends mainly on the heat dissipation area, profile, natural convection, and thermal radiation of the heat sink; it almost has nothing to do with the thermal conductivity of materials. Therefore, as long as the thermal conductivity is adequate, heat-dissipating coating can still be used as good heat dissipation modules for electronic products. Proper structural design of product or module can easily achieve a large heat dissipation surface area for convection. However, to achieve high heat dissipation efficiency through radiation, high thermal radiation coefficient is necessary [3].

Graphene is a nanomaterial with only one layer of carbon atoms. It features low density, low chemical activity, high thermal conductivity, large specific surface area, and high infrared emissivity. Graphene has superior heat conduction characteristics and its thermal radiation coefficient is greater than 0.95 [4]. Balandin et al. reported the thermal conductivity of suspended single-layer graphene measured near 5000 W m^{-1} K^{-1}, which is one of the highest thermal conductivity of the currently known materials [5]. Therefore, from the perspective of heat conduction, heat dissipation, or thermal management, graphene can effectively improve the heat dissipation performance of existing thermal dissipation products for electronic components, assemblies, and LEDs as long as graphene products can be configured to meet the design requirements. However, the stacking tendency of graphene led to poor dispersion and greater post-processing difficulties, thereby preventing graphene from exhibiting its superior characteristics [6,7].

Thermoset powder coating comprises thermoset resin, hardener, dye, filler, and additives. There are several types of thermoset powder coatings: epoxy resin, polyester, and acrylic resin. Table 1 compares the pros and cons of these three types of powder coatings. The constituents are first mixed according to a specific ratio, followed by hot extrusion and crushing and other preparation processes. The coating is then applied by an electrostatic spray or friction spray (a thermoset method) at ambient temperature. It is then baked, melted, and cured to form a shiny permanent coating for heat dissipation and corrosion prevention. [8,9]. Powder coating generally has a better thermal conductivity than solvent coating due to the better binding between the coating and the substrate. More thoroughly cured coating leads to more stable crosslinking and hence denser and tighter coating [10,11]. This favors the reduction of scattering in the "lattice vibration" of the thermal dissipation mechanism.

Table 1. Types and surface characteristics of resins.

	Epoxy	Epoxy Polyester	Polyester
Hardness	Excellent	Very good	Very good
Softness	Excellent	Excellent	Excellent
Baking resistance	Very Poor	Very good	Excellent
Weatherability	Poor	Poor	Excellent
Corrosion resistance	Excellent	Very good	Very good
Chemical resistance	Excellent	Good	Very good
Operability	Very good	Excellent	Excellent

The discussion of radiation and convection is rare. This study is aiming to investigate the enhancing effect of graphene-loading on the thermal dissipation performance of aluminum plate. The aluminum plate was attached to a heater as the heat source. The heat was transferred through the Al plate to the ambient atmosphere via convection and radiation. The plate was either bare or coated with a thin layer of polymer filled with graphene nanoflakes or boron nitride nanoparticles. The performance of the heat dissipation was evaluated by measuring the surface temperature on the plates with or without coating at a constant heat flux under forced convection or natural convection conditions. This study will demonstrate the significance of radiation heat transfer in the heat dissipation.

2. Materials and Methods

2.1. Materials

Graphene (AG05, grain size 5 μm, thickness 3.5 nm, aspect ratio 1429) was supplied by Allightec Co., Taichung, Taiwan. Aluminum plates (AL101001, Kuopont Chemical, Taoyuan, Taiwan) were used as the substrate for coating. The dimensions of the plate were $10 \times 10 \times 0.1$ cm^3. Epoxy resin (E12(604), Dow Chemical, Midland, MI, USA) and polyester (SJ4ET, Shenjian New Materials, Wuhu, China) were used as the matrix of the coating. Furthermore, hardener (HR0001, Kuopont Chemical, Taiwan) and additive (AD0001 Chemical, Kuopont, Taoyuan, Taiwan) were employed to give the coating (Table 2) both chemical resistance and weather resistance.

Table 2. Composition of powder coatings.

Ingredient		Product ID	Content, wt%	Manufacturer
Epoxy resin		E12–604	33	Shang-shan, Dow Chemical, Midland, MI, USA
Polyester resin		SJ4ET	35	Shen-Jian, Wuhu, China
Curing agent		KPC–03	6	Kuopont, Taoyuan, Taiwan
Auxiliary		KPA–01	5	Kuopont, Taoyuan, Taiwan
TiO$_2$		BLR-698	18	Lomon Billions, Jiaozuo, China
Filler	Graphene	AG05	3	Allightec, Taichung, Taiwan
	Boron nitride	TSD–03	3	Topspin, Kaohsiung, Taiwan

2.2. Preparation of Powder Coating

All the ingredients were blended using a single-screw extruder (PK–55, Pinying Machine Co., Kaohsiung, Taiwan) at 85–90 °C and a screw speed of 60 rpm. The resultant blend was pressed into sheets using roller miller and ground into powder (diameter: 0.1–2 μm) using a milling machine (SFM–22, Shehui Co., Taoyuan, Taiwan). The powder was deposited directly onto the substrate surface through electrostatic spraying using a sprayer (PEM–X1, Wagner, Markdorf, Germany) before curing at 160–200 °C.

2.3. Measurement of Thermal Conductivity

The thermal conductivity was determined using a thermal conductivity meter (LFA447 NanoFlash, Netzsch, Selb, Germany). Thermocouples were attached to the surface of the specimens. The coating contained 3 wt% of either multilayer graphene, boron nitride, or without additive as the control. The thermal conductivity of the coating was calculated according to the following equation:

$$\frac{L_T}{k_T} = \frac{L_1}{k_1} + \frac{L_2}{k_2} \tag{1}$$

where L_1, L_2 and L_T are the thicknesses of the coating, the substrate and the total thickness, respectively, and k_1, k_2 and k_T are the thermal conductivities of the coating, the substrate, and the overall thermal conductivity, respectively. The thickness of the aluminum plate was 1 mm, whereas that of the coating was measured using a coating thickness meter (Qnix Qua Nix 4200P, Automation Dr. Nix GmbH & Co. KG, Cologne, Germany). The coating thickness was 40 μm.

2.4. Measurement of Thermal Emissivity

The thermal emissivity was measured using an infrared emissivity detector (ED01, Conjutek Co., New Taipei City, Taiwan) in the wavelength range of 2 to 22 μm.

2.5. Forced Convective Heat Transfer

The forced convective heat transfer of the coated and bare plates was performed according to the standard of AMCA 210-07. Figure 1 depicts the experimental setup for conducting forced convection. The heat supply was set either 8 W or 16 W. The plate was placed horizontally under a flow rate of 2 m/s. Temperatures were measured at four points on the bottom surface of the plate.

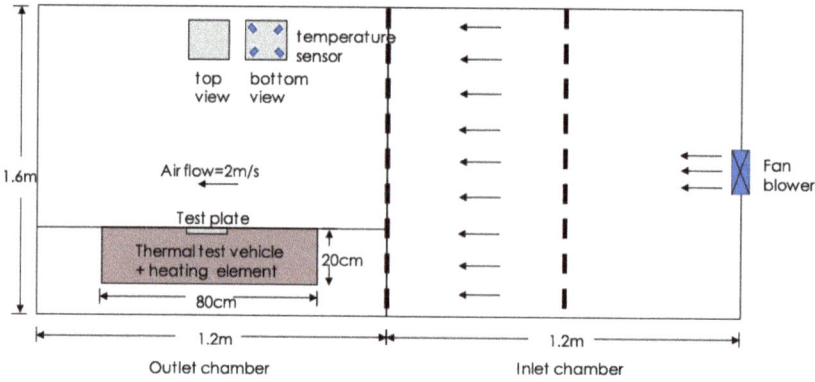

Figure 1. The experimental setup for conducting forced convection.

2.6. Natural Convective Heat Transfer

The natural convective heat transfer was performed by placing the plate horizontally as illustrated in Figure 2. The temperature was monitored until reaching steady state.

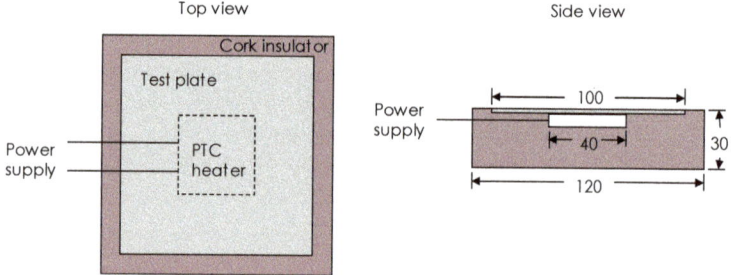

Figure 2. The experimental setup for conducting natural convection.

3. Results and Discussion

3.1. Characteristics of Graphene and Powder Coating

Table 3 and Figure 3 show the characteristics of the graphene obtained from the supplier. From the Raman spectrum of graphene, there are three distinct absorption peaks: D peak at 1353 cm^{-1}, G peak at 1581 cm^{-1}, and 2D peak at 2720 cm^{-1}. The I_D/I_G is about 0.05 and the I_{2D}/I_G is about 0.36, indicating that this is multilayer graphene. The AFM image shows that the horizontal dimension of the graphene sheet is between 3–25 µm.

Table 3. Characteristics of graphene nanoparticles.

Item	Properties	Test Method
appearance	Black Granules	visual
lateral size (µm)	3–25	particle analyzer
number of layers	6–10	AFM
carbon content (%)	>99.5	X-ray photo-electronic spectroscopy
oxygen content (%)	<0.1	X-ray photo-electronic spectroscopy
water adsorption content (%)	≤0.5	ASTM D570–2005
bulk density (g/cm^3)	0.03–0.1	powder densitometer
true density (g/cm^3)	2.25	density tester
specific surface area (m^2/g)	25–50	specific surface area tester

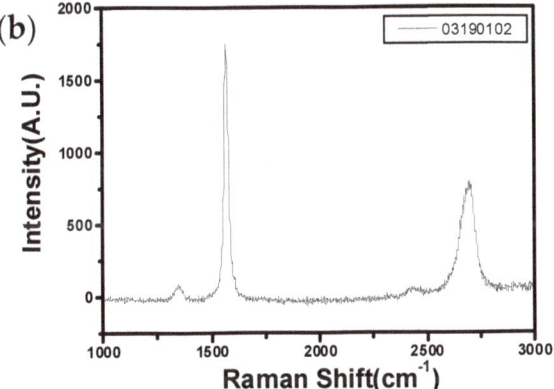

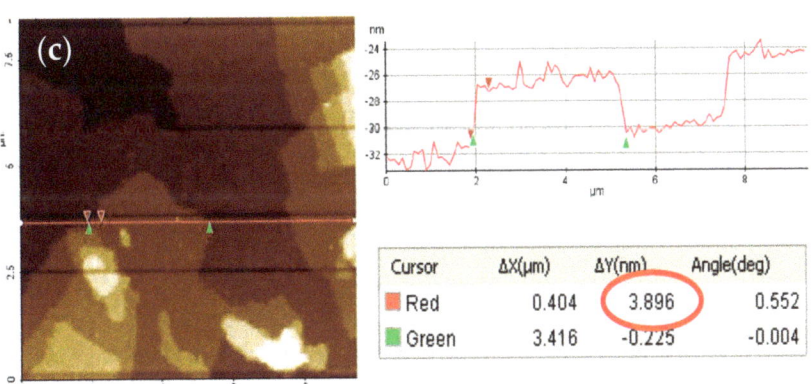

Figure 3. The characteristics of graphene nanoparticles. (**a**) SEM image; (**b**) Raman spectrum; (**c**) AFM image.

Figure 4 shows the SEM image of the cross section of graphene-loaded coating as well as the EDS images of carbon and oxygen. These images indicated that graphene nanoparticles were well distributed in the coating matrix. Furthermore, Table 4 shows that the carbon content in the coating with graphene was slightly higher than that in the pristine coating, indicating the presence of graphene. Some micro-scale aggregates were observable in Figure 4a. Similar observation was also reported in the literature [12]. This may affect the thermal conductivity of the coating, however, it is out of scope of this study.

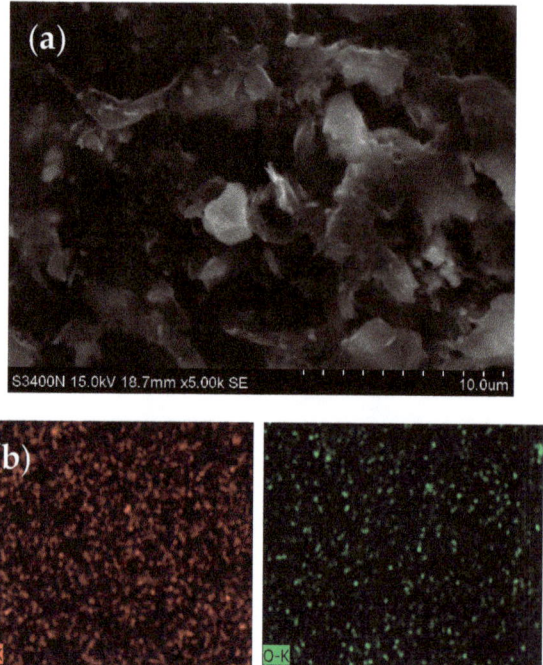

Figure 4. The SEM image of the graphene-loaded coating. (a) The micrograph of cross-section; (b) EDS images of carbon and oxygen.

Table 4. Atomic compositions of the coatings with or without graphene from EDS results.

Element	Pristine Coating	Graphene-Loaded Coating
C (mol%)	75.7	77.4
O (mol%)	24.3	22.6

Graphene loaded nanocomposites have been considered for thermal managements. There are several reviews regarding the thermal conductivity of graphene-polymer composites [13–15]. In recent years, graphene and expanded graphite have been widely studied as nanofillers for polymer composites, as thermal interface materials and heat sinks [16–19]. In addition to the extremely high thermal conductivity of single-layer graphene, two-dimensional morphology also makes graphene more conducive, thus improving heat transfer performance. The thermal conductivity of graphene-polymer composites is affected by factors including loading, graphene orientation, and interface [20]. Graphene exhibits a very high specific surface area leading to large interface with the polymer chains, and causing phonon scattering and hence ultra-high interface thermal resistance. Therefore, heat is difficult to transfer through the graphene-polymer interface. In addition, when the loading of graphene is above the percolation threshold, the thermal conductivity of this composite would be increased significantly. When the orientation of graphene is in the direction of heat flow, facilitating the formation of thermal conductive channel and hence improve the thermal conductivity. However, in this study, the powder was deposited onto the substrate through electrostatic spraying, thus these graphene nanosheets were randomly oriented.

This study chose thermoset powder coating as the research object. A thermoset resin is used as the film forming material, and a hardener with a crosslinking reaction is added to form an insoluble, non-melting hard coating after heating. Such a coating would not soften like thermoplastic coating even at elevated temperatures; it can only fracture. Since the resin used in the thermoset powder coating is a low molecular weight pre-polymer with a low degree of polymerization, it has good leveling and decorative properties. Moreover, this low molecular weight pre-polymer can be crosslinked into 3D network after curing, endowing the coating good corrosion resistance and mechanical properties. This has led to rapid development of the thermoset powder coating.

3.2. Thermal Conductivity

Table 5 shows the thermal conductivities of the coated and uncoated aluminum plates. The overall thermal conductivity was reduced from 196.7 W/m-K of the bare aluminum to 88.2 W/m-K of the epoxy/BN coated aluminum plate. This indicates that the coating on the surface can impair the heat conduction. This may appear to violate the purpose of improving thermal dissipation. However, the heat generated from the electronic elements dissipates to the ambient through not only conduction but also convection and radiation. In the subsequent sections, the coating actually did facilitate the dissipation of the heat.

Table 5. Thermal conductivity of the aluminum plates with or without coating *.

Sample K (W/m-K)	Al (Bare Aluminum Plate)	EPC (Epoxy-Polyester Coating)	EBN (Boron Nitride-Loaded Coating)	EGR (Graphene-Loaded Coating)
Overall	196.7	79.5	88.2	165.0
Coating	-	5.0	6.0	33.3

* $T = 25\ °C$, Light voltage = 250 V, pulse width = 0.02 ms, model = Cowan.

The thermal conductivity of the coating in Table 5 was calculated from the overall thermal conductivity according to Equation (1). Three types of coating were measured: pristine epoxy-polyester coating (EPC), BN-loaded (EBN) and graphene-loaded (EGR) epoxy-polyester coating. The thermal conductivity of the BN-loaded coating was slightly higher than that of the pristine epoxy coating. On the other hand, the loading of graphene improved the thermal conductivity of the coating to above 6 folds. This is reasonable since graphene is well-known for high thermal conductivity. Because the pristine epoxy-polyester coating exhibited low thermal conductivity, this coating was not studied further in the subsequent heat transfer experiments. Only Al, EBN and EGR were employed in the heat transfer tests.

3.3. Thermal Emissivity

Table 6 shows the emissivity of the samples in the wavelength range from 2 to 22 µm. In general, the emissivity values of metals are low while those of polymers are much higher. In this study, EBN coating appears white, whereas EGR coating appears black.

Table 6. The emissivities of aluminum plate and two types of coatings *.

Test Item	Al	EBN	EGR
Emissivity, ε	0.07	0.40	0.88

* $T = 25\ °C$, test time = 3 s.

3.4. Forced Convective Heat Transfer

In order to investigate the role of radiation heat transfer in the thermal dissipation performance of coating, the aluminum plates were subject to heat transfer experiments under natural convection and forced convection.

Table 7 summarizes the results of heat transfer under forced convection. For a small object in a big room, the radiative heat flux was calculated according to the Stefan-Boltzmann Law: [21]

$$q_r = \varepsilon\sigma(T_s^4 - T_a^4) \tag{2}$$

where q_r is the radiative heat flux from the sample to the ambient, ε is the emissivity of the surface, σ is the Stefan-Boltzmann constant (5.67 × 10⁸ W/m/K⁴), and T_s and T_a are respectively the surface temperature and ambient temperature (in K). The convective heat flux (q_c) equals the total heat flux (q_t) minuses the radiative heat flux. The radiative heat transfer ratio is q_r/q_t. For bare aluminum plate, because of low emissivity, the radiative heat transfer ratio was 1.7–1.9%. However, for aluminum plates coated with epoxy-polyester resin loaded with BN or graphene, the radiative heat transfer ratio increased to 8.9–9.4% and 15.9–16.6%, respectively. These additional heat flux would improve the heat dissipation, making the surface temperature lower, thus the heating source (e.g., IC or LED) would be cooler. Indeed, the surface temperature for EGR were 7 °C and 13 °C lower than those for bare aluminum when the heat flux was respectively 800 and 1600 W/m².

Figure 5 shows that the convective heat flux depends linearly with the temperature difference. The slope (28.456 W/m²K) is the convective heat transfer coefficient under this specific test condition. The coefficient of determination (R^2) was 0.996, indicating that this correlation fits very well to the experimental results. We can use this value to predict the heat dissipation rate at other heat flux at the same air flow speed. Furthermore, the heat transfer coefficient is independent on the substrate, whether it is bare aluminum or coated with a layer of polymer coating.

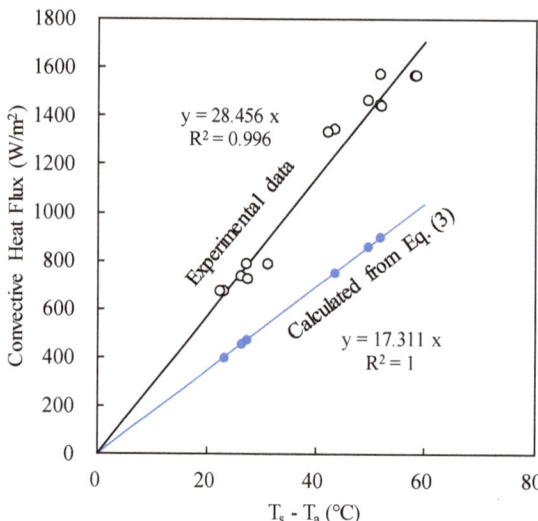

Figure 5. The linear correlation between convective heat flux and temperature difference.

The Reynolds number Re (= uL/ν) for this test condition was around 1.2 × 10⁴, less than 5 × 10⁵, suggesting the air flow was laminar. For laminar forced convection, the heat transfer coefficient based on boundary layer model is as follows

$$h_L = 0.664 Pr^{1/3} Re_L^{1/2} \left(\frac{k}{L}\right) \qquad (3)$$

where L is the length of the plate, k is the thermal conductivity of air, Pr is the Prandtl number of the air, Re is the Reynolds number of the air stream, u is the speed of the air stream, and v is the kinematic viscosity the air. The resultant convective heat flux was then calculated as

$$q_{fc} = h_L(T_s - T_a) \qquad (4)$$

The calculated results were presented in Figure 5 as well. However, the heat transfer coefficient (the slope) was only 60% of the experimental results. This probably is due to the turbulence in the actual measuring environment, which would accelerate heat transfer.

3.5. Natural Convective Heat Transfer

In addition to forced convection, natural convection is the other path for heat dissipation. Table 8 summarizes the results of heat transfer under natural convection. All the conditions were the same as in Section 3.4, except there was no air flowing on the surface. The temperature difference was higher that its counterpart in Table 7, suggesting that natural convection is slower than forced convection in heat dissipation. Furthermore, because of higher surface temperature, the radiative heat flux in natural convection was higher than in forced convection. Consequently, the convective heat flux in natural convection was lower than in forced convection, reflecting the slower heat dissipation in natural convection. The order of the radiative heat transfer ratio was the same as in Table 7, that is, EGR > EBN > Al. This order is the same as that of the emissivity, suggesting that graphene-loaded coating can enhance heat dissipation.

Table 7. The heat transfer rates by convection and radiation under forced convection *.

Surface	Al	Al	EBN	EBN	EGR	EGR
Total heat flux, q_T (W/m^2)	800	1600	800	1600	800	1600
Temperature difference, ΔT (°C)	29.9 ± 2.3	56.3 ± 3.8	27.2 ± 0.8	51.2 ± 1.3	22.6 ± 0.4	42.7 ± 0.7
Radiative heat flux, q_r (W/m^2)	14 ± 2	30 ± 3	71 ± 4	151 ± 10	127 ± 2	266 ± 4
Convective heat flux, q_c (W/m^2)	786 ± 2	1570 ± 3	728 ± 4	1450 ± 11	673 ± 2	1336 ± 6
Radiative heat transfer ratio, %	1.7 ± 0.2	1.9 ± 0.2	8.9 ± 0.6	9.4 ± 0.6	15.9 ± 0.2	16.6 ± 0.2

* RH = 76.2%, P_{amb} = 747.5 mm Hg, air flow rate = 2 m/s.

Table 8. The heat transfer rates by convection and radiation under natural convection.

Surface	Al	Al	EBN	EBN	EGR	EGR
Total heat flux, q_T (W/m^2)	800	1600	800	1600	800	1600
Temperature difference, ΔT (°C)	55.4 ± 1.0	98.5 ± 1.2	45.2 ± 2.6	79.0 ± 2.1	41.0 ± 3.2	69.6 ± 3.2
Radiative heat flux, q_r (W/m^2)	31 ± 1	67 ± 1	135 ± 8	278 ± 8	263 ± 22	515 ± 29
Convective heat flux, q_c (W/m^2)	767 ± 2	1538 ± 2	673 ± 1	1321 ± 5	538 ± 13	1086 ± 25
Radiative heat transfer ratio, %	3.8 ± 0.1	4.1 ± 0.1	16.7 ± 0.9	17.4 ± 0.5	32.8 ± 2.3	32.2 ± 1.7

Natural convection is a result of the motion of the fluid due to density changes arising from the heating. In this study, the heated plate was placed horizontally, inducing an upward air stream. The flow pattern is complicate. No reliable empirical correlation is capable to predict the heat transfer. Therefore, we construct an empirical correlation of convective heat flux vs temperature difference. Because the aluminum plate has a low emissivity, the aluminum plate was used to measure the surface temperature for a series of total heat fluxes. The convective heat flux was obtained by subtracting the radiative heat flux from the total heating flux. Figure 6 shows that the convective heat flux depends on the temperature difference. Linear regression yielded a quadratic correlation with R^2 equals to 0.981.

$$q_c = 0.0369(\Delta T)^2 + 12.27\Delta T \tag{5}$$

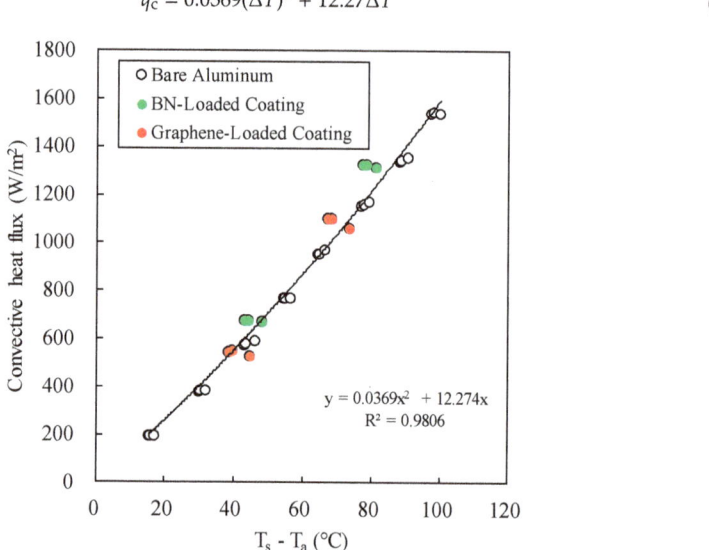

Figure 6. The convective heat fluxes of coated and uncoated aluminum plates under natural convection.

3.6. Heat Transfer Coefficients

Table 9 summarizes heat transfer coefficients calculated from the experimental results in Tables 7 and 8. Heat transfer coefficient is the measure of heat dissipation. Among these heat transfer coefficients, the total heat transfer coefficient (h_T) was calculated as follows:

$$h_T = q_T/\Delta T \tag{6}$$

and the convective heat transfer coefficient (h_c) and the radiative heat transfer coefficient (h_r) were calculated respectively as follows:

$$h_c = q_c/\Delta T \tag{7}$$

$$h_r = q_r/\Delta T \tag{8}$$

where ΔT is the temperature difference between the surface temperature and the ambient temperature.

These three heat transfer coefficients are affected by three factors: type of convection, surface coating, and total heat flux. The weight of each factor on each coefficient can be evaluated statistically with analysis of variance (ANOVA).

Table 9. Comparison of heat transfer coefficients in forced and natural convection.

Surface	Forced Convection								Natural Convection							
	AL	EBN	EGR	AL	EBN	EGR	AL	EBN	EGR	AL	EBN	EGR	AL	EBN	EGR	
q_T (W/m^2)	800	800	800	1600	1600	1600	800	800	800	1600	1600	1600				
h_T (W/m^{2}K)	26.8 ± 2.1	29.4 ± 0.8	35.4 ± 0.6	28.5 ± 2.0	31.3 ± 0.8	37.5 ± 0.6	14.4 ± 0.2	17.9 ± 0.9	19.6 ± 1.3	16.3 ± 0.2	20.2 ± 0.5	23.0 ± 1.0				
h_c (W/m^{2}K)	26.4 ± 2.1	26.8 ± 0.9	29.7 ± 0.4	28.0 ± 2.0	28.3 ± 0.9	31.3 ± 0.4	13.8 ± 0.2	14.9 ± 0.9	13.2 ± 1.3	15.6 ± 0.2	16.7 ± 0.5	15.6 ± 1.0				
h_r (W/m^{2}K)	0.46 ± 0.02	2.62 ± 0.09	5.63 ± 0.17	0.53 ± 0.02	2.95 ± 0.11	6.22 ± 0.18	0.55 ± 0.00	2.98 ± 0.01	6.41 ± 0.03	0.68 ± 0.00	3.52 ± 0.01	7.40 ± 0.07				
h_r/h_T (%)	1.7 ± 0.2	8.9 ± 0.6	15.9 ± 0.2	1.9 ± 0.2	9.4 ± 0.6	16.6 ± 0.2	3.8 ± 0.1	16.7 ± 0.9	32.8 ± 2.3	4.1 ± 0.1	17.4 ± 0.5	32.2 ± 1.7				

3.6.1. Total Heat Transfer Coefficient

Table 10 presents the results of ANOVA for total heat transfer coefficient. The results show that all three factors significantly affect h_T. Among these factors, the type of convection was the most influential while q_T was the least.

Table 10. Results of ANOVA for total heat transfer coefficient.

Source	SS	df	MS	F	p-Value	sig
Convection	1504.15	1	1504.15	835.96	5.5×10^{-24}	yes
Surface	326.84	2	163.42	90.82	1.1×10^{-13}	yes
q_T	45.11	1	45.11	25.07	2.1×10^{-5}	yes
Error	55.78	31	1.8			
Total	1931.88	35	55.20			

Figure 7 shows that the total heat transfer coefficient of the forced convection was about twice of that of the natural convection. This reflects the fact that forced convection can remove heat faster than natural convection. Furthermore, bare aluminum surface exhibited lower h_T and h_r than the other two coated surfaces. This can be attributed to the faster radiative heat transfer from coated aluminum plates, and that graphene-loaded coating exhibited higher h_T than other surfaces, since the emissivity of EGR was much higher than others. Figure 7 also shows that higher total heat flux (q_T) led to higher h_T for each surface. In forced convection, the increase was at most 6%, whereas in natural convection, the increase jumped to 17%. However, the effect of q_T was less than the effect of the surface, which is consistent with ANOVA.

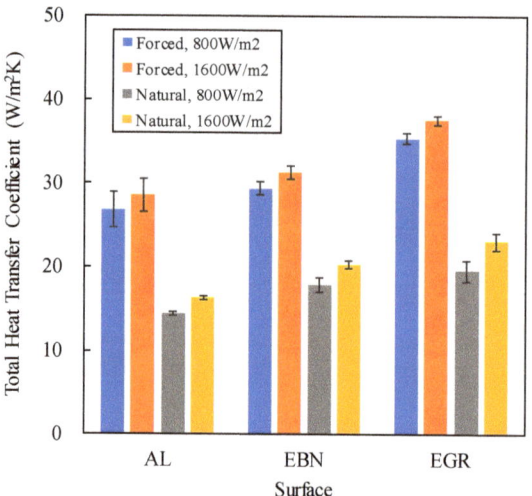

Figure 7. Effect of surface type on the total heat transfer coefficient.

3.6.2. Convective Heat Transfer Coefficient

Table 11 shows that the major factor affecting h_c was the type of convection. Figure 8 also shows that the h_c of forced convection was about twice of that of natural convection This is expected because h_c is the "convective" heat transfer coefficient. The type of surface coating affects less significantly to h_c. This is obvious because thermal radiation depends only on the temperature difference and would not affect the air flow.

The ANOVA results indicated that q_T was the minor factor for h_c. This is supported in Figure 8 that higher q_T led to slightly higher h_c. In forced convection, according to Equation (3), the convective heat transfer coefficient is proportional to the thermal conductivity of the air, which increases with the temperature. Because the surface temperature increased with the total heat flux, leading to higher thermal conductivity and hence higher h_c. However, the increase in h_c was small, thus the slope of q_c in Figure 5 was a constant, suggesting a constant h_c.

In natural convection, Figure 6 shows that q_c is a quadratic function of ΔT, thus h_c is a linear function of ΔT:

$$h_c = 0.0369(\Delta T) + 12.27 \tag{9}$$

However, the prefactor 0.0369 was small, making a weak dependency of h_c on ΔT.

Table 11. Results of ANOVA for convective heat transfer coefficient.

Source	SS	df	MS	F	p-Value	sig
Convection	1620.06	1	1620.06	809.34	0.0000	yes
Surface	13.35	2	6.68	3.34	0.0487	yes
q_T	28.98	1	28.98	14.48	0.0006	yes
Error	62.05	31	2			
Total	1724.45	35	49.27			

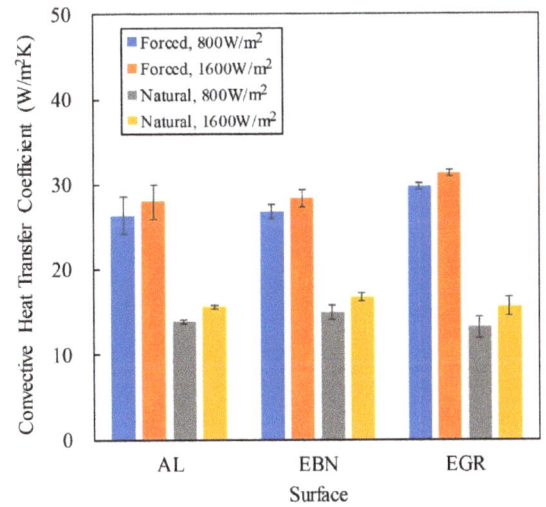

Figure 8. Effect of surface type on the convective heat transfer coefficient.

3.6.3. Radiative Heat Transfer Coefficient

Table 12 shows the ANOVA results and that for h_r, the major factor is the surface coating and the minor factor is the type of convection. The total heat flux affected the least the radiative heat transfer. The radiative heat transfer increased with the emissivity of the surface. In this study, the emissivity varied greatly, ranging from 0.07 for aluminum, 0.4 for BN-loaded coating, to 0.88 for graphene-loaded coating. Thus, the effect of emissivity on h_r is significant. Figure 9 also shows this effect. The type of convection affected h_r through T_s and T_a, because h_r can be calculated as follows:

$$h_r = \sigma \varepsilon \left(T_s^2 + T_a^2 \right) (T_s + T_a) \tag{10}$$

The surface temperature was lower for forced convection because of higher h_c.

Figure 9 summarizes the effect of surface coating on h_r. The difference between convection types was less than that between surfaces. The effect of q_T was further lower than the effect of convection.

Table 12. Results of ANOVA for radiative heat transfer coefficient.

Source	SS	df	MS	F	p-Value	sig
Convection	2.439	1	2.439	34.99	1.6×10^{-6}	yes
Surface	207.8	2	103.9	1490.46	1.6×10^{-31}	yes
q_T	1.756	1	1.756	25.18	2.0×10^{-5}	yes
Error	2.161	31	0.07			
Total	214.154	35	6.119			

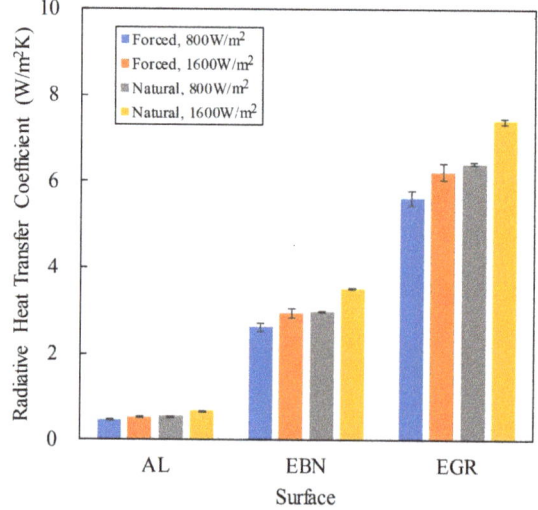

Figure 9. Effect of surface type on the radiative heat transfer coefficient.

4. Conclusions

The transfer of heat from the source (IC, LED, etc.) to the sink (ambient) involves both heat convection and heat conduction. There is another route for heat dissipation occurring in the ambient, that is, radiation, as long as the surface temperature is different to the ambient temperature. In nature and in engineering, the natural cooling or heating of objects is achieved by natural convection heat transfer. The intensity of natural convection heat transfer is weak, especially in the air environment, with radiation heat transfer of the same order of magnitude. At relatively high temperatures, the intensity of radiative heat transfer is much stronger than that of natural convective heat transfer. Therefore, in the actual calculation of natural convective heat transfer, radiative heat transfer should not be neglected.

In this study, graphene nanoparticles were blended into epoxy-polyester powder. Aluminum plate was then coated with aforementioned powder blends. For comparison, BN-loaded coating plates were also prepared. The thermal conductivity of the coating was improved from 5 W/m·K to 6 and 33.3 W/m·K for the BN- and graphene-loaded coating, respectively. The performance of heat dissipation of the resulting plates was further investigated under forced and natural convection. Under the forced convection, the radiative heat transfer coefficient (h_r) of the bare Al plate took about 1.8% of the total heat transfer coefficient (h_T), whereas for the graphene-loaded coating, h_r took about 16% of h_T. Therefore, radiative heat transfer is not negligible in heat dissipation through forced convection.

Under the natural convection, the h_r of bare Al plate was about 4% of h_T, while the h_r/h_T of graphene-loaded coating was about 33%, indicating that the thermal radiation cannot be ignored in the dissipation through natural convection.

The heat dissipation in this study showed that thermal radiation is a non-negligible route under either forced convection or natural convection. Based on this finding, a thin layer of graphene-loaded coating with a high emissivity can improve the heat dissipation performance of metal substrate.

Author Contributions: F.K., prepared experiments, as well as wrote the original draft; M.-C.Y., supervised the research project and finalized the manuscripts. All authors have read and agreed to the published version of the manuscript.

Funding: This research received no external funding.

Conflicts of Interest: The authors declare no conflict of interest.

References

1. Ganguli, S.; Roy, A.K.; Anderson, D.P. Improved thermal conductivity for chemically functionalized exfoliated graphene/epoxy composites. *Carbon* **2008**, *46*, 806–817. [CrossRef]
2. Maasilta, I.J.; Minnich, A.J. Heat under the microscope. *Phys. Today* **2014**, *67*, 27–32. [CrossRef]
3. Matsumoto, T.; Koizumi, T.; Kawakami, Y.; Okamoto, K.; Tomita, M. Perfect blackbody radiation from a graphene nanostructure with application to High-Temperature spectral emissivity measurements. *Opt. Express* **2013**, *21*, 30964–30974. [CrossRef] [PubMed]
4. Lim, M.; Lee, S.S.; Lee, B.J. Near-Field thermal radiation between Graphene-Covered doped silicon plates. *Opt. Express* **2013**, *21*, 22173–22185. [CrossRef] [PubMed]
5. Balandin, A.A.; Ghosh, S.; Bao, W.; Calizo, I.; Teweldebrhan, D.; Miao, F.; Lau, C.N. Superior thermal conductivity of Single-Layer Graphene. *Nano Lett.* **2008**, *8*, 902–907. [CrossRef] [PubMed]
6. Subrina, S.; Kotchetkov, D.; Balandin, A.A. Graphene heat spreaders for thermal management of nanoelectronic circuits. *IEEE Electron Device Lett.* **2009**, *30*, 1281–1284. [CrossRef]
7. Serov, A.Y.; Ong, Z.-Y.; Pop, E. Effect of grain boundaries on thermal transport in graphene. *Appl. Phys. Lett.* **2013**, *102*, 33104. [CrossRef]
8. Yan, J.; Yang, Y.; Chen, J.; Yu, K.; Yan, J.; Cen, K. Plasma-Enhanced chemical vapor deposition synthesis of vertically oriented graphene nanosheets. *Nanoscale* **2013**, *5*, 5180. [CrossRef]
9. Li, B.; Li, Z.; Zheng, B.; Sun, B.; Dai, G.C. Properties and interfacial treatment effect on thermal conductivity and electrical insulativity of the polymer composites. *J. East China Univ. Sci. Technol.* **2008**, *34*, 219–224.
10. Li, C.; Shi, G. Three-Dimensional graphene architectures. *Nanoscale* **2012**, *4*, 5549. [CrossRef] [PubMed]
11. Nardeccia, S.; Carriazo, D.; Ferrer, M.L.; Gutiérrez, M.C.; del Monte, F. Three dimensional macroporous architects and aerogels built of carbon nanotubes and/or graphene synthesis and applications. *Chem. Soc. Rev.* **2013**, *42*, 794–830. [CrossRef] [PubMed]
12. Marra, F.; D'Aloia, A.G.; Tamburrano, A.; Ochando, I.M.; De Bellis, G.; Ellis, G.; Sarto, M.S. Electromagnetic and dynamic mechanical properties of epoxy and Vinylester-Based composites filled with graphene nanoplatelets. *Polymers* **2016**, *8*, 272. [CrossRef] [PubMed]
13. Chen, H.; Ginzburg, V.V.; Yang, J.; Yang, Y.; Liu, W.; Huang, Y.; Du, L.; Chen, B. Thermal conductivity of Polymer-Based composites: Fundamentals and applications. *Prog. Polym. Sci.* **2016**, *59*, 41–85. [CrossRef]
14. Huang, X.; Jiang, P.; Tanaka, T. A review of dielectric polymer composites with high thermal conductivity. *IEEE Electr. Insul. Mag.* **2011**, *27*, 8–16. [CrossRef]
15. Han, Z.; Fina, A. Thermal conductivity of carbon nanotubes and their polymer nanocomposites: A review. *Prog. Polym. Sci.* **2011**, *36*, 914–944. [CrossRef]
16. Lin, Y.-F.; Hsieh, C.-T.; Wai, R.-J. Facile synthesis of graphene sheets for heat sink application. *Solid State Sci.* **2015**, *43*, 22–27. [CrossRef]
17. Im, H.; Kim, J. Thermal conductivity of a graphene Oxide–Carbon nanotube hybrid/epoxy composite. *Carbon* **2012**, *50*, 5429–5440. [CrossRef]
18. Li, Q.; Guo, Y.; Li, W.; Qiu, S.; Zhu, C.; Wei, X.; Chen, M.; Liu, C.; Liao, S.; Gong, Y.; et al. Ultrahigh thermal conductivity of assembled aligned multilayer graphene/epoxy composite. *Chem. Mater.* **2014**, *26*, 4459–4465. [CrossRef]
19. Song, S.H.; Park, K.H.; Kim, B.H.; Choi, Y.W.; Jun, G.H.; Lee, N.J.; Kong, B.-S.; Paik, K.-W.; Jeon, S. Enhanced thermal conductivity of Epoxy-Graphene composites by using Non-Oxidized graphene flakes with Non-Covalent functionalization. *Adv. Mater.* **2012**, *25*, 732–737. [CrossRef] [PubMed]

20. Li, A.; Zhang, C.; Zhang, Y.-F. Thermal conductivity of Graphene-Polymer composites: Mechanisms, properties, and applications. *Polymers* **2017**, *9*, 437.
21. Holman, J.P. *Heat Transfer*, 10th ed.; McGraw Hill Higher Education: Boston, MA, USA, 2010.

© 2020 by the authors. Licensee MDPI, Basel, Switzerland. This article is an open access article distributed under the terms and conditions of the Creative Commons Attribution (CC BY) license (http://creativecommons.org/licenses/by/4.0/).

Article

Synergistic Effects of Black Phosphorus/Boron Nitride Nanosheets on Enhancing the Flame-Retardant Properties of Waterborne Polyurethane and Its Flame-Retardant Mechanism

Sihao Yin [1,2,3], Xinlin Ren [1,2,3,*], Peichao Lian [1,2,3], Yuanzhi Zhu [1,2,3] and Yi Mei [1,2,3,*]

1. Faculty of Chemical Engineering, Kunming University of Science and Technology, Kunming 650500, China; yinsihao1213@126.com (S.Y.); lianpeichao@126.com (P.L.); yuanzhi_zhu@kust.edu.cn (Y.Z.)
2. The Higher Educational Key Laboratory for Phosphorus Chemical Engineering of Yunnan Province, Kunming University of Science and Technology, Kunming 650500, China
3. Yunnan Provincial Key Laboratory of Energy Saving in Phosphorus Chemical Engineering and New Phosphorus Materials, Kunming 650500, China
* Correspondence: ren8877@126.com (X.R.); meiyi412@kmust.edu.cn (Y.M.); Tel.: +86-138-8855-1958 (Y.M.)

Received: 4 February 2020; Accepted: 2 July 2020; Published: 3 July 2020

Abstract: We applied black phosphorene (BP) and hexagonal boron nitride (BN) nanosheets as flame retardants to waterborne polyurethane to fabricate a novel black phosphorus/boron nitride/waterborne polyurethane composite material. The results demonstrated that the limiting oxygen index of the flame-retarded waterborne polyurethane composite increased from 21.7% for pure waterborne polyurethane to 33.8%. The peak heat release rate and total heat release of the waterborne polyurethane composite were significantly reduced by 50.94% and 23.92%, respectively, at a flame-retardant content of only 0.4 wt%. The superior refractory performances of waterborne polyurethane composite are attributed to the synergistic effect of BP and BN in the gas phase and condensed phase. This study shows that black phosphorus-based nanocomposites have great potential to improve the fire resistance of polymers.

Keywords: black phosphorene; boron nitride; flame retardant; waterborne polyurethane

---

1. Introduction

Polymeric materials have been widely used in electronic devices, construction, and transportation. However, most of the polymeric materials have been intrinsically inflammable [1,2]. Therefore, flame-retardant additives are important to mitigate the risk of fire [3,4]. Some halogen-based flame retardants have been banned because they form carcinogens during combustion [5–7]. Compared to traditional halogenated flame retardants, phosphorus-containing flame retardants have attracted much attention due to the advantages of having low smoke and low halogen content as well as being non-toxic [8–10]. Phosphorus flame retardants can be divided into inorganic phosphorus flame retardants and organophosphorus flame retardants. Organophosphorus flame retardants have low cost and good compatibility with polymers, but they have high volatility and poor thermal stability [11]. Inorganic phosphorus flame retardants have high phosphorus content, high flame retardant efficiency and low toxicity, but their particles are usually large, thus resulting in poor compatibility with polymer materials and uneven dispersion [12]. For example, red phosphorus needs to be modified or coated to increase its compatibility with polymer materials [13,14].

Black phosphorus (BP) is a new kind of 2D layered material that is only composed of the phosphorus element [15–17]. Recently, BP has been demonstrated to be a good flame retardant [18–20].

The high specific surface area of the layered structure can result in an efficient barrier effect in the process of polymer combustion. Compared with volatile white phosphorus and amorphous red phosphorus, BP also exhibits higher thermal stability and phase-dispersion, which could improve the flame-retardant performance and reduce damage to the mechanical properties of the polymers. In previous research, we reported that BP can effectively enhance the thermal stability and fire resistance of polymers [21]. Yuan Hu, et al. synthesized the BP/carbon nanotube composite and demonstrated its synergistic flame retardant performance for epoxy resin [22]. However, some key indicators of flame-retardant property, such as the limit oxygen index (LOI), for the reported BP-based composite materials need to be further improved.

Hexagonal boron nitride (BN) nanosheet is a widely studied 2D material with high thermal stability, good mechanical strength, and large surface area. It has been found to be a good flame retardant for several polymers [23–25]. Considering the fact that the flame-retardant mechanism of BP materials has been mainly attributed to the formed radicals in the gas phase while the BN nanosheet mainly works in the condensed phase, the combination of BP and BN may have a good synergistic effect to further improve the flame-retardant performance and decrease the additive amount of the flame retardants.

Thus, in this paper, we designed a series of experiments to analyze the synergistic effect and flame-retardant mechanism of BP and BN by filling them into waterborne polyurethane (WPU). Through systematic characterization, we found that adding only 0.4 wt% BP/BN nanosheet could significantly raise the LOI of WPU from 21.7% to 33.8%. The improvement of flame retardancy involves the combination of the condensed phase and gas phase effects during combustion. In addition, such a small additive amount of BP/BN showed negligible color effect on the WPU (Figure 1), which broadens its real application range.

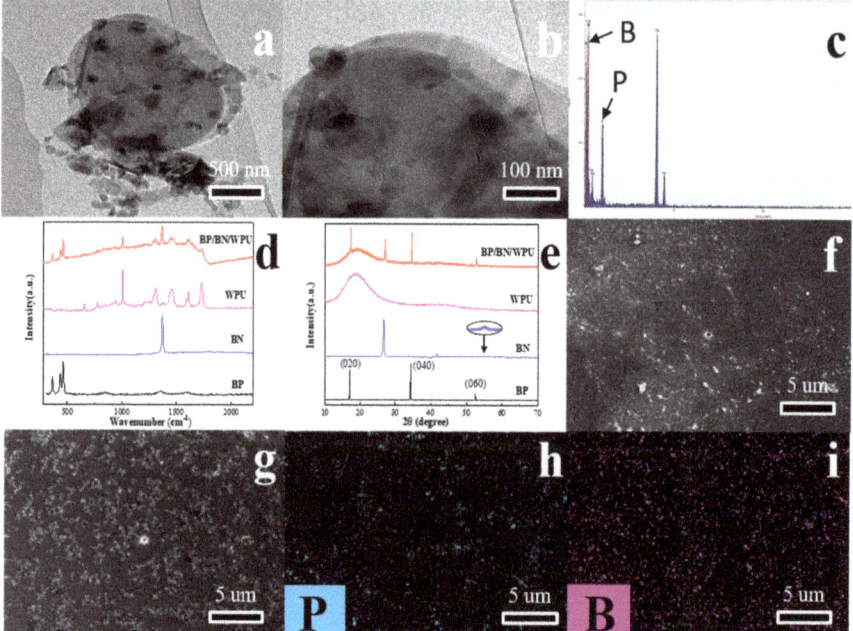

Figure 1. (**a**,**b**) TEM image of BP/BN nanosheets; (**c**) The EDS of BP/BN nanosheets; (**d**) Raman spectra of BP, BN, WPU and BP/BN/WPU; (**e**) XRD spectra of BP, BN, WPU and BP/BN/WPU; (**f**) and (**g**) SEM image of WPU and BP/BN/WPU; (**h**) phosphorus mapping in BP/BN/WPU and (**i**) boron mapping in BP/BN/WPU.

2. Materials and Methods

2.1. Materials

In this study, BP was prepared by a mineralization transformation method from red phosphorus with the help of iodine and tin in a quartz tube under argon atmosphere. The prepared BP crystals were washed with toluene to remove residual mineralizer, followed by washing with water and acetone. The utilized red phosphorus, iodine, tin, toluene and acetone were analytically pure. The BN was purchased from Shanghai Huayi Group Huayuan Fine Chemicals Co., Ltd., Shanghai, China with a particle size lower than 30 μm. The WPU latex with a solid content of 25 wt% was purchased from Anhui Huatai New Material co. Ltd., Hefei, China The water used was deionized water.

2.2. Fabrication of BP and BN Nanosheets

BP was ground for 2 h into powder, and then 0.5 g of the powder was added into a conical flask with 500 mL of deionized water. The conical flask was sealed with argon gas to stop oxidation of the BP. Then the dispersion was added to a working ultrasonic device (50 Hz, 200 W) for 24 h with the temperature controlled below 30 °C. Afterwards, the dispersed liquid was centrifuged at 3500 rpm for 15 min by a centrifugal machine (TGL 16C, Shanghai Anting Scientific Instrument Factory, Shanghai, China). Finally, the supernatant liquid was collected and condensed by suction filtration. An argon atmosphere should be used to prevent the oxidation of phosphorene throughout the whole experimental process. The dispersion of boron nitride was also obtained by liquid phase stripping. In order to calculate the concentration of the dispersion, freeze drying was used to remove the moisture. Then the remaining solid was weighed.

2.3. Preparation of BP/BN/WPU Composite Materials

The obtained BP and h-BN dispersion were added into a beaker with WPU. After stirring for a few minutes, the beaker was sealed by filling with an argon atmosphere. The mixture was ultrasonicated for 2 h at low temperature with an ice bath. Then the obtained suspension was poured onto a plate of the size of 120 × 120 mm and dried under vacuum at 22 °C. After it was dried completely, the BP/BN/WPU material was formed. The additive amount of BP/BN is 0.2%/0.2%. For comparison, the BP/WPU with 0.4% BP and BN/WPU with 0.4% BN were synthesized using the same method (Table 1).

Table 1. The additive amount of the BP and BN nanosheets in the samples.

Samples	Weight (g)	Percentage of BP (%)	Percentage of BN (%)	Content of BP (g)	Content of BN (g)
WPU	22.1	0	0	0	0
0.4BP/WPU	22	0.40	0	0.0880	0
0.4BN/WPU	22.4	0	0.40	0	0.0896
0.2/0.2BP/BN/WPU	22.2	0.20	0.20	0.0444	0.0444

2.4. Analytical Test

2.4.1. Structure Characterizations

Transmission electron microscopy (TEM, Philips CM100, Amsterdam, The Netherlands) was conducted to observe the BP and BN nanosheets at an acceleration voltage of 100 kV.

An X-ray diffraction device (XRD, PANalytical Empyrean, Almelo, The Netherlands) was used to analyze the BP, BN powders, as well as the WPU and its composite materials, respectively.

Raman spectra were obtained on a LabRAM-HR Confocal Raman Microprobe (HORIBA Scientific Co., Palaiseau, France) with excitation provided in backscattering geometry by a 633 nm argon laser line.

Scanning electron microscopy (SEM, Bruker Nano, Bruker, Karlsruhe, Germany) was used to analyze the microstructure of the materials and the charred residue. The WPU and its composite materials with BP and BN nanosheets were fractured by liquid nitrogen, then the fracture surfaces of the samples were uniformly coated with a layer of gold and then observed by SEM. In order to present the distribution of phosphorene in the polymers, elemental mapping tests were also conducted on another two specimens without a coating of a gold layer. In addition, An EDS test of the charred residue was applied to determine the content of BP and BN in the condensed phase.

X-ray photoelectron spectroscopy (XPS) analysis was performed using Al radiation as a probe (K-alpha, Thermo Fisher Scientific, Waltham, MA, USA) to measure the valence states and chemical composition of the residue of BP/BN/WPU.

2.4.2. Thermal Properties Measurement

Thermogravimetric analysis (TG) of the materials was performed on a thermal analyzer (NETZSCH STA449F3, NETZSCH, Selb, Germany) with a gas flow rate of 80 mL/min under nitrogen atmosphere. The heating rate was 10 °C/min and the temperature ranged from 40 °C to 800 °C.

Thermogravimetric analysis–Fourier transform infrared spectroscopy (TG–FTIR) was performed via a TGA/DSC3 thermogravimetric analyzer (METTLER TOLEDO, Greifensee, Switzerland) linked with a Nicolet FTIR IS50 spectrometer (Thermo Fisher) at a heating rate of 10 °C/min within a temperature range of 40 °C to 800 °C under a N_2 flow of 50 mL/min. The temperature of the transfer line in the TG–FTIR system was 200 °C.

2.4.3. Flammability Property Measurement

The limiting oxygen index (LOI) values were measured according to the standard oxygen index test ASTM D2863-77 by using the device of COI from Motis combustion technology co. LTD (Kunshan, China).

In order to study the combustion behavior, cone calorimetry (CC, PX-07-007, Suzhou phoenix quality inspection instrument co. LTD, Suzhou, China) was performed at a heat flux of 35 kW/m^2. The material specimens were made into a square with the size of 100 × 100 × 3 mm. After wrapping in a piece of aluminum film, the specimens were set on fire on the CC.

3. Results and Discussion

3.1. Characterization of BP/BN/WPU

The TEM images reveal the micromorphology of BP and BN nanosheets prepared by liquid phase stripping. As shown in Figure 1a,b, it can be clearly seen that the bulk black phosphorus and BN were peeled into a few layers of nanoflakes. We suggest that the BP nanosheets are arranged on the BN nanosheets surface, which were fully stripped with clear edges with a length of several micrometers. The Energy Dispersive Spectrometer (EDS) results (Figure 1c) confirmed the elements' distribution in the BP/BN nanosheets. Raman spectra and X-ray diffraction (XRD) were carried out for BP, BN powder, pure WPU and BP/BN/WPU composite, respectively. As shown in the Raman spectra (Figure 1d), the BP nanosheets show three characteristic peaks of A_g^1, B_g^2, and A_g^2, corresponding to the crystal orientation [26], thickness [27], and angle [28]. The BN shows the in-plane ring vibration peak of BN (E2g vibration mode) at 1365 cm^{-1} [29]. The BP/BN/WPU composite shows four new peaks in comparison with pure WPU, indicating the successful introduction of BP and BN nanosheets into the WPU matrix. The XRD spectrum of BP reveals three obvious diffraction peaks (Figure 1e), corresponding to the (020), (040), and (060) plane, respectively [30,31]. The spectrum of BN also has three obvious diffraction peaks, indicating the good crystallinity of BN. The XRD spectrum of BP/BN/WPU retains the main peaks of BP and BN, which is consistent with the Raman results.

The SEM images of pure WPU (Figure 1f) are observed on the fractured surface. It is shown to be homogeneous and without any additive particles due to the typical fracture behavior of a homogeneous material. Compared with the pure WPU, BP/BN/WPU composite shows several white nanosheets (Figure 1g), which could be attributed to the additives BP and BN. As shown in Figure 1h–i, both the P and B elements distribute uniformly over a large region for the BP/BN/WPU sample. These results indicated that the BP and BN nanosheets were uniformly distributed into the WPU.

3.2. Thermal Stability of BP/BN/WPU and Its Nanocomposite

The TG analysis of the specimens was carried out in a N_2 atmosphere to analyze the decomposition behavior in a real fire scenario. As shown in Figure 2a, the TG curves of pure BP have an obvious mass loss starting at 420 °C, while the BN only shows negligible mass loss over the whole temperature range. These results show that the BP turns into the gas phase after 420 °C, and the BN is very stable even at higher temperature. From Figure 2b,c, we can see that the WPU shows two major mass loss stages over the selected temperature range. This is caused by the difference in thermal stability between the hard segment and soft segment of WPU. The first mass loss stage is mainly caused by the breakage of urethane bonds in the amine and isocyanate while the second stage is assigned to the decomposition of residual polyols. In contrast to pure WPU, the BP/WPU, BN/WPU and BP/BN/WPU exhibited three main mass loss stages. The third stage of the composite occurred in the range of 440–500 °C, which indicates that the mass loss became slower at this temperature after the addition of BP and BN nanosheets. The residue char of BP/BN/WPU reached 10.11%, which is also higher than that of WPU (0.93%), BP/WPU (8.11%), and BN/WPU (6.44%). These results revealed that the BP and BN nanosheets can synergistically improve the thermal stability of WPU and promote the formation of residue char [32].

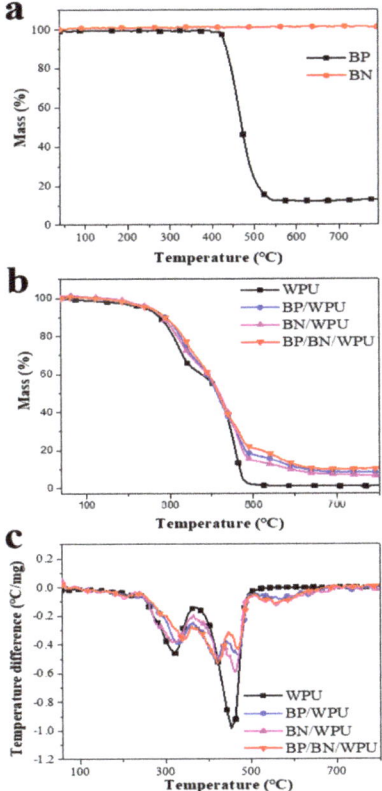

Figure 2. (a) the TGA curves of BP and BN; (b) the TGA curves of WPU, BP/WPU, BN/WPU and BP/BN/WPU; and (c) the resulted DTG curves of WPU, BP/WPU, BN/WPU and BP/BN/WPU.

3.3. Fire Safety Properties of BP/BN/WPU

The limiting oxygen index (LOI) test is used to determine the flammability of the samples. As depicted in Figure 3, the LOI of pure WPU is 21.8%, indicating that the polymeric matrix is a flammable material. Adding 0.4 wt% of BP or BN increased the LOI to 24.5% for 0.4%BP/WPU and 26.7% for 0.4%BN/WPU, respectively. Interestingly, simultaneous addition of 0.2% of BP and BN into WPU significantly increase the LOI to 33.8% and reached the V0 of fire resistance. That is to say, BP and BN have a synergistic effect in enhancing the self-extinguishing ability of WPU in a real fire scenario.

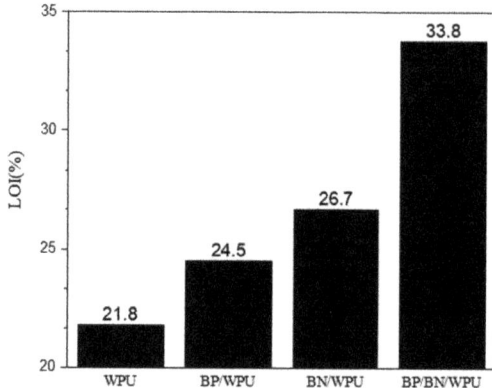

Figure 3. The additive amount of the flame retardant and values of the limit oxygen index (LOI) of the samples.

The cone calorimeter (CC) tests were applied to simulate a real fire scenario. The results are shown in Figure 4 and detailed data are given in Table 2. The total heat release (THR) and the peak of heat release rate (PHRR) of pure WPU are 84.20 MJ/m^2 and 452.5 kW/m^2, respectively, indicating that pure WPU released a large amount of heat and made it easy to cause the "flashover" phenomenon. The BP/BN/WPU exhibits the best flame-retardant properties, with a 50.94% decrease in PHRR (Figure 4a) and 23.92% decrease in THR (Figure 4b), respectively. The introduction of BP and BN nanosheets into WPU significantly reduced heat release and restricted the "flashover" phenomenon. The time to PHRR (TPHRR) and time to ignition (TTI) values of the materials are also shown in Table 2. The TPHRR of BP/BN/WPU arrived last of all. That is to say the burning time of the material was increased, which is a benefit for escaping, rescuing, and firefighting. The TTI of pure WPU and BN/WPU have bigger values than pure BN/WPU and BP/BN/WPU, which indicates that the addition of BP makes the polymer matrix easy to ignite. This phenomenon is a common feature of phosphorus-based flame retardants [33,34]. The reason may be that the addition of phosphorus results in the composite polymers having a lower decomposition temperature, leading them to release inflammable gases and ignite earlier. The average of the effective heat of combustion (av-EHC) reveals the volatiles dilution in the gas phase, and usually also discloses the inhibition of gas phase combustion by flame retardants. As shown in Table 2, the av-EHC of WPU and BN/WPU are not much different. This indicates that BN mainly plays a role in the condensed phase rather than the gas phase. Compared with WPU, the av-EHC of BP/WPU and BP/BN/WPU is much lower. The low av-EHC is due to the capture of gas phase radicals by black phosphorus during combustion.

The CO_2 release (Figure 4c) and the CO release (Figure 4d) curves indicate that CO_2 is the main released gas, with little difference in release amount during the burning process. However, the CO release amount of BP/WPU is the highest, and the CO/CO_2 ratio was increased due to added BP, which indicates that BP works in the gas phase with the ability to restrict the complete combustion of polymer.

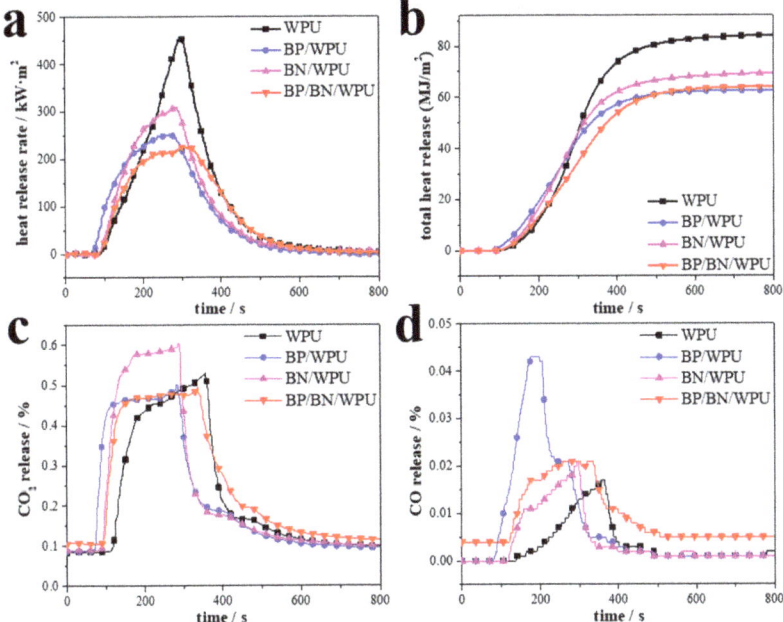

Figure 4. Cone calorimeter test of the samples: (**a**) HRR curves; (**b**) THR curves; (**c**) CO_2 release curves; (**d**) CO release curves.

Table 2. The cone calorimeter test data of WPU, G/WPU, and BP/G/WPU.

Sample	TTI (s)	TPHRR (s)	PHRR (kW/m^2)	THR (MJ/m^2)	Av-EHC (MJ/kg)	CO/CO$_2$	EFF	SE
WPU	65	297	452.5	84.20	31.07	0.021		
BP/WPU	40	257	252.7	62.57	29.29	0.056	499.5	1.15
BN/WPU	62	264	308.3	72.25	31.30	0.027	360.5	1.5
BP/BN/WPU	51	305	222.0	64.06	27.9	0.037	576.25	

The flame retardant effectivity (EFF) and synergistic effectivity (SE) are used to numerically evaluate the synergistic effect of multi-component flame retardant systems [35,36]. Flame retardant effectivity (EFF) and synergistic effectivity (SE) were calculated from CC data as follows: EFF = (PHRR$_{polymer}$ − PHRR$_{composite}$)/Flame-retardant content; SE = EFF$_{Flame-retardant}$ + synergists/EFF$_{Flame-retardant}$. BP/BN/WPU has the highest EEF value, indicating that compared with the single flame retardant, BP and BN synergistic flame retardant has the highest flame retardant efficiency. The SE values of BP/WPU and BN/WPU materials are 1.15 and 1.5, respectively, indicating that BP and BN can synergistically improve the flame retardancy of WPU.

3.4. Flame-Retardant Mechanism of BP/BN/WPU

3.4.1. Products Analysis of Carbon Residue

The residues after the CC tests were used to analyze the products after combustion, and the corresponding results are summarized in Table 3. The SEM images were used to observe the microstructure of the residues (Figure 5). The pure WPU left scarcely any residue (Table 3). Both the addition of BP and BN could enhance the residues of WPU, and the BP/BN/WPU further increase the residues up to 10.34%. This result was also confirmed by the digital images of residues in Figure 5a1–d1. Figure 5a2 shows the residue of the pure WPU presents a lot of large cracks and ~10 μm holes.

The enlarged image (Figure 5a3) shows that the residue of WPU is shaggy in shape with discontinuous particles. This structure usually facilitates heat transformation and the release of inflammable gas. After the addition of BP and BN, the BP/WPU and BN/WPU show obvious structure difference with relatively smaller holes and smoother surface (Figure 5b3,c3). The residue of BP/BN/WPU exhibited a completely dense surface without small holes, which indicates the good synergistic effect between the BP and BN nanosheets in triggering the catalytic carbonization and constructing a barrier.

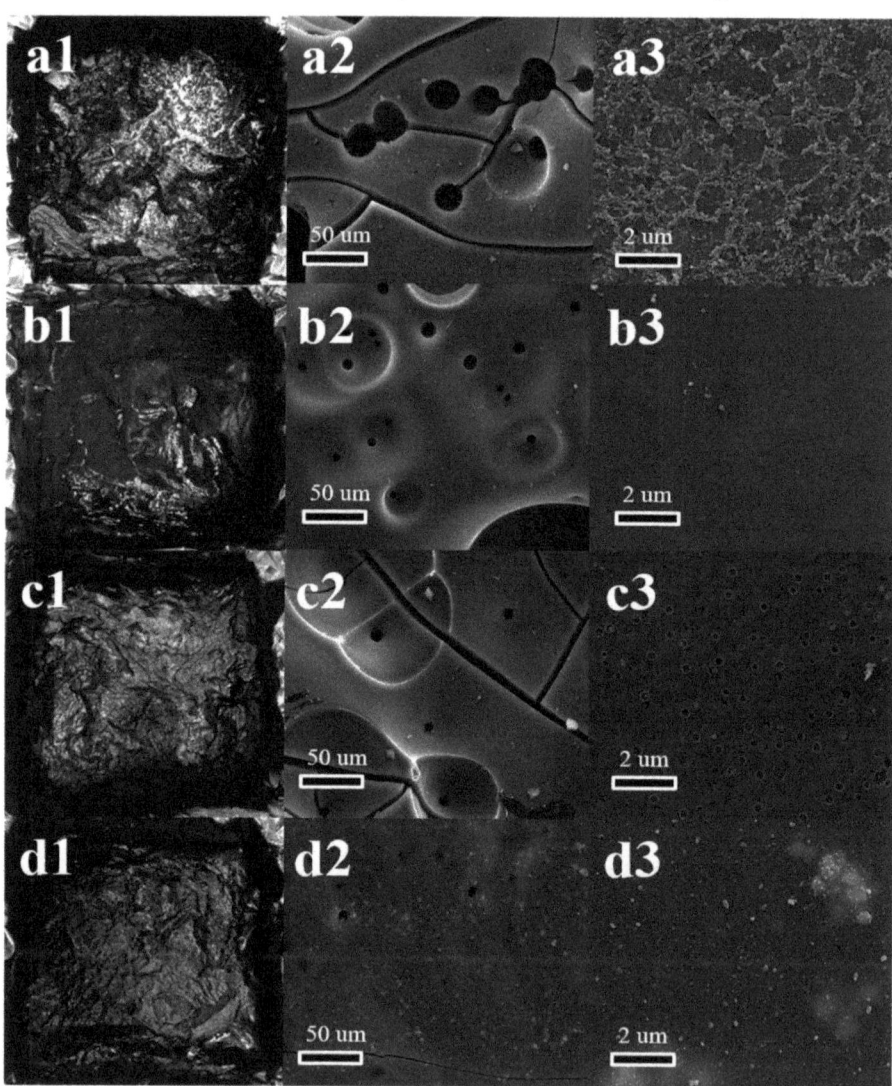

Figure 5. The images of the residues after the CC test: (**a1,b1,c1,d1**) the real picture of the pure WPU, BP/WPU, BN/WPU, BP/BN/WPU, respectively; (**a2,b2,c2,d2**) SEM image of the pure WPU, BP/WPU, BN/WPU, BP/BN/WPU, respectively; (**a3,b3,c3,d3**) the enlarged SEM image of the pure WPU, BP/WPU, BN/WPU, BP/BN/WPU.

Table 3. The content of BP and BN nanosheets in the residues after the CC test.

Residue Samples	Weight (g)	Total Residues (%)	Condensed Phase of P (%)	Condensed Phase of BN (%)	Gas Phase of P (%)	Gas Phase of BN (%)
WPU	0.20	0.1	0	0	0	0
0.4%BP/WPU	1.3	6.34	9.75	0	90.25	0
0.4%BN/WPU	2.1	7.38	0	66.96	0	33.04
0.2%/0.2%BP/BN/WPU	2.4	10.34	19.46	68.69	80.54	31.31

EDS analysis was carried out to confirm the different roles of BP and BN in the flame retardant. As shown in Table 3, the element P content in the residue of BP/WPU was only 0.0086 g, indicating that 90.25% of BP is consumed during combustion. The element B in the residue of BP/BN/WPU was 0.060 g. which means most BN was retained and functioned in the condensed phase. According to a previous study on the flame retardant properties of phosphorus [37], BP may also play a key role in the gas phase because it can be converted into P–O radicals and diffuse in the surrounding gas, which can then react with the H or OH radicals generated by polymers under the burning conditions, consequently reducing the energy of the flame. BN is stable and will accumulate in the condensed phase, which can form a physical barrier to reduce heat transfer and release of combustible gases [38–40].

In order to further analyze the component of BP in the residue after the CC test, XPS was conducted on the residue powder of BP/BN/WPU. The high-resolution P1s XPS of BP/BN/WPU (Figure 6a) can be deconvoluted into three peaks with binding energies at 132.3, 134.0 and 134.8 eV, corresponding to P–C, P–O–C and phosphoric anhydride (P_2O_5) [41], respectively. Figure 6b shows the high-resolution XPS spectra of B 1s, which was deconvoluted into two peaks with binding energies at 190.7 and 192.2 eV corresponding to the B–N, B–O–C. Boron element mainly exists in BN form in the residue, and a small amount of B forms chemical bonds with C. This result indicates that BP can be converted into phosphoric acid and phosphoric anhydride at high temperature, which can promote the polymers to produce a carbon layer. In addition, BN also play a catalytic role in carbon formation. The generated carbon layer and the amount of BN left behind will form an insulating layer to prevent the materials from contacting with oxygen and heat transferring, consequently weakening the fire [42,43].

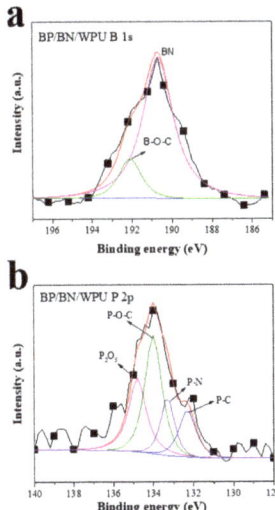

Figure 6. XPS spectra of the residue of BP/BN/WPU composite material after CC test; (a) high-resolution XPS spectra of B 1s of the residue of BP/BN/WPU composite material; (b) high-resolution XPS spectra of P 2p of the residue of BP/BN/WPU composite material.

3.4.2. The Product Analysis of the Gas Phase

In order to determine the pyrolytic mechanism, the decomposition process of BP/BN/WPU was analyzed by Thermogravimetric analysis–Fourier transform infrared spectroscopy (TG–FTIR). The 3D map of the FTIR spectra shows BP/BN/WPU releases CO_2 earlier than pure WPU (Figure 7a,b). The results indicated that the addition of BP/BN can release non-combustible gas in the initial thermal decomposition of WPU, which can dilute the oxygen from the air and dilute combustible gases from the material.

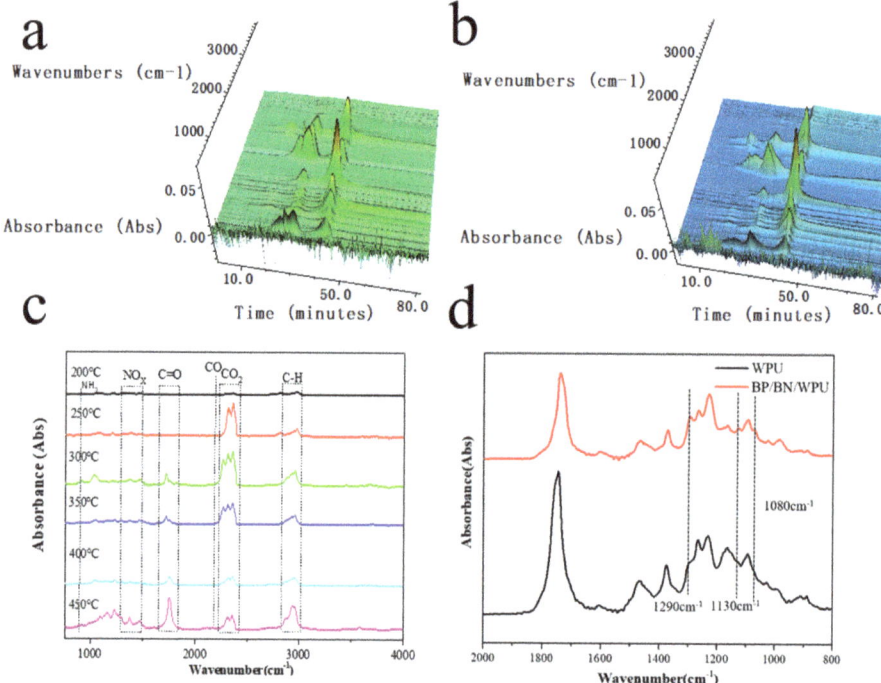

Figure 7. (a) 3D image of FTIR spectra of the WPU; (b) 3D image of FTIR spectra of the BP/BN/WPU; (c) The FTIR spectra of BP/BN/WPU obtained at different temperatures; (d) The enlarged image for comparison of the biggest mass loss in the second TG stages of the pure WPU and BP/BN/WPU.

To further study the thermal degradation process, the chemical structure changes of the BP/BN/WPU at different pyrolysis temperatures were investigated (Figure 7c). The changes in characteristic peaks reflect the detailed decomposition process of WPU and WPU composites. The peak at 2800–3100 cm^{-1} could be attributed to the stretching vibration of C–H [44,45]. The peaks at 2250–2400 cm^{-1}, 2190 cm^{-1} and 1650–1810 cm^{-1} are related to CO_2, CO, and carbonyl compound, respectively. The peak at 915 cm^{-1} corresponds to the bending vibration of the NH_3, and peaks at 1320–1550 cm^{-1} refer to the stretching vibration of NO_x (such as N_2O, NO and NO_2) [46], indicating WPU combustion has two main stages: Before the combustion, the WPU decomposed to give flammable gases such as hydrocarbon and carbonyl compounds. With the increase of temperature, a large amount of combustible gas was ignited and the WPU began to burn violently. The FT-IR spectra of the BP/BN/WPU sample (Figure 7d) show a significant enhancement at 1290, 1130 and 1080 cm^{-1} compared to WPU, which could be explained by the P=O, PO_2- and P–O groups [47,48]. The presence of phosphorus-containing gas should be attributed to the formation of phosphorus-containing radicals from the decomposition of BP, which demonstrates the significant role of BP in the gas-phase flame retardancy.

3.4.3. Synergistic Flame-Retardant Mechanisms

Based on the above flame-retardant performances and analyses, a possible flame retardant mechanism is proposed in Figure 8. At high temperature or in a real fire scenario, BP starts to decompose at about 240 °C. When the temperature reaches 420 °C or higher, most of the BP begins to enter the gas phase. The BP nanosheets work in both of the gaseous and condense phases for fire restriction. On the one hand, most of the BP will form radicals in the gas phase by absorbing surrounding oxygen and hydrogen atoms, which react with the pyrolytic radicals of the matrix polymers to inhibit the chain reaction. At the same time, the liberated incombustible gases will dilute the combustible gases and reduce their contact with oxygen. On the other hand, the residual BP will capture the oxygen in the polymers and transform it into phosphoric anhydride, which will promote the formation of a char layer and prevent the release of carbon-containing gas. BN nanosheets mainly remain in the condensed phase, synergistically catalyze carbon, and build a nano-barrier formation with black phosphorus due to the fact that they have good thermo-stability and a peculiar spatial three-dimensional network structure. Therefore, the high efficiency flame retardant performance of BP/BN can be attributed to the synergy effects of the gas phase mechanism of black phosphorus and the condensation phase mechanism of boron nitride during the combustion process.

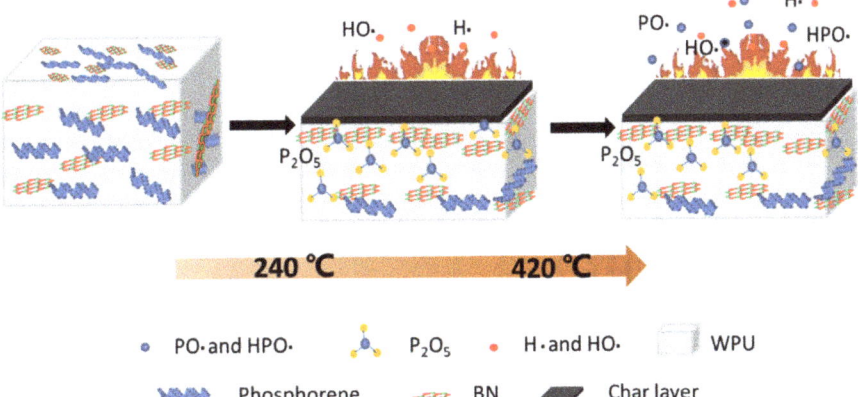

Figure 8. Schematic illustration of the flame-retardant mechanism.

4. Conclusions

BP and BN nanosheets were used as a flame retardant for WPU. The SEM and mapping results indicated that the BP and BN nanosheets were distributed uniformly in the matrix WPU. The flame-retardant tests demonstrated that the PHRR of WPU decreases by 50.94% and the THR decreases by 23.92% at a BP/BN content of only 0.4 wt%. The LOI of the BP/BN/WPU composite increased from 21.7% to 33.8%, compared with pure WPU. The residue of BP/BN/WPU after the CC test was denser and approximately 10 times more than the residue of pure WPU, according to TG, SEM, XPS, TG-IR analysis of WPU during the combustion process. Efficient flame retardancy is due to the synergistic effects of BP and BN. To conclude, BP/BN nanosheets can provide a good flame retardant effect with very low addition amount, and have a good application prospect in the field of flame retardancy.

Author Contributions: Conceptualization: X.R., Y.M., P.L., and S.Y.; formal analysis: X.R.; investigation, S.Y. and X.R.; project administration: S.Y.; resources: Y.M.; supervision: Y.M.; writing—original draft: S.Y. and X.R.; writing—review and editing: Y.Z. and P.L. All authors have read and agreed to the published version of the manuscript.

Funding: This work was supported by the Cultivation Fund for Yunling (No. 10978195, Yunnan, China), the Higher Educational Key Laboratory for Phosphorus Chemical Engineering of Yunnan Province (No. 14058501, Yunnan, China), Yunnan Provincial Key Laboratory of Energy Saving in Phosphorus Chemical Engineering and New Phosphorus Materials (No. 20190002, Yunnan, China), Natural Science Foundation of China (No. 21968012).

Conflicts of Interest: The authors declare no conflict of interest.

References

1. Laoutid, F.; Bonnaud, L.; Alexandre, M.; Lopez-Cuesta, J.M.; Dubois, P. New prospects in flame retardant polymer materials: From fundamentals to nanocomposites. *Mater. Sci. Eng. R Rep.* **2009**, *63*, 100–125. [CrossRef]
2. Chattopadhyay, D.K.; Webster, D.C. Thermal stability and flame retardancy of polyurethanes. *Prog. Polym. Sci.* **2009**, *34*, 1068–1133. [CrossRef]
3. Chattopadhyay, D.K.; Raju, K.V.S.N. Structural engineering of polyurethane coatings for high performance applications. *Prog. Polym. Sci.* **2007**, *32*, 352–418. [CrossRef]
4. Engels, H.W.; Pirkl, H.G.; Albers, R.; Albach, R.W.; Krause, J.; Hoffmann, A.; Casselmann, H.; Dormish, J. Polyurethanes: Versatile materials and sustainable problem solvers for todays challenges. *Angew. Chem.* **2013**, *52*, 9422–9441. [CrossRef]
5. de Wit, C.A. An overview of brominated flame retardants in the environment. *Chemosphere* **2002**, *46*, 583–624. [CrossRef]
6. Alaee, M.; Arias, P.; Sjodin, A.; Bergman, A. An overview of commercially used brominated flame retardants, their applications, their use patterns in different countries/regions and possible modes of release. *Environ. Int.* **2003**, *29*, 683–689. [CrossRef]
7. Birnbaum, L.S.; Staskal, D.F. Brominated flame retardants: Cause for concern? *Environ. Health Perspect.* **2004**, *112*, 9–17. [CrossRef]
8. Tabatabaee, F.; Khorasani, M.; Ebrahimi, M.; Gonzalez, A.; Irusta, L.; Sardon, H. Synthesis and comprehensive study on industrially relevant flame retardant waterborne polyurethanes based on phosphorus chemistry. *Prog. Org. Coat.* **2019**, *131*, 397–406. [CrossRef]
9. Yan, Y.; Liang, B. Synthesis of phosphorus-nitrogen flame retardant and its application in epoxy resin. *Fine Chem.* **2019**, *36*, 316–321.
10. Lu, S.Y.; Hamerton, I. Recent developments in the chemistry of halogen-free flame retardant polymers. *Prog. Polym. Sci.* **2002**, *27*, 1661–1712. [CrossRef]
11. Zhang, S.; Horrocks, A.R. A review of flame retardant polypropylene fibres. *Prog. Polym. Sci.* **2003**, *28*, 1517–1538. [CrossRef]
12. van der Veen, I.; de Boer, J. Phosphorus flame retardants: Properties, production, environmental occurrence, toxicity and analysis. *Chemosphere* **2012**, *88*, 1119–1153. [CrossRef] [PubMed]
13. Liu, Y.; Wang, Q. Melamine cyanurate-microencapsulated red phosphorus flame retardant unreinforced and glass fiber reinforced polyamide 66. *Polym. Degrad. Stabil.* **2006**, *91*, 3103–3109. [CrossRef]
14. Wang, B.B.; Sheng, H.B.; Shi, Y.Q.; Hu, W.Z.; Hong, N.N.; Zeng, W.R.; Ge, H.; Yu, X.J.; Song, L.; Hu, Y. Recent advances for microencapsulation of flame retardant. *Polym. Degrad. Stabil.* **2015**, *113*, 96–109. [CrossRef]
15. Liu, H.; Neal, A.T.; Zhu, Z.; Luo, Z.; Xu, X.F.; Tomanek, D.; Ye, P.D. Phosphorene: An unexplored 2d semiconductor with a high hole mobility. *ACS Nano* **2014**, *8*, 4033–4041. [CrossRef]
16. Li, L.; Yu, Y.; Ye, G.J.; Ge, Q.; Ou, X.; Wu, H.; Feng, D.; Chen, X.H.; Zhang, Y. Black phosphorus field-effect transistors. *Nat. Nanotechnol.* **2014**, *9*, 372–377. [CrossRef]
17. Qiao, J.; Kong, X.; Hu, Z.-X.; Yang, F.; Ji, W. High-mobility transport anisotropy and linear dichroism in few-layer black phosphorus. *Nat. Commun.* **2014**, *5*, 1–7. [CrossRef]
18. Qiu, S.L.; Zhou, Y.F.; Zhou, X.; Zhang, T.; Wang, C.Y.; Yuen, R.K.K.; Hu, W.Z.; Hu, Y. Air-stable polyphosphazene-functionalized few-layer black phosphorene for flame retardancy of epoxy resins. *Small* **2019**, *15*, 13. [CrossRef]
19. Qu, Z.; Wu, K.; Jiao, E.; Chen, W.; Hu, Z.; Xu, C.; Shi, J.; Wang, S.; Tan, Z. Surface functionalization of few-layer black phosphorene and its flame retardancy in epoxy resin. *Chem. Eng. J.* **2020**, *328*, 122991. [CrossRef]

20. Ren, X.; Mei, Y.; Lian, P.; Xie, D.; Deng, W.; Wen, Y.; Luo, Y. Fabrication and application of black phosphorene/graphene composite material as a flame retardant. *Polymers* **2019**, *11*, 193. [CrossRef]
21. Ren, X.; Mei, Y.; Lian, P.; Xie, D.; Yang, Y.; Wang, Y.; Wang, Z. A novel application of phosphorene as a flame retardant. *Polymers* **2018**, *10*, 227. [CrossRef]
22. Zou, B.; Qiu, S.; Ren, X.; Zhou, Y.; Zhou, F.; Xu, Z.; Zhao, Z.; Song, L.; Hu, Y.; Gong, X. Combination of black phosphorus nanosheets and mcnts via phosphoruscarbon bonds for reducing the flammability of air stable epoxy resin nanocomposites. *J. Hazard. Mater.* **2020**, *383*, 121069. [CrossRef] [PubMed]
23. Qiu, S.L.; Hou, Y.B.; Xing, W.Y.; Ma, C.; Zhou, X.; Liu, L.X.; Kan, Y.C.; Yuen, R.K.K.; Hu, Y. Self-assembled supermolecular aggregate supported on boron nitride nanoplatelets for flame retardant and friction application. *Chem. Eng. J.* **2018**, *349*, 223–234. [CrossRef]
24. Zhang, Q.R.; Li, Z.W.; Li, X.H.; Yu, L.G.; Zhang, Z.J.; Wu, Z.S. Zinc ferrite nanoparticle decorated boron nitride nanosheet: Preparation, magnetic field arrangement, and flame retardancy. *Chem. Eng. J.* **2019**, *356*, 680–692. [CrossRef]
25. Feng, Y.Z.; Han, G.J.; Wang, B.; Zhou, X.P.; Ma, J.M.; Ye, Y.S.; Liu, C.T.; Xie, X.L. Multiple synergistic effects of graphene-based hybrid and hexagonal born nitride in enhancing thermal conductivity and flame retardancy of epoxy. *Chem. Eng. J.* **2020**, *379*, 13. [CrossRef]
26. Wu, J.X.; Mao, N.N.; Xie, L.M.; Xu, H.; Zhang, J. Identifying the crystalline orientation of black phosphorus using angle-resolved polarized raman spectroscopy. *Angew. Chem. Int. Ed.* **2015**, *54*, 2366–2369. [CrossRef]
27. Favron, A.; Gaufres, E.; Fossard, F.; Phaneuf-L'Heureux, A.L.; Tang, N.Y.W.; Levesque, P.L.; Loiseau, A.; Leonelli, R.; Francoeur, S.; Martel, R. Photooxidation and quantum confinement effects in exfoliated black phosphorus. *Nat. Mater.* **2015**, *14*, 826–832. [CrossRef]
28. Zhang, S.; Yang, J.; Xu, R.J.; Wang, F.; Li, W.F.; Ghufran, M.; Zhang, Y.W.; Yu, Z.F.; Zhang, G.; Qin, Q.H.; et al. Extraordinary photoluminescence and strong temperature/angle-dependent raman responses in few-layer phosphorene. *ACS Nano* **2014**, *8*, 9590–9596. [CrossRef] [PubMed]
29. Yu, X.X.; Cai, R.R.; Jiao, J.L.; Fan, Y.L.; Gao, Q.; Li, J.W.; Pan, N.; Wu, M.Z.; Wang, X.P. Enhanced thermal stability of boron nitride-coated au nanoparticles for surface enhanced raman spectroscopy. *J. Alloys Compd.* **2018**, *730*, 487–492. [CrossRef]
30. Chen, L.; Zhou, G.M.; Liu, Z.B.; Ma, X.M.; Chen, J.; Zhang, Z.Y.; Ma, X.L.; Li, F.; Cheng, H.M.; Ren, W.C. Scalable clean exfoliation of high-quality few-layer black phosphorus for a flexible lithium ion battery. *Adv. Mater.* **2016**, *28*, 510–517. [CrossRef]
31. Zhao, M.; Qian, H.L.; Niu, X.Y.; Wang, W.; Guan, L.; Sha, J.; Wang, Y.W. Growth mechanism and enhanced yield of black phosphorus microribbons. *Cryst. Growth Des.* **2016**, *16*, 1096–1103. [CrossRef]
32. Yu, B.; Xing, W.; Guo, W.; Qiu, S.; Wang, X.; Lo, S.; Hu, Y. Thermal exfoliation of hexagonal boron nitride for effective enhancements on thermal stability, flame retardancy and smoke suppression of epoxy resin nanocomposites via sol–gel process. *J. Mater. Chem. A* **2016**, *4*, 7330–7340. [CrossRef]
33. Zhou, F.; Zhang, T.; Zou, B.; Hu, W.; Wang, B.; Zhan, J.; Ma, C.; Hu, Y. Synthesis of a novel liquid phosphorus-containing flame retardant for flexible polyurethane foam: Combustion behaviors and thermal properties. *Polym. Degrad. Stabil.* **2020**, *171*, 109029. [CrossRef]
34. Rao, W.-H.; Liao, W.; Wang, H.; Zhao, H.-B.; Wang, Y.-Z. Flame-retardant and smoke-suppressant flexible polyurethane foams based on reactive phosphorus-containing polyol and expandable graphite. *J. Hazard. Mater.* **2018**, *360*, 651–660. [CrossRef]
35. Lewin, M. Synergistic and catalytic effects in flame retardancy of polymeric materials—An overview. *J. Fire Sci.* **1999**, *17*, 3–19. [CrossRef]
36. Lewin, M. Synergism and catalysis in flame retardancy of polymers. *Polym. Adv. Technol.* **2001**, *12*, 215–222. [CrossRef]
37. Pecht, M.; Deng, Y.L. Electronic device encapsulation using red phosphorus flame retardants. *Microelectron. Reliab.* **2006**, *46*, 53–62. [CrossRef]
38. Cai, W.; Guo, W.W.; Pan, Y.; Wang, J.L.; Mu, X.W.; Feng, X.M.; Yuan, B.H.; Wang, B.B.; Hu, Y. Polydopamine-bridged synthesis of ternary h-BN@PDA@SNO2 as nanoenhancers for flame retardant and smoke suppression of epoxy composites. *Compos. Pt. A Appl. Sci. Manuf.* **2018**, *111*, 94–105. [CrossRef]
39. Qu, T.; Yang, N.; Hou, J.; Li, G.; Yao, Y.; Zhang, Q.; He, L.; Wu, D.; Qu, X. Flame retarding epoxy composites with poly(phosphazene-co-bisphenol a)-coated boron nitride to improve thermal conductivity and thermal stability. *RSC Adv.* **2017**, *7*, 6140–6151. [CrossRef]

40. Dong, W.; Xiaowei, M.; Wei, C.; Lei, S. Constructing phosphorus, nitrogen, silicon-co-contained boron nitride nano-sheets to reinforce flame retardant properties of unsaturated polyester resin. *Compos. Part A Appl. Sci. Manuf.* **2018**, *109*, 546–554.
41. Ryder, C.R.; Wood, J.D.; Wells, S.A.; Yang, Y.; Jariwala, D.; Marks, T.J.; Schatz, G.C.; Hersam, M.C. Covalent functionalization and passivation of exfoliated black phosphorus via aryl diazonium chemistry. *Nat. Chem.* **2016**, *8*, 597–602. [CrossRef] [PubMed]
42. Kuo, P.L.; Chang, J.M.; Wang, T.L. Flame-retarding materials—I. Syntheses and flame-retarding property of alkylphosphate-type polyols and corresponding polyurethanes. *J. Appl. Polym. Sci.* **1998**, *69*, 1635–1643. [CrossRef]
43. Horold, S. Phosphorus flame retardants in thermoset resins. *Polym. Degrad. Stabil.* **1999**, *64*, 427–431. [CrossRef]
44. Hedrick, S.A.; Chuang, S.S.C. Temperature programmed decomposition of polypropylene: In situ ftir coupled with mass spectroscopy study. *Thermochim. Acta* **1998**, *315*, 159–168. [CrossRef]
45. Yu, B.; Wang, X.; Qian, X.D.; Xing, W.Y.; Yang, H.Y.; Ma, L.Y.; Lin, Y.; Jiang, S.H.; Song, L.; Hu, Y.; et al. Functionalized graphene oxide/phosphoramide oligomer hybrids flame retardant prepared via in situ polymerization for improving the fire safety of polypropylene. *RSC Adv.* **2014**, *4*, 31782–31794. [CrossRef]
46. Zhang, P.K.; He, Y.Z.; Tian, S.Q.; Fan, H.J.; Chen, Y.; Yan, J. Flame retardancy, mechanical, and thermal properties of waterborne polyurethane conjugated with a novel phosphorous-nitrogen intumescent flame retardant. *Polym. Compos.* **2017**, *38*, 452–462. [CrossRef]
47. Yuan, B.H.; Fan, A.; Yang, M.; Chen, X.F.; Hu, Y.; Bao, C.L.; Jiang, S.H.; Niu, Y.; Zhang, Y.; He, S.; et al. The effects of graphene on the flammability and fire behavior of intumescent flame retardant polypropylene composites at different flame scenarios. *Polym. Degrad. Stabil.* **2017**, *143*, 42–56. [CrossRef]
48. Zhou, S.; Song, L.; Wang, Z.Z.; Hu, Y.; Xing, W.Y. Flame retardation and char formation mechanism of intumescent flame retarded polypropylene composites containing melamine phosphate and pentaerythritol phosphate. *Polym. Degrad. Stabil.* **2008**, *93*, 1799–1806. [CrossRef]

© 2020 by the authors. Licensee MDPI, Basel, Switzerland. This article is an open access article distributed under the terms and conditions of the Creative Commons Attribution (CC BY) license (http://creativecommons.org/licenses/by/4.0/).

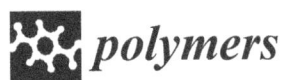

Article

The Effect of OMMT on the Properties of Vehicle Damping Carbon Black-Natural Rubber Composites

Wei Liu [1,2], Lutao Lv [1], Zonglin Yang [1], Yuqing Zheng [1] and Hui Wang [1,3,*]

[1] School of Polymer Science and Engineering, Qingdao University of Science and Technology, Qingdao 266042, China; qdlorin@qust.edu.cn (W.L.); tao19961225@sina.com (L.L.); gaofenzi@qust.edu.cn (Z.Y.); min@qust.edu.cn (Y.Z.)
[2] Key Laboratory of Rubber-Plastics, Ministry of Education/Shandong Provincial Key Laboratory of Rubber and Plastics, Qingdao University of Science and Technology, Qingdao 266042, China
[3] Department of Chemical Engineering, University of Waterloo, 200 University Avenue West, Waterloo, ON N2L 3G1, Canada
* Correspondence: hwang@qust.edu.cn

Received: 20 July 2020; Accepted: 25 August 2020; Published: 31 August 2020

Abstract: In this study, the filled natural rubber (NR) was prepared with organic montmorillonite (OMMT) and carbon black (CB). The effects of the amount of OMMT on the properties of CB/NR composites were investigated by measuring the physical and mechanical properties, compression set and compression heat properties, processing properties and damping properties. The formulation was optimized depending on the different conditions of end applications and the damping properties of rubber were maximized without affecting the other properties of the rubber. The results showed that the rubber composite system filled with 2 phr (parts per hundreds of rubber) OMMT had better mechanical properties and excellent damping performance.

Keywords: organic montmorillonite; natural rubber; damping properties; mechanical properties

1. Introduction

With the development of rail transit, it has become a major problem to prepare high-performance (high strength, high elasticity, high flexibility and high damping) rubber shock absorbers to improve NVH (noise, vibration and harshness) performance. Rubber can transform the mechanical energy to heat energy, and so can reduce or even eliminate the vibration, and has been widely used as a damping material [1,2]. In recent years, the improvement of damping properties of natural rubber (NR) has become a hot topic [3]. It was found that damping vibration system can be obtained by changing the rubber shock absorber damping performance to improve the performance of rubber, which provides a theoretical support for later research and development of rubber damping products. Gao et al. [4] blended butadiene rubber (BR) with NR to design the new formulation of automotive shock absorber rubber products, and obtained better damping performance of composite rubber materials. Some researchers [5,6] used a grafting method to improve the rubber macromolecule chain branching degree, by which the molecular chain of internal friction was raised and then the performance of damping materials was increased.

Wang [7] reported that when increasing the content of the short fiber filling, the peak value of the loss factor Tanδ of the NR composites increased from 0.2 to 0.44. The effects of sulfur and accelerator, rubber alloy and carbon black (CB) on damping rubber were also studied extensively [8–12]. The amount of carbon black and the dispersion of carbon black in the material can impose significant impact on the damping properties of the material. It was also found that graphite can make a great effect on the damping performance [13]. When 5 phr graphite was added to nitrile rubber to replace the same weight of carbon black, the storage modulus of nitrile rubber increased correspondingly, and the effective damping temperature range expanded, which are beneficial to improve the damping performance. Other studies showed that CNFs provide much stronger reinforcement than carbon black [14]. Noor Azammi [15] filled kenaf fiber into TPU-NR composites and got a sample that would shift the damping temperature range up to 135 °C. Therefore, the wider the applicable temperature range, the better the damping property of the material. Joseph et al. [16] studied the changes of the damping parameters when adding palm oil microfibers and long fibers to acrylonitrile butadiene rubber (NBR). The damping parameters includes the loss factor peak, glass transition temperature (T_g), storage modulus, and the damping properties of composite materials. The results showed that the storage modulus increased with the increase of the dosage and the loss factor decreases gradually. Guo et al. [17] reported that when filled with micro glass flake, the storage modulus of NBR increased in low-temperature stage and the mechanical properties and high-temperature performances almost kept well. Perera et al. [18] grafted methyl methacrylate (MMA) on the molecular chain of NR, which could greatly improve the damping performance of rubber. Xu et al. [19] prepared a rubber composite with excellent properties of improved mechanical property, high damping value and wide damping temperature range by using multilayered structure material and revealing the damping mechanism at the molecular level and in a quantitative manner.

Organic montmorillonite (OMMT) has attracted extensive attention due to its good reinforcing effect, low price, abundant reserves, simple preparation and environmental protection. As a result, it is used as a new reinforcing filler in rubber. The influence of OMMT content on swelling behavior was investigated, showing that it can significantly improve the delayed expansion property of rubber [20]. Chen et al. [21,22] found that the physical properties of rubber composites could be enhanced by OMMT, which shed light on the application of structural damping materials in the future.

It is rarely reported that OMMT was used to improve the vibration damping performance of rubber, while the layered structure of it is worth studying. At the same time, it can be found from the previous research that the damping performance of rubber is mainly affected by the structure of rubber and its filler. In this study, the processing performance, mechanical properties and damping properties of rubber composites were all improved only with little amount of OMMT. To characterize the damping performance of rubber material with OMMT (widely used as nanofillers in polymeric composites), the OMMT/CB/NR composites were obtained by mechanical mixing reaction intercalation method. In addition, studies on the influence of OMMT content and the composition of rubber components on its mechanical, dynamic mechanical properties and damping properties were performed. The characterization of the samples was done by rubber processing analysis (RPA), dynamic mechanical analysis (DMA), scanning electron microscope (SEM) and the mechanical analysis instruments.

2. Experiment Section

2.1. Materials and Preparation

Natural rubber (SCR, Hainan Natural Rubber Company's product, Haikou, Hainan, China) was used as basis material. Sulfur (99.5% purity) was purchased from Jinchangsheng Co., Guangzhou, Guangdong, China. Organic montmorillonite (OMMT, 85%, organic modified by quaternary ammonium salt, particle size 44 μm, colloidal viscosity 4.0 mPas·s) was purchased from Zhejiang Fenghong clay chemical Co., Anji, Zhejiang, China. Cabon Black (N330) was purchased from Cabot Corporation, Tianjin, China. N-cyclohexyl thio-phthalimide (CTP-70GR, a kind of scorch retarder), N-1,3-dimethylbutyl-N'-phenyl-p-phenylenediamine (4020, a kind of antioxidant), Poly (1,2-dihydro-2,2,4-trimethyl-quinoline) (RD, a kind of antioxidant) and Zinc oxide (ZnO-80) were purchased from Ningbo Actmix Rubber Chemicals Co., Ltd, Ningbo, China The remaining ingredients were provided by Jinchangsheng Co., Guangzhou, Guangdong, China. Table 1 shows the specific formula proportion of samples.

Table 1. Formula of natural rubber (NR) with organic montmorillonite (OMMT).

Ingredient	NR	Sulfur	ZnO	Stearic Acid	Microcrystalline Wax	CTP	4020	RD	N330	OMMT
Phr	100	2	5	1	1	1	3	1	50	0/2/4/6/8

The formula of this study was listed in Table 1, on the basis of weight. In the preparation process, all raw materials were weighed according to the formula ratio, and the actual weight of NR was 200 g. Firstly, NR was dried at 60 °C for 4 h before use. Then, it was processed with stearic acid and microcrystalline wax in sk-160B two-roll mill (Shanghai Plastics and Rubber Machinery Co., Shanghai, China) at room temperature, the initial distance between the two rollers was 2 mm, completely encased on the roller for 2 min. Next, the CTP (N-cyclohexyl thio-phthalimide, a kind of scorch retarder), 4020 (N-1,3-dimethylbutyl-N'-phenyl-p-phenylenediamine, a kind of antioxidant), RD (poly (1,2-dihydro-2,2,4-trimethyl-quinoline), a kind of antioxidant), mixed for 3 min. The following were carbon black N330 and OMMT, mixed for 3 min. After that, sulfur and ZnO (Zinc oxide) were added into the blended compound. Next, set the distance between the two rollers to 1mm, made the "triangle bag" in the process for 5 times. Lastly, the compound obtained was molded into sheets in a Type Plate Vulcanizer (XLB-400 × 400 × 2 50 T) (Qingdao Yadong Machinery, Qingdao, China) at 150 °C for 15 min.

2.2. Measurements

2.2.1. Processing Performance

Mooney viscosity test was carried out on a CL-2000G Mooney Viscosity Meter at 100 °C (Jiangdu Rectify Test Machine Factory, Yangzhou, China) according to the standard of ASTM-D1646. Vulcanization performance was conducted using a MDR2000 Rotor Rheometer at 150 °C (GOTECH Testing Machines Inc., Dongguan, guangzhou, China) according to the standard of ASTM-GB/T 16584-1996. Processing performance of each content were tested with three samples.

2.2.2. Mechanical Tests

Tensile properties, tear strength and elasticity modulus were determined using AT-7000M Universal Electronic Tensile Machine (GOTECH Testing Machines Inc., Shanghai, China) at a tensile speed of 500 mm·min^{-1} according to HG/T 3849-2008, GB/T 529-2008 and HG/T 3321-2012. Hardness was measured by GT-GS-MB Shore A Hardness Tester (GOTECH Testing Machines Inc., China) according to GB/T 531.2-2009. Compression set was tested according to GB/T 7759.1-2015 (homemade equipment, type A sample, the compression ratio of sample of 25%, time for 72 h). Compression heat was tested by GT-RH-200 (GOTECH Testing Machines Inc., China) according to GB/T1687.3-2016. (Resilience was carried out using a CJ-6A Rubber Rebound Testing Machine (Shanghai Chemical Machinery No. 4 Factory, Shanghai, China) at an impact speed of 14 m·s^{-1} according to GB/T 1681-2009. Five samples of each content were tested for mechanical properties.

2.2.3. Rubber Processing Analysis (RPA)

Dynamic rheological measurements of unvulcanized rubber compounds were carried out on a RPA8000 Rubber Processing Analyzer (GOTECH Testing Machines Inc., China) by a shear mode at a temperature of 60 °C, a frequency of 1 Hz, and a strain amplitude in the range of 0.28–200%, according to ASTM D 6204. Three samples of each content were tested for data of RPA.

2.2.4. Dynamic Mechanical Analysis (DMA)

The DMA was carried out using a DMA 242 Dynamic Mechanical Analyzer (NETZSCH Company, Selb, Germany) in a double cantilever beam deformation mode within a temperature range of −80 to

80 °C with a heating rate of 3 °C/min and at a fixed frequency of 1 Hz. The experiment process was carried out in the atmosphere of nitrogen. Three samples of each content were tested for data of DMA.

2.2.5. Scanning Electron Microscope (SEM)

The micromorphology of OMMT in rubber matrix was investigated using a JEOL JSM-7500F scanning electron microscope (SEM) (JEOL Ltd., Tokyo, Japan). Resolution: 1.0 nm at 15 kV; 1.4 nm at 1 kV. Acceleration voltage: 0.1–30 kV. One sample for each content was tested for SEM.

3. Results and Discussion

3.1. Vulcanization Performance

Normally, the change in torque during vulcanization is proportional to the density of cross-linking including the physical and chemical cross-linking. The vulcanization properties of each content were tested. It can be seen from Figure 1 that the minimum torque (M_L) increased gradually with an increase in the content of OMMT. It was determined by the viscosity of the rubber compound, which indicates the strength of physical interaction when the rubber is not cross-linked. Therefore, when the content of OMMT was increased, the physical interactions of the rubber gradually increased. The maximum torque (M_H) increased appreciably with addition of 2 phr OMMT. However, the increment was not obvious with more amounts of OMMT addition. M_H depends on the filler and the vulcanization system, and it reflected the superposition of physical interactions and chemical interactions after cross-linking. When OMMT was used in rubber, the MH-ML of compound increased, and the viscosity became larger thereby producing the agglomeration effect. Therefore, the larger content of OMMT, the stronger the agglomeration effect, and thus it damaged the mechanical properties of the material.

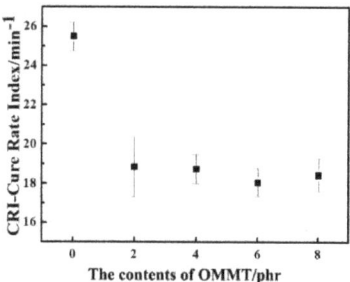

Figure 1. The effects of the contents of OMMT on the cure rate index of carbon black (CB)/natural rubber (NR) composites.

3.2. Mechanical Properties

Table 2 shows the effect of OMMT addition amount (on the basis of weight) on the mechanical properties of vulcanized CB/NR polymer composites. It can be seen that the OMMT content has an adverse effect on the physical and mechanical properties of the material, especially on the tearing strength. However, the tensile strength decreased by less than 11.5%, which still met the requirements of automobile vibration reduction products. According to the tensile strength at different elongations, it can be seen that the modulus had a decreasing trend with increasing of the OMMT contents. It is worth mentioning that the composite obtained has a greater resilience by adding 2 phr OMMT, which is similar to the variation trends of M_H and MH-ML as discussed in the above analysis.

Table 2. The effects of OMMT contents on mechanical properties of vulcanized rubber.

The Content of OMMT/phr	0	2	4	6	8
Tensile Strength/MPa	27.5	27.6	25.2	25.6	24.4
Elongation/%	616	502	519	515	488
Tensile Strength at 100%/MPa	2.1	3.2	2.7	2.8	2.8
Tensile Strength at 300%/MPa	11.0	15.0	12.4	12.9	13.2
Elasticity Modulus/MPa	2.6	3.4	2.9	3.1	3.2
Tear Strength/kN/m	131.9	111.9	111.3	96.6	87.0
Resilience Property/%	47.7	51.9	51.4	52.0	51.4

Hardness was measured at five different locations of each sample. Figure 2 shows the hardness of composite material changing with the amounts of OMMT. It can be seen that the hardness values of vulcanized rubber filled different contents of OMMT were all within 65 ± 2 and the general trend of it is increase slightly with more contents, but not all of it fits this pattern due to experimental mistakes. It also reflects its effect on the packing-packing network of vulcanized rubber. These data indicate that OMMT has minor influence on the vulcanized rubber hardness, which is of great importance to the formulation design of shock absorbing products.

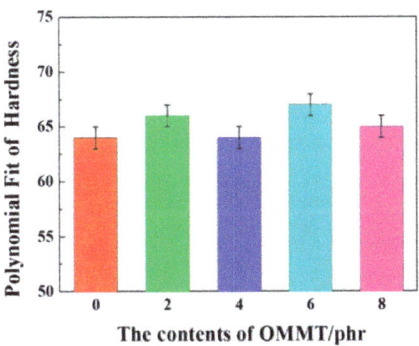

Figure 2. The effects of the contents of OMMT on hardness of vulcanized rubber.

It can be seen from Table 3 that the final temperature rise of compression heat generation was the lowest with addition of 2 phr OMMT, and it increased gradually with increasing of OMMT amounts. When the rubber endures the cyclic stress, the macromolecular chain segment can generate relative motion, which then converts the mechanical energy into the thermal energy. It was preliminarily expected that 2 phr OMMT will reduce the agglomeration effect and improve the dispersion of CB. When excessive content of OMMT was added, the fillers were more unevenly dispersed. The filler-filler network thus became stronger and the friction between them caused the heat of rubber increased. At the same time, the static compression permanent deformation of polymer composites reduced gradually. This may be a result of the insertion of the rubber macromolecular segment into the OMMT sheet structure, which then increases the physical cross-linking and reduces the static compression permanent deformation.

Table 3. The effects of OMMT content on the compressed heat generation and compression permanent deformation of vulcanized rubber.

OMMT/phr	Final Temperature Rise/°C	Dynamic Compression Permanent Deformation/%	Static Compression Permanent Deformation/%
0	15.85	0.060	0.093
2	13.90	0.040	0.091
4	14.25	0.051	0.083
6	15.30	0.055	0.081
8	15.80	0.045	0.072

3.3. Dynamic Mechanical Performance

In general, polymers are not used in their pure form, which means that the CB, silica and OMMT play an important role in the polymer reinforcement. The fillers in polymer strongly modifies the viscoelastic behavior, especially the dynamic mechanical properties of the composite. In dynamic mechanics measurements of shear modulus, the Payne effect is often referred to as the progressive breakdown of packing-packing interactions, where G' is used commonly [23–25].

RPA8000 was used to conduct three-dimensional strain scanning on CB/NR compounds filled with different contents of OMMT, and the interaction between packing-packing and the interaction between packing-rubber were accurately characterized. Each content was tested with three samples and the experimental results were shown in Figure 3, which indicates that the storage modulus curves of the rubber compound changed significantly with different contents of OMMT. The last two storage modulus curves were greatly lower than the first one under low strain conditions with respect to each different OMMT contents of rubber and the latter two curves were basically coincident. There were packing-packing networks in the composite which was destroyed during the first strain sweep and cannot be instantaneously recovered, and the last two measurements were performed on the basis of the completely destroyed packing-packing network.

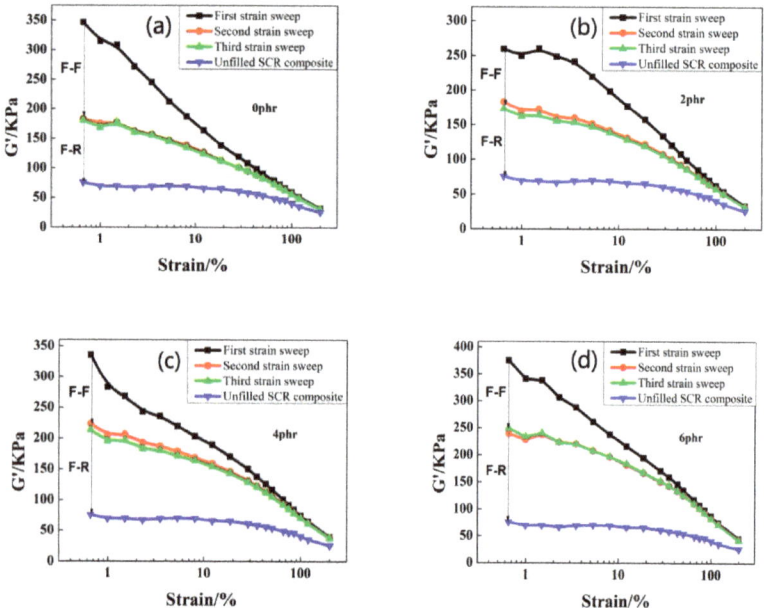

Figure 3. *Cont.*

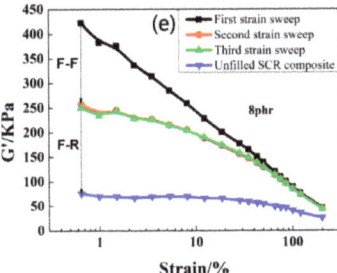

Figure 3. The three times G'-strain curves of the CB/NR rubber composites with different contents of OMMT. The content of OMMT: (**a**) 0 phr, (**b**) 2 phr, (**c**) 4 phr, (**d**) 6 phr, (**e**) 8 phr.

In order to further study the dispersion of fillers in the CB/NR composites, we marked it in the RPA scans. The difference values of G'(1)-G'(3) at low strain could indicate the strength of the packing network, and the different values of G'(3)-G'(0) without adding filler at low strain could indicate the strength of the filler and the rubber [26]. It can be seen from Figure 4 that the value of G'(1)-G'(3) with 2 phr OMMT was the lowest in all composites. With adding more OMMT, the value of G'(1)-G'(3) increased progressively. When the content of OMMT was added more than 8 phr, it becomes larger than that of without OMMT. It shows that a small amount (2 and 4 phr) of OMMT induces the weaker filler-filler network, which is helpful for the dispersion of carbon black. If the content is too large, the OMMT agglomeration effect becomes stronger and the material properties decreased. The order of G'(3)-G'(0) value influenced by OMMT was 10 > 8 > 6 > 4 > 0 > 2 phr. When the content of OMMT was 2 phr, the filler-rubber interaction was lowest; the interaction of the filler was also minimized; and the MH-ML was the largest. It can be preliminarily concluded that when 2 phr OMMT added, the rubber-rubber interaction was the strongest and the performance of rubber was good.

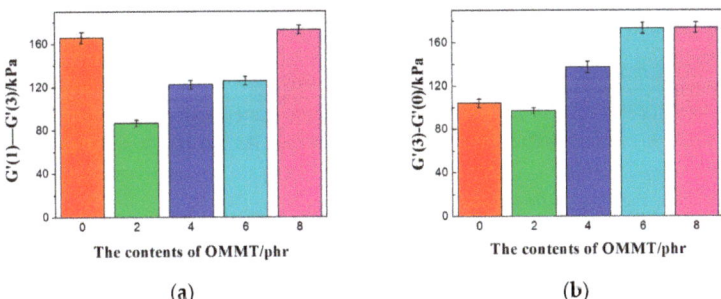

Figure 4. (**a**) G' (1)-G' (3) and (**b**) G' (3)-G' (0) of CB/NR composites with different contents of OMMT.

Figure 5 shows the storage modulus of the rubber compound filled with different contents of OMMT under small strain, which exhibited a nonlinear decrease with the increase of strain. This was the so-called Payne effect. Normally, the better the fillers dispersion, the lower the storage modulus under the same condition of the strain sweep.

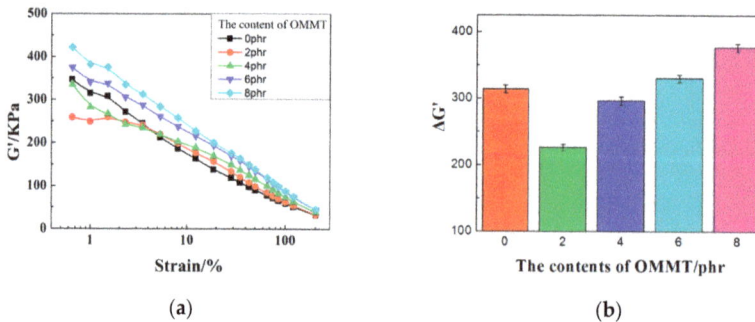

Figure 5. (a) G′-strain and (b) ΔG′ of CB/NR composites with different contents of OMMT.

It can be seen from Figure 5 that the rubber composites with 2 phr OMMT had a lowest storage modulus, and the storage modulus was found increase gradually with increasing of OMMT amount. Therefore, a small content of OMMT can improve the dispersibility of filler. In order to analyze the effect of it, ΔG′ (G′0-G′∞) was presented with the variation of OMMT content. It was found that the ΔG′ with 2 phr OMMT is the lowest and more amounts of OMMT made it gradually stronger, which was similar to the results of three consecutive strain scans on it. Since the ΔG′ is analyzed under large strain conditions, the filler network can be completely broken, and its accuracy should be higher than that obtained by the RPA8000 three-strain scan. Based on the results of two strain scans, the order of the filler network strength was obtained as 8 > 6 > 0 > 4 > 2 phr.

As showed in Figure 6, the T_g of the OMMT/CB/NR composites slightly shifts toward the high temperature direction for the addition amount of 2, 4, and 6 phr OMMT. The slight increase of T_g is due to the fact that the rubber macromolecular segment is restricted by the sheet structure of OMMT and the exercise capacity is thus reduced. In the case of the 8 phr content, the agglomeration was produced and T_g was found decrease slightly, which is consistent with the analysis results of RPA. The peak value of Tanδ increased after OMMT was added, which was because the density of cross-linking within the vulcanized rubber increased. In the glass transition zone, the rubber was deformed by the external force and the flaky filler OMMT was oriented subsequently, which would generate friction with the rubber. This will transform the mechanical energy generated by the vibration into the heat energy loss, and then improve the damping of the material. As a result, the damping value of the material increased and the damping temperature range (Tanδ ≥ 0.3) was widened (Figure 7). Finally, the damping performance was increased.

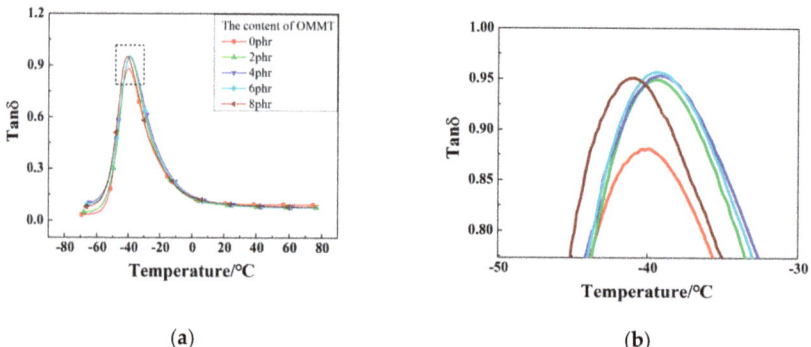

Figure 6. The loss factor-temperature curves of the CB/NR composites with different contents of OMMT. (Note: (**b**) is a local enlarged view of (**a**).)

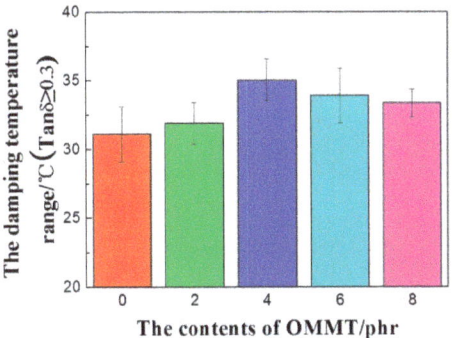

Figure 7. Comparison the damping temperature of CB/NR with different contents of OMMT.

3.4. Microstructural Characterization

Figure 8 shows the SEM images of frozen fracture surface (sprayed with gold) of the composites with and without OMMT. The SEM image of the sample without OMMT in Figure 8a shows that the morphology of fractured surface exhibits a relatively smooth surface, except for the presence of small particles of CB. Compared with the surface morphology of the composites sample (0 phr), OMMT particles were occasionally observed on the surface of the composites sample (2 phr OMMT) in the rubber matrix and carbon black, as shown in Figure 8b. Figure 8c,d shows that layered OMMT particles were more to be found in the cross-section of the OMMT/CB/NR composites. With an increase of OMMT in content, the agglomerated OMMT particles were easily observed on the surface of OMMT/CB/NR composite sample (8 phr OMMT) seen in Figure 8e. The lamellar structure of OMMT which may helpful to the damping property of the composite, but the agglomeration phenomenon of OMMT particles was not beneficial to the improvement of mechanical properties and damping performance, as shown in the previous data.

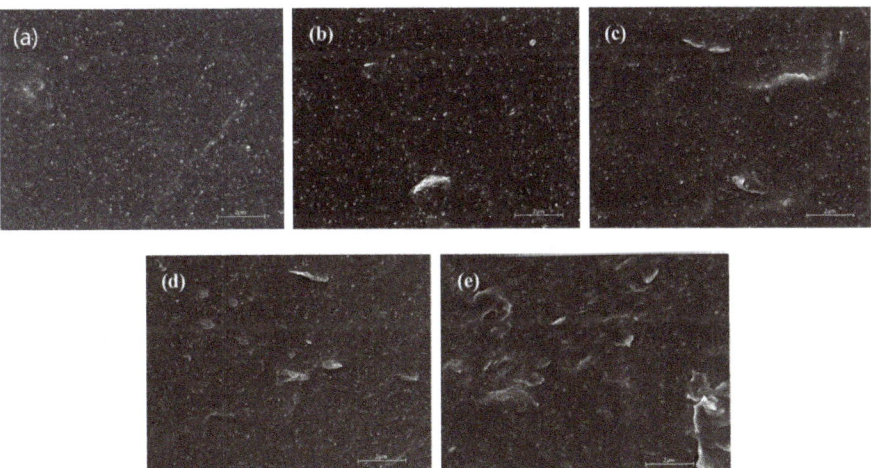

Figure 8. SEM images of rubber sections with various OMMT contents: (**a**) 0 phr, (**b**) 2 phr, (**c**) 4 phr, (**d**) 6 phr, and (**e**) 8 phr.

4. Conclusions

The CB/NR composite filled with 2 phr OMMT was found to have excellent mechanical properties and was up to the standard of the performance requirements of automobile damping vibration absorbing parts. Compared with the composite without OMMT, the crosslink density of the rubber was increased; the scorch time was extended; and the vulcanization rate was decreased. The OMMT is beneficial to improve the performance of the anti-vibration parts of the automobile and the processing safety. Meanwhile, the addition of OMMT could increase the dispersibility of CB and reduce the final temperature rise of compression heat generation. Both dynamic and static compression permanent deformations were quite small. After filling the OMMT, the system dynamic mechanical and damping properties were enhanced and the effective damping temperature range was broadened.

Author Contributions: Conceptualization, W.L. and H.W.; methodology, W.L.; software, Z.Y.; validation, W.L. and H.W.; formal analysis, L.L.; investigation, L.L.; resources, Y.Z.; data curation, L.L.; writing—original draft preparation, W.L.; writing—review, editing and visualization, W.L.; supervision, Y.Z.; project administration, H.W.; funding acquisition, H.W. All authors have read and agreed to the published version of the manuscript.

Funding: This research received no external funding.

Acknowledgments: The authors thank the financial support from the Natural Science Foundation of Shandong Province (ZR2019MB001), the Taishan Scholar Project of Shandong Province (ts201712047), and the Special Fund Project to Guide Development of Local Science and Technology by Central Government.

Conflicts of Interest: The authors declare no conflict of interest.

References

1. Essawy, H.A.; El-Sabbagh, S.H.; Tawfik, M.E. Assessment of provoked compatibility of NBR/SBR polymer blend with montmorillonite amphiphiles from the thermal degradation kinetics. *Polym. Bull.* **2018**, *75*, 1417–1430. [CrossRef]
2. Fleischmann, D.D.; Ayalur, K.S.; Arbeiter, F. Influence of crosslinker and water on mechanical properties of carboxylated nitrile butadiene rubber (XNBR). *Polym. Test.* **2018**, *66*, 24–31. [CrossRef]
3. Lang, Z.Q.; Jing, X.J.; Billings, S.A. Theoretical study of the effects of nonlinear viscous damping on vibration isolation of SDOF systems. *J. Sound Vib.* **2009**, *323*, 352–365. [CrossRef]
4. Gao, F.N. The application of silane coupling agent in rubber sealing products and shock absorbing products. *Rubber Technol. Mark.* **2009**, *23*, 11–13.
5. Urayama, K.; Miki, T.; Takigawa, T.; Shinzo, K. Damping elastomer based on model irregular networks of end-linked poly (dimethylsiloxane). *Chem. Mater.* **2004**, *16*, 173–178. [CrossRef]
6. Yamazaki, H.; Takeda, M.; Kohno, Y.; Ando, H.; Urayama, K.; Takigawa, T. Dynamic viscoelasticity of poly (butyl acrylate) elastomers containing dangling chains with controlled lengths. *Macromolecules* **2011**, *4*, 8829–8834. [CrossRef]
7. Wang, Z.R. Japan patent collection of shock absorbing rubber products. *World Rubber Ind.* **2001**, *28*, 27–32.
8. Zhang, X.L.; Liang, D.; Zhao, S.G. Study on vulcanization kinetics and properties of natural rubber damping damping materials by vulcanization system. *Rubber Ind.* **2018**, *65*, 845–849.
9. Liu, C.; Fan, J.F.; Chen, Y.K. Design of regulable chlorobutyl rubber damping materials with high damping value for a wide temperature range. *Polym. Test.* **2019**, *79*, 106003. [CrossRef]
10. Wang, J.G.; Ren, H.; Wang, S.J.; Wei, J.J.; Ren, H.; Xue, M.L. Effect of rubber alloy SG-301 on properties of natural rubber/nitrile rubber damping material. *China Synth. Rubber Ind.* **2018**, *41*, 420–424.
11. Li, B.; Zhang, B.S.; Yang, L.B.; Tang, J.H.; Liu, H.P. Effect of carbon black on the properties of high damping natural rubber/bromobutyl rubber and rubber. *Rubber Ind.* **2018**, *65*, 665–668.
12. Litvinov, V.M.; Orza, R.A.; Kluppel, M.; Van Duin, M.; Magusin, P.C.M.M. Rubber-filler interactions and network structure in relation to stress-strain behavior of vulcanized, carbon black filled EPDM. *Macromolecules* **2011**, *44*, 4887–4900. [CrossRef]
13. Liu, Q.X.; Ding, X.B.; Zhang, H.P.; Yan, X. Preparation of high-performance damping materials based on carboxylated nitrile rubber: Combination of organic hybridization and fiber reinforcement. *J. Appl. Polym. Sci.* **2009**, *114*, 2655–2661. [CrossRef]
14. Sinclair, A.; Zhou, X.Y.; Tangpong, S.; Bajwa, D.S.; Quadir, M.; Jiang, L. High-Performance styrene-butadiene rubber nanocomposites reinforced by surface-modified cellulose nanofibers. *ACS Omega* **2019**, *4*, 13189–13199. [CrossRef] [PubMed]

15. Noor Azammi, A.M.; Sapuan, S.M.; Ishak, M.R.; Sultan, M.T. Physical and damping properties of kenaf fibre filled natural rubber/ thermoplastic polyurethane composites. *Def. Technol.* **2020**, *16*, 29–34. [CrossRef]
16. Joseph, S.; Sreekumar, P.A.; Kenny, J.M.; Puglia, D.; Thomas, S.; Joseph, K. Dynamic Mechanical Analysis of Oil Palm Microfibril-Reinforced Acrylonitrile Butadiene Rubber Composites. *Polym. Compos.* **2010**, *31*, 236–244. [CrossRef]
17. Guo, Y.B.; Tan, H.; Gao, Z.Q.; Wang, D.; Zhang, S. Mechanical and unlubricated sliding wear properties of nitrile rubber reinforced with mico glass flake. *Polymers* **2018**, *10*, 705. [CrossRef]
18. Perera, M.C.S.; Wen, C.C. Radiation degradation of MG rubber studied by dynamic mechanical analysis and solid state NMR. *Polymers* **2000**, *41*, 323. [CrossRef]
19. Xu, K.M.; Hu, Q.M.; Wu, H.; Guo, S.; Zhang, F. Designing a polymer-based hybrid with simultaneously improved mechanical and damping properties via a multilayer structure construction: Structure evolution and a damping mechanism. *Polymers* **2020**, *12*, 446. [CrossRef]
20. Zhang, Z.L.; Wang, P.; Cheng, K.; Xin, C.; Zhang, Z.; Li, Z. Delayed expansion behavior and mechanical properties of water-swelling rubber/OMMT composites. *J. Macromol. Sci. Part A* **2020**, *57*, 610–617. [CrossRef]
21. Chen, S.B.; Wang, Q.H.; Wang, T.M. Damping, thermal, and mechanical properties of carbon nanotubes modifed castor oil-based polyurethane/epoxy interpenetrating polymer network composites. *Mater. Des.* **2012**, *38*, 47–52. [CrossRef]
22. Chen, S.B.; Wang, Q.H.; Wang, T.M. Damping, thermal, and mechanical properties of montmorillonite modifed castor oil-based polyurethane/epoxy graft IPN composites. *Mater. Chem. Phys.* **2011**, *130*, 680–684. [CrossRef]
23. Payne, A.R. *Reinforcement of Elastomers*; Kraus, G., Ed.; Inter Science Publishers: New York, NY, USA, 1965; Volume 3, p. 69.
24. Payne, A.R. The dynamic properties of carbon black-loaded natural rubber vulcanizates. *J. Appl. Polym. Sci.* **1962**, *6*, 57–63. [CrossRef]
25. Payne, A.R. Effect of dispersion on the dynamic properties of filler-loaded rubbers. *J. Appl. Polym. Sci.* **1965**, *9*, 2273–2284. [CrossRef]
26. Choi, S.S. Filler-polymer interactions in both silica and carbon black-filled styrene-butadiene rubber compounds. *J. Polym. Sci. Part B Polym. Phys.* **2001**, *39*, 439–445. [CrossRef]

© 2020 by the authors. Licensee MDPI, Basel, Switzerland. This article is an open access article distributed under the terms and conditions of the Creative Commons Attribution (CC BY) license (http://creativecommons.org/licenses/by/4.0/).

MDPI
St. Alban-Anlage 66
4052 Basel
Switzerland
Tel. +41 61 683 77 34
Fax +41 61 302 89 18
www.mdpi.com

Polymers Editorial Office
E-mail: polymers@mdpi.com
www.mdpi.com/journal/polymers

www.ingramcontent.com/pod-product-compliance
Lightning Source LLC
LaVergne TN
LVHW070138100526
838202LV00015B/1842